Springer Series in Materials Science

Volume 327

The Springer Series in Materials Science covers the complete spectrum of materials research and technology, including fundamental principles, physical properties, materials theory and design. Recognizing the increasing importance of materials science in future device technologies, the book titles in this series reflect the state-of-the-art in understanding and controlling the structure and properties of all important classes of materials.

Colin Tong

Advanced Materials and Components for 5G and Beyond

Springer

Colin Tong
Bolingbrook, IL, USA

ISSN 0933-033X ISSN 2196-2812 (electronic)
Springer Series in Materials Science
ISBN 978-3-031-17206-9 ISBN 978-3-031-17207-6 (eBook)
https://doi.org/10.1007/978-3-031-17207-6

This Springer imprint is published by the registered company Springer Nature Switzerland AG
The registered company address is: Gewerbestrasse 11, 6330 Cham, Switzerland

This book is dedicated to my wife Dali and to our family.

Preface

The rapidly increasing number of mobile devices, voluminous data, and higher data rate is pushing the development of fifth-generation (5G) and beyond wireless communications. The 5G networks are broadly characterized by ubiquitous connectivity, extremely low latency, and very high-speed data transfer via the adoption of new technology to equip future millimeter band wireless communication systems down to nanoscale and massive multi-input multi-output (MIMO) with extreme base station and device densities, as well as unprecedented numbers of antennas. 5G has become a novel driving force for leading innovation and stimulating new types of information technology, as well as an emerging engine for promoting industrial upgrading and driving sustained economic growth. 5G is not an incremental improvement over previous generations and thus requires a new set of materials, including plastics, metals, ceramics, composites, nanomaterials, and functional materials. Due to the construction of 5G network systems and the popularization of 5G terminals, demand for components such as base station antennas, filters, EMI shields, protective films and sealants, thermal management solutions, and high-frequency printed circuit boards (PCBs) has increased. Clearly, 5G and beyond technology is still under development, with improvements to well meet future unknown requirements and the formulation of better strategies for its utilization in the fully connected world. Therefore, there is much work to do for scientists and engineers to achieve breakthroughs in material development, component design, and related technological innovations.

To meet the demands of students, scientists, engineers, and marketing technologists for a systematic reference source, ***Advanced Materials and Components for 5G and Beyond*** introduces the current status and future trends of material advancement and component design in technology development for 5G and beyond wireless communications. Coverage includes semiconductor solutions for 5G electronics, design and performance enhancement for 5G antennas, high frequency PCB materials and design requirements, materials for high frequency filters, EMI shielding materials and absorbers for 5G systems, thermal management materials and components, and protective packaging and sealing materials for 5G devices. It explores fundamental physics, design, and engineering aspects, as well as the full array of state-of-the art applications of 5G and beyond wireless communications. Future challenges and potential trends of 5G and beyond applications and related materials technologies are also addressed.

It is a great pleasure to acknowledge the help and support I have received from my colleagues and friends. I would like to express my sincere gratitude to Dr. Sam Harrison and all other editors who have done a fantastic job on the publication of this book.

Bolingbrook, IL, USA Colin Tong

Contents

Abbreviations

1G	First generation
2G	Second generation
3G	Third generation
3GPP	3rd Generation Partnership Project
4G	Fourth generation
5G	Fifth generation
6G	Sixth generation
AAS	Active antenna system
ADAS	Advanced driver assistance system
ADC	Analog-to-digital converter
AESA	Active electronically scanned array
AI	Artificial intelligence
AiP	Antenna-in-package
AOI	Automated optical inspection
AR	Augmented reality
ATMS	Arc thermal metal spray
AVA	Antipodal Vivaldi Antenna
BAW	Bulk-acoustic-wave
BPF	Band pass filter; Berkeley Packet Filter
BT	Bismaleimide triazine
BTS	Base transmit system
CBCPW	Conductor-backed coplanar waveguide
CCAM	Connected, cooperative and automated mobility
CCL	Copper-clad laminate
CDMA	Code division multiple access
CF	Cell-free
CMOS	Complementary metal-oxide semiconductor
CNT	Carbon nanotube
COTS	Commercial off-the-shelf
CPC	Conductive polymer-based composite
CPE	Customer Premise Equipment

CPS	Cyber-physical system
CPW	Coplanar waveguide
CR	Cognitive radio
CSA	Cognitive service architecture
CSRR	Complementary Split Ring Resonator
CTE	Coefficient of thermal expansion
CVD	Chemical vapor deposition
DAC	Digital-to-analog data converter
DGS	Defected ground structure
DRL	Deep reinforcement learning
DSC	Differential scanning calorimetry
DSP	Digital signal processing
DRA	Dielectric resonator antenna
EBG	Electromagnetic bandgap
ECC	Envelop correlation coefficient
ECG	Electrocardiogram
ED	Electrodeposited
EHD	Electrohydrodynamic
EM	Electromagnetic
EMI	Electromagnetic interference
eWLB	Embedded wafer-level ball grid array
FBAR	Film bulk acoustic resonator
FDD	Frequency division duplex
FEM	Finite element method
FLG	Few-layer graphene
FOWLP	Fan-out wafer-level packaging
FSS	Frequency selective surface
FTTH	Fiber to the home
GCPW	Grounded coplanar-waveguide
GSM	Global system for mobile communication
HPC	High-performance computing
HVAC	Heating, ventilation, and air conditioning
IC	Integrated circuit
ICP	Intrinsically conducting polymer
IDT	Interdigital transducer
IoE	Internet of everything
IoMT	Internet of Medical Things
IoT	Internet of Things
IIoT	Industrial Internet-of-Things
IPD	Integrated passive device
IT	Information technology
IZO	Indium–zinc oxide
LCP	Liquid crystal polymer
LNA	Low noise amplifier
LO	Local oscillator

LTCC	Low temperature co-fired ceramics
LTE	Long Term Evolution
MC	Mutual coupling
ME	Magneto-electric
MEC	Multi-access edge computing; Mobile edge computing
MEMS	Micro-electro-mechanical system
MI	Machine intelligence
MIM	Metal–insulator-metal
MIMO	Multi-input multi-output
ML	Machine learning
MLO	Multilayer organic
mMTC	Massive machine-type communication
M2M	Machine-to-machine
MNO	Mobile network operator
MOM	Metal oxide metal
MOS	Metal oxide semiconductor
MPA	Monopolar Patch Antenna
MRC	Maximum-ratio-combining (receiver)
MZI-PLC	Mach-Zehnder interferometer/planar lightwave circuit
NF	Noise figure
NFV	Network functions virtualization
NGN	Next generation network
NS	Network slicing
NTT	Nippon Telegraph and Telephone Public Corporation
OT	Operational technology
PA	Power amplifier
PC	Polycarbonate
PCB	Printed circuit board
PCM	Phase-change material
PDK	Process development kit
PECVD	Plasma-enhanced chemical vapor deposition
PEEK	Polyether ether ketone
PMMA	Poly(methyl methacrylate)
PP	Polypropylene
PPDR	Public protection and disaster relief
PPE	Polyphenol ether
PPI	Pores per inch
PTFE	Polytetrafluoroethylene
PTH	Plated through hole
PVA	Poly(vinyl alcohol)
PVC	Polyvinyl chloride
PVD	Physical vapor deposition

QoE	Quality of experience
QoS	Quality of service
RAN	Radio access network
RDL	Redistribution layer
RF	Radio frequency
RFFE	Radio-frequency front-end
RIE	Reactive ion etching
ROF	Radio-over-fiber
SAW	Surface acoustic wave
SBA	Service-based architecture
SBD	Schottky barrier diode
SDR	Software-defined-radio
SDN	Software defined network
SE	Shielding effectiveness
SFDR	Spurious-free dynamic range
SISO	Single input single output
SIW	Substrate-integrated waveguide
SMC	Sheet molding compound; surface-mount component
SMR	Solidly mounted resonator
SMT	Surface-mount technology
SoC	System on chip
SOI	Silicon on insulator
SoP	System-on-package
SPI	Solder paste inspection
SRR	Split-ring resonator
STAR	Simultaneous transmit and receive
TC	Temperature compensated
TDD	Time division duplex
TEC	Thermoelectric cooler
TEM	Transverse electromagnetic
TIM	Thermal interface material
TPV	Through package via
TSN	Time-sensitive networking
UAV	Unmanned aerial vehicle
UE	User equipment
URLLC	Ultra-reliable low-latency communication
UV	Ultraviolet light
UVA	Ultraviolet light absorber
V2X	Vehicle-to-everything
VCO	Voltage-controlled oscillator
VGA	Variable gain amplifier
VNA	Vector network analyzer
VR	Virtual reality
WBG	Wide-bandgap

About the Author

Colin Tong is a materials expert with considerable professional experience in the past two decades. His research and development activities and industrial practices cover a broad range of different fields with a special focus on materials testing and characterization, component design and processing of advanced composite materials, metallurgy, thermal management of electronic packaging, electromagnetic interference shielding, integrated optical waveguides, functional metamaterials and metadevices, energy materials, 5G materials, as well as flexible and printed electronics. He holds a PhD in materials science and engineering, and a master's as well as a bachelor's degree in materials and mechanical engineering. Dr. Tong has published 6 books and more than 30 peer-reviewed papers, and he holds 10 patents. He is a senior member of IEEE (Institute of Electrical and Electronics Engineers). He received the Henry Marion Howe Medal from ASM International for his contribution to research and development on advanced aluminum composite materials in 1999.

Chapter 1
5G Technology Components and Material Solutions for Hardware System Integration

Abstract 5G technologies and applications transform mobile broadband to an enhanced wideband millimeter spectrum for a ubiquitous ultrafast experience, enable massive machine-type communications, and empower network applications that require ultrahigh reliability and ultralow latency. 5G capabilities will also be foundational to support a plethora of use cases, for instance, scientific experimentation in extreme environments, consumer applications (e.g., autonomous vehicles, cellular networks), artificial intelligence (AI) and machine learning (ML) applications, simulation and data analytics, and software-defined or cognitive radio. This chapter will give a brief review of 5G technology components and material solutions to provide ubiquitous connectivity, extremely low latency, and very high-speed data transfer via the adoption of innovative technology to equip millimeter band wireless communication systems at the nanoscale and massive multi-input multi-output (MIMO) with extreme base station and device densities, as well as unprecedented numbers of antennas.

1.1 Evolution of 5G Technology

Since the Nippon Telegraph and Telephone Public Corporation (NTT) initiated the world's first cellular mobile communication service in December 1979, the technology of mobile communications has continued to evolve with the progress of technology. From the first generation (1G) to the second generation (2G) voice calls were the main means of communication, and simple e-mail was possible. However, from the third generation (3G), data communications such as "i-mode" and multimedia information, including photos, music, and video, could be communicated using mobile devices. From the fourth generation (4G), smartphones have been explosively popularized by high-speed communication technology exceeding 100 Mbps using the Long Term Evolution (LTE), and a wide variety of multimedia communication services have appeared. 4G technology continues to evolve in the

C. Tong, *Advanced Materials and Components for 5G and Beyond*, Springer Series in Materials Science 327, https://doi.org/10.1007/978-3-031-17207-6_1

form of LTE-Advanced and has now reached a maximum communication speed close to 1 Gbps. The evolution made possible through 5G and future 6G technology is expected to provide more services for industry and society, along with artificial intelligence (AI) and the Internet of Things (IoT), as well as further upgrading of the multimedia communication services with its technical features such as high speed, high capacity, low latency, and massive connectivity. As shown in Fig. 1.1, the mobile communication system has evolved technically every decade, while the services of mobile communications have changed greatly in cycles of approximately 20 years. Therefore, the "Third Wave" initiated by 5G is expected to become a larger wave through 5G evolution and sixth generation (6G) technology and will support industry and society in the 2030s (Nakamura 2020).

5G radio represents a major step in mobile network capabilities. To date, mobile networks have mainly provided connectivity for smartphones, tablets, and laptops for consumers. 5G will take the traditional mobile broadband to the extreme in terms of data rates, capacity, and availability. In addition, 5G will enable new services, including industrial Internet of Things (IoT) connectivity and critical communication. 5G targets are set very high with data rates up to 20 Gbps and capacity increases of up to 1000 times with flexible platforms for device connectivity, ultralow latency, and high reliability. It is expected that 5G can fundamentally impact all sections of society by improving efficiency, productivity, and safety. 4G networks were designed and developed mainly by telecom operators and vendors for the smartphone use case. Today, other parties, including different industries and communities, are taking a lot of interest in 5G networks; they want to understand 5G capabilities and take full advantage of 5G networks. 4G was about connecting people. 5G is about connecting everything (Holma et al. 2019).

Figure 1.2 shows the requirements for wireless technology to be actualized by 6G through 5G evolution. In addition to the higher requirements of 5G, new requirements that were not considered in 5G have been added, and they have been expanded

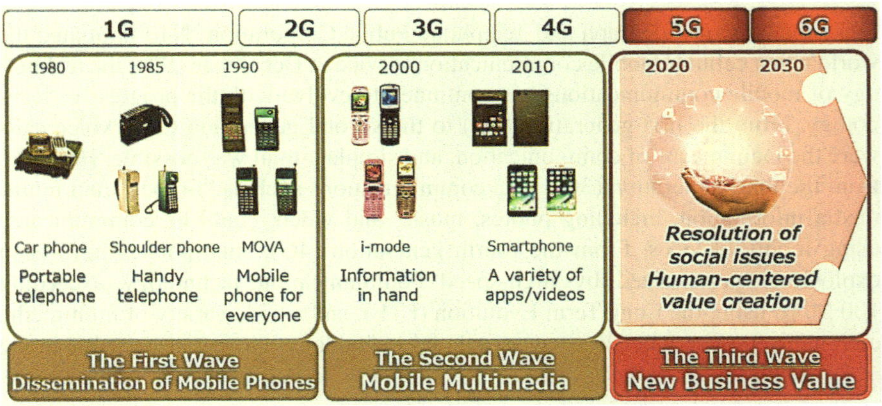

Fig. 1.1 Evolution of technologies and services in mobile communications. (Adapted from Nakamura 2020 with permission from IEEE)

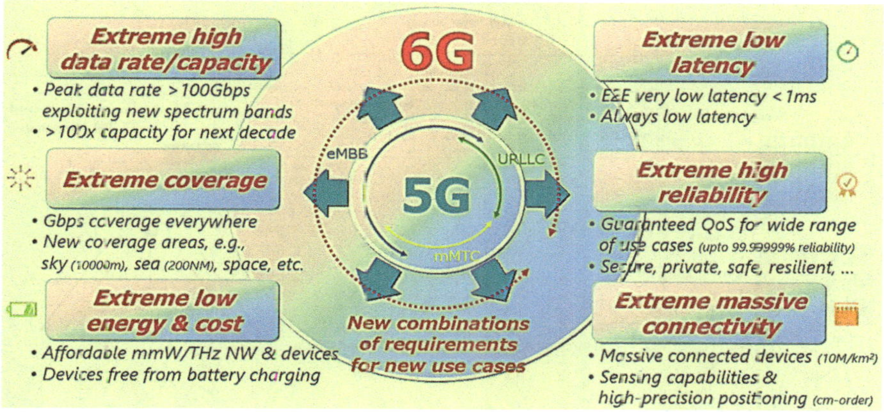

Fig. 1.2 Requirements for 6G wireless technology. (Adapted from Nakamura 2020 with permission from IEEE)

more widely. Moreover, as with 5G, not all requirements need to be met at the same time, but new combinations of requirements will be required for future new use cases (Nakamura 2020).

1.2 5G Technology Components

The 5G wireless communications are a converged system with multiple radio access technology components integrated. The main technology components are shown in Fig. 1.3 (Hao et al. 2020).

1.2.1 5G Spectrum

Compared with existing 4G, 5G supports high frequencies and spectra. 5G mobile radio technology is designed to operate on any frequency band between 400 MHz and 90 GHz. The low bands are needed for coverage, and the high bands are needed for high data rates and capacity. The initial 5G deployments use Time Division Duplex (TDD) between 2.5 and 5.0 GHz, Frequency Division Duplex (FDD) below 2.7 GHz, and TDD at millimeter wave at 24–39 GHz. The three main spectrum options are illustrated in Fig. 1.4. The millimeter wave spectrum above 20 GHz can provide a wide bandwidth up to 1–2 GHz, which ramps up the data rate to a very high 5–20 Gbps for extreme mobile broadband capacity (Holma et al. 2019).

Millimeter waves are mainly suitable for short-range transmission compared with traditional bandwidths. The greater the attenuation that describes the

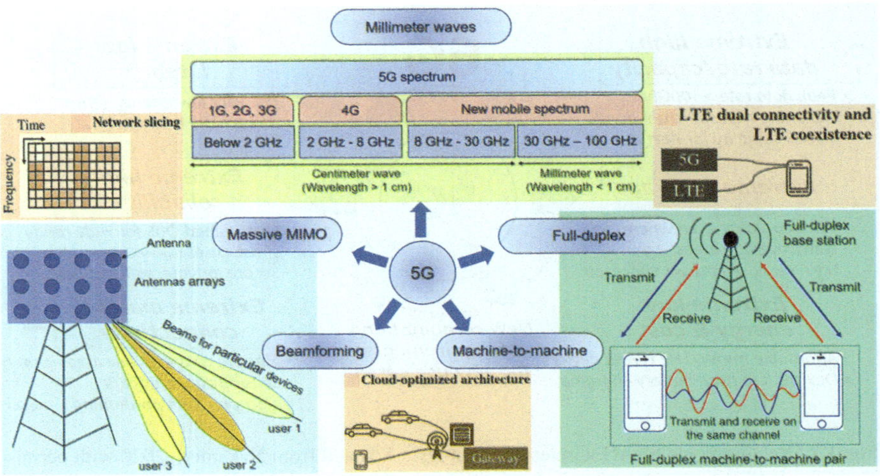

Fig. 1.3 Key 5G technology components including millimeter waves, network slicing, massive MIMO, Beamforming, full-duplex, and machine-to-machine cloud-optimized architecture applied in 5G wireless communications making it smarter, faster, and more efficient. (Modified from Hao et al. 2020 with permission from Walter de Gruyter and Company)

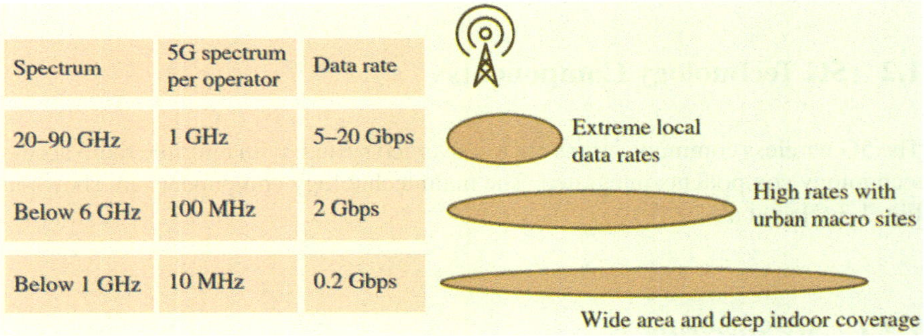

Fig. 1.4 Utilization of all spectrum options by 5G. (Adapted from Holma et al. 2019 with permission from John Wiley and Sons)

amplitude of a signal decays over distance, the shorter the waves travel. Furthermore, millimeter wave signals exhibit reduced diffraction and more specular propagation than their microwave counterparts; hence, they are much more susceptible to blockages. Millimeter waves are more easily obstructed by the walls of buildings, trees, and other foliage and even inclement weather. The signal of mobile terminals in 5G wireless communications is easily interfered with by surrounding metal devices. Macro base stations spaced many miles apart are sufficient for signal transmission in 4G wireless communications; however, in 5G wireless communications, the range of base station towers is far lower, and the size of the base station can be reduced in 5G wireless communications. Considering the short travel distance for

5G, macro base stations still connect in the frequency range of less than 6 GHz, and numerous miniature base stations composed of nanodevices that operate at millimeter wave frequencies are distributed and connected. The 5G wireless communications no longer depend on the construction of large-scale base stations, as 3G and 4G did, but instead use many miniature base stations to complement traditional cellular towers. Such highly densified miniature base stations provide an efficient solution to the short transition of millimeter waves (Hao et al. 2020).

Accordingly, 5G wireless communications can offer tremendous data capabilities, unrestricted call volumes, and infinite data broadcasts. There is a significant improvement in terms of signaling, management, and accounting procedures in 5G communications; therefore, they can accommodate needs from diverse applications outside the traditional mobile broadband category (Hu 2015).

1.2.2 Massive Multiple-Input Multiple-Output (MIMO) Antennas

Massive MIMO systems where macro base stations are equipped with antenna arrays can accurately concentrate transmitted energy to mobile devices. A massive MIMO system comprises an array of hundreds of antennas simultaneously serving multiple user terminals. Each single-antenna user in a massive MIMO system can scale down its transmit power proportional to the number of antennas at the macro base stations to achieve the same performance as a corresponding single-input single-output system. Massive MIMO has allowed multiple orders of magnitude improvements in energy efficiency, data speed, and capacity as well as enhanced link reliability and coverage. For example, numerical averaging over random terminal locations has shown that approximately 95% of terminals can receive a throughput of 21.2 Mb/s per terminal, whereas the array antennas offer a downlink throughput of approximately 20 Mb/s for 1000 terminals (Larsson et al. 2014; Hao et al. 2020).

Compared to a single antenna transmitting and receiving the signal in all directions for 4G, the massive MIMO in the 5G network enables energy radiation in the intended directions through beamforming and beamsteering, which can reduce intercell interference. A directional beam can reduce power consumption as all radio frequency signals are targeted toward a receiving unit instead of being scattered in all directions. The directional beam is obtained using an array of antennas allowing the beam to be guided through a combination of constructive and destructive interference and to focus the signal on a specific device. As there are arrays of multiple antennas embedded in multiple dispersed base stations, a larger number of individual antennas are required for the 5G network. Moreover, the performance of massive MIMO systems is generally less sensitive to the propagation environment than in point-to-point MIMO. Beamforming can help massive MIMO arrays make more efficient use of the spectrum around them with reduced latency. The reduced latency, high data speed, ultrareliability, and massive connectivity of 5G wireless communications can drastically change human lives and enable a new generation of applications, services, and business opportunities. Other advanced technologies include beamforming, which focuses on a wireless signal in a specific direction rather than broadcasting to a wide area; full duplex, which is able to transmit and receive data at the same time; and machine-to-machine, which enables the handling of billions of nodes and can substantially increase spectral efficiency and network coverage (Jang et al. 2016; Hao et al. 2020).

However, the adoption of millimeter waves and massive MIMO antenna arrays for beamsteering unavoidably engenders substantial challenges for antenna systems. Conventional antennas in portable devices, such as those found in 4G terminals, are not suitable for millimeter waves. Antennas for 5G wireless communications are easily affected by surrounding components such as batteries and shielding cases when they are integrated into a real terminal such as a mobile phone. The size of antennas used for 5G is down to micrometers and even nanometers at frequencies from the low band to the high band, and thus, very large numbers of antennas can conceivably fit into portable devices. The antennas cannot be fabricated simply by reducing the size of classical metallic antennas down to nanometers because the low mobility of electrons in nanoscale metallic structures and the high resonant frequencies of small antennas result in large channel attenuation and difficulty in implementing transceivers at such a high frequency. The use of traditional metallic materials for nanoantennas to implement wireless communications has become impossible. Identifying the best materials for nanoantennas to be applied in 5G is a challenge. For instance, the efficiency and bandwidth of an individual nanoantenna are a function of its dielectric constant, so nanoantenna materials for the 5G network require a lower dielectric constant. Since each nanoantenna element acts as both a transmitter and a receiver, it is critical to isolate the elements from each other to prevent the leaking of the transmitted signal from one element into the receiving portion of an adjacent element. Nanoantenna materials with the correct properties are ideal for reducing this crosstalk and for eliminating reflections from other parts of the device that can interfere with the desired signal (Hao et al. 2020).

1.2.3 Network Slicing

Physical and protocol layers in 5G need flexible designs to support different use cases and different frequency bands and to maximize the energy and spectral efficiency. Network slicing will create virtual network segments for the different services within the same 5G network. This slicing capability allows operators to support different use cases and enterprise customers without having to build dedicated networks.

More specifically, 5G networks are designed to support very diverse and extreme requirements for latency, throughput, capacity, and availability. Network slicing offers a solution to meet the requirements of all use cases in a common network

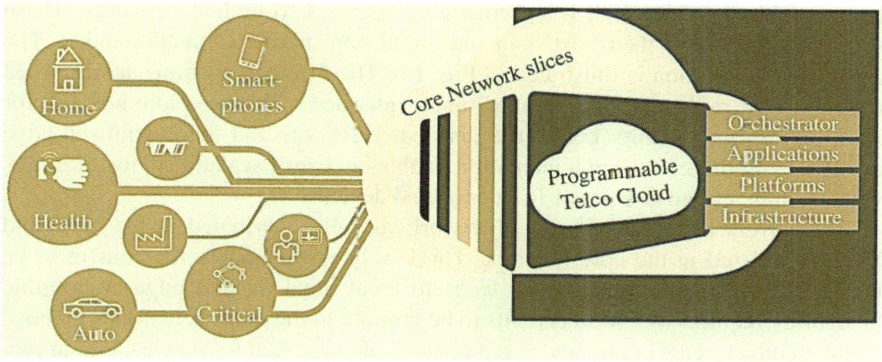

Fig. 1.5 Network slicing concept. (Adapted from Holma 2019 with permission from John Wiley and Sons)

infrastructure. The concept of network slicing is illustrated in Fig. 1.5. The same network infrastructure can support, for example, smartphones, tablets, virtual reality connections, personal health devices, critical remote control, or automotive connectivity (Holma 2019).

1.2.4 Dual Connectivity and Long Term Evolution (LTE) Coexistence

5G is the first radio solution that is closely integrated with the legacy radio network for a smooth rollout and seamless experience. The solution is called dual connectivity, where a 5G UE (user equipment) can have simultaneous connections to a 5G radio network and an LTE radio network. 5G can be deployed as a stand-alone system, but more typically, 5G will be deployed together with LTE in the early phase. A 5G device can have simultaneous radio connections to 5G and to LTE. Dual connectivity can make the introduction of 5G simpler, increase the user data rate, and improve reliability. 5G is also designed for LTE coexistence, which makes spectrum sharing feasible and simplifies spectrum refarming (Holma 2019).

1.2.5 Support for Cloud Implementation and Edge Computing

The radio network architecture has typically been distributed, where all radio processing is performed close to the antenna in the base station. The core network architecture has been highly centralized, with only a small number of core sites. Future architecture will be different: radio processing will become more centralized for better scalability, and core processing will become more distributed for lower latency. This evolution will bring edge cloud servers to mobile networks. These server locations can then host both radio and core network functionalities. The architecture evolution is illustrated in Fig. 1.6. The 5G specifications are designed to support the radio cloud by defining a new interface within the radio network for splitting the functionality between a distributed RF site and the centralized edge cloud site. Radio cloud implementation enables network scalability, for example, when adding a large number of IoT-connected devices.

The current architecture in LTE networks is fully distributed in the radio and fully centralized in the core network. The low latency requires the content to be brought close to the radio, which leads to local breakout and edge computing. Scalability requires the cloud benefits to be brought to the radio networks with edge cloud architecture. 5G radio and core networks are specified for native cloud implementation, including new interfaces inside the radio network (Holma 2019).

Edge computing refers to the practice of moving computing resources closer to the edge of a network. Mobile or multiaccess edge computing can be thought of as the hosting of a miniature data center much closer to the user, perhaps in the central office of a telco provider or even on the premises in the case of sophisticated manufacturing or enterprise use cases. Edge computing is widely expected to bring large benefits to data and processing-intensive applications that benefit from very low latency, such as virtual and augmented reality, robotics control, and other industrial uses. Applications enabled by edge computing are expected to benefit producers in the manufacturing sector through digital twin technologies (Dong et al. 2019).

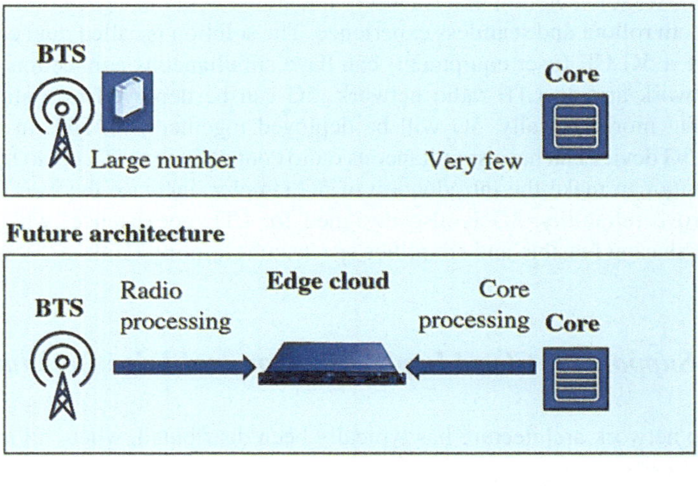

Fig. 1.6 Network architecture evolution. (Adapted from Holma 2019 with permission from John Wiley and Sons)

1.3 Materials Solutions for 5G Hardware System Integration

Hardware systems that operate in 5G are classified as User Equipment (UE), Customer Premise Equipment (CPE), and Base Station or infrastructure. They have varying needs, sizes, and power constraints. UE is driven by miniaturization and reduced power, while infrastructure equipment is geared toward high gain, communication range, massive MIMO needs, and broadband. The technology metrics for these classes are compared in Table 1.1 (Watanabe et al. 2021).

The landscape of base transmit systems (BTSs) has changed as they evolve from early 2/3G systems into existing 4G LTE and now 5G. Supporting 5G involves reallocation of the existing spectrum along with allocation of the new spectrum at frequencies below 6 GHz as well as the spectrum in the millimeter-wave (mm-wave) frequency region. The RF hardware supporting 5G BTSs also has quite different requirements compared to earlier generations. The adoption of higher-order active antenna systems (AASs) composed of many (32, 64, and higher) transmit and receive paths connected to an array of antenna elements is typical in 5G, whereas earlier 4G deployments focused largely in the range of two to eight transmit paths

Table 1.1 Technology metrics for 5G communication systems (Watanabe et al. 2021)

Metrics	Downlink (Base station)	Uplink (CPE)	User equipment (UE)
Antenna and module size	70 × 70 × 2.7 mm^3	450–1400 mm^2 Substrate thickness: 1.5 mm	20 × 5 × 2 mm^3 (Qualcomm, QTM052)
Number of antennas	64–256	16–32	4–8
PA power	33 dBm	19 dBm	10–15 dBm (6–8 dBm usually) 28 nm CMOS DC power for four elements: 360–380 mW Power consumption per channel <100 mW
Antenna gain	27 dBi EIRP[a] of 50 dBm (IBM)	18–21 dBi for 8 × 4 array patch antenna with grounded rings	20 dBi for 2 × 2 antenna array
End-to-end loss	–	–	2.5 dB
Pathloss	135 dB	135 dB	–
Received power	−75 dBm	−90 dBm	−140 dBm[b]
SNR per RX element	5 dB	−15 dB	6.2 dB[c]
Rx gain	21 dBi	27 dBi	~10 dB
Rx SNR after gain	26 dB	12 dB	–

[a]Effective isotropic radiated power (EIRP)
[b]Minimum reference signal received power (RSRP)
[c]Minimum performance requirements for 16-QAM modulation format and 0.48 code rate for Rank 1 (beamforming), defined by 3GPP

per sector. It is envisioned that the antennas will be connected to RF front-end units that support from 4 to 24 or more antennas each and ideally provide a power amplifier (PA), low noise amplifier (LNA), switch, phase shifter, and variable gain amplifier (VGA) for each antenna. A number of these RF front-end components will then connect with RF transceivers, responsible for signal generation, modulation, and demodulation. Therefore, 5G has placed significantly new and different requirements on RF front-end electronics and is driving industry to bring new technologies and solutions to realize RF transmit amplification and receiver (Rx) hardware functionality (Asbeck et al. 2019; Westberg et al. 2019).

Many characteristic benefits promised by 5G will operate at high frequency (above 26 GHz), i.e., mmWave 5G. Such high frequency requires new materials and different device designs. On the one hand, high frequency leads to more significant transmission loss, which offers opportunities for low-loss materials with a small dielectric constant and small tan loss. Advanced packaging designs also aim to reduce signal loss by integrating passive components into the whole package. On the other hand, high frequency requires high power to drive and will generate more heat. Power amplifiers with higher power density and higher gain will be essential, as well as thermal management. Figure 1.7 shows common requirements for 5G materials and hardware design (Jiang et al. 2020). Compared to earlier generations, AASs and 5G are driving both opportunities and challenges, including semiconductor technologies, levels of integration and functionality, packaging, cost, and size, and other RF BTS front-end electronics.

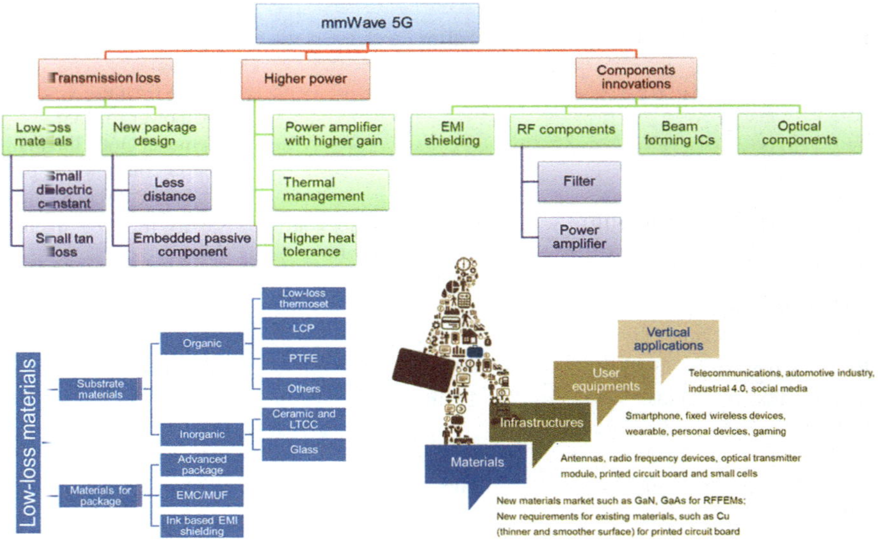

Fig. 1.7 Common requirements for 5G materials and hardware design. (Modified from Jiang et al. 2020. Credit: IDTechEX)

1.3.1 Evolution of the Cellular Base Station and Its Construction Materials

The evolution of BTSs is depicted in Fig. 1.8, which can be divided into three separate stages. First, Fig. 1.8a is representative of previous BTS deployments, where the transceiver station is located on the ground. The RF signal is then routed from ground level up to the masthead with high-level power amplification driving antenna elements. This architecture still plays a large role in many of today's infrastructure systems. However, a lot has changed as shown in Fig. 1.8b, a representation is more in line with many of the systems being deployed today and during the past couple of years. The base station gateway remains located on the ground, and it contains channel cards. The transition to RF, however, is now completed at the top of the mast in the remote radio heads, with antennas located just above the radio heads. This solution is far more modular and has significant benefits for the power efficiency of the system compared to earlier approaches. It also makes the architecture more modular so that network providers have more flexibility to adapt to their specific needs and requirements (Westberg et al. 2019).

The transition to real 5G is shown in Fig. 1.8c, where AASs come on top of the remote radio heads. Here, the full RF electronics, consisting of RF power amplification, Rx functionality, filtering, and so on, are all integrated into an AAS. This opens exciting possibilities for such system architectures as mMIMO, which will enable much higher data rates. In addition, as shown in Fig. 1.8c, a large increase in small-cell deployments at both sub6-GHz and mm-wave frequencies will increase density and allow operators to serve even more users. Another use case is fixed wireless access. With 5G, this is moving toward a phased array to enable point-to-point

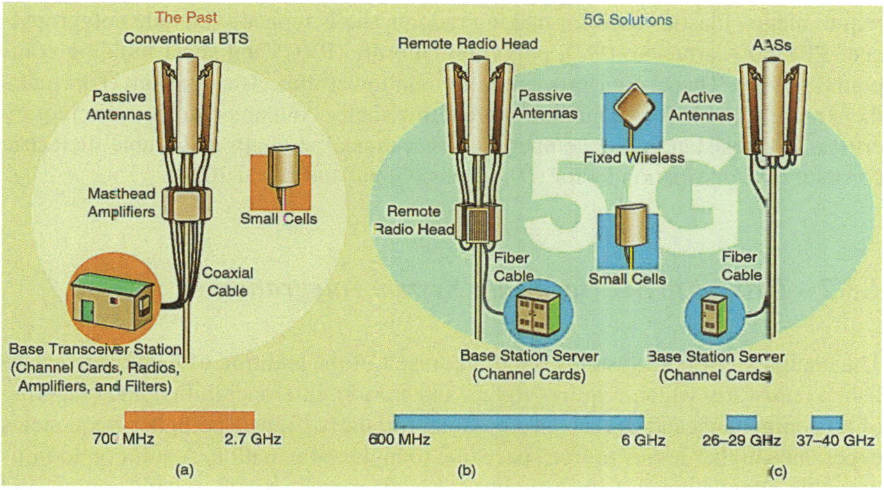

Fig. 1.8 The evolution of the cellular base station (Adapted from Westberg et al. 2019 with permission from IEEE): (**a**) Conventional BTS, (**b**) remote radio head, and (**c**) AASs

solutions. Furthermore, the frequency range also increases. Sub6 GHz is moving from 2.7 to approximately 6 GHz. The entrance of mm-wave frequencies at 28 GHz and near 40 GHz is also coming online (Westberg et al. 2019).

The base station that uses radio waves to relay communications between the mobile device and the core network is commonly mounted outside a building. Previously, such outdoor base stations were made of metal. Usually, it is made of polycarbonate, which undergoes various degradation reactions when exposed to sunlight and must therefore be light stabilized. With Tinuvin® 360 as an additive to the polycarbonate material, the outdoor 5G base stations can withstand weathering and degradation by intense sunlight, thus maintaining stable service with an extended life span. Tinuvin® 360 is a very low volatile ultraviolet light absorber (UVA) of the hydroxyphenyl benzotriazole class that imparts outstanding light stability to a variety of polymers and will withstand weathering and degradation by intense sunlight, thus maintaining stable service with an extended life span. In addition, low penetration loss in 5G base stations is managed using special brick material consisting of coal residue, phosphogypsum, desulfurization gypsum, quartz sand and clay (Gopal and Desai 2020).

A base station in the wireless world is a device that connects other wireless devices to a central hub. It is a wireless receiver and short-range transceiver that consists of an antenna and analog-to-digital converters (ADCs) to convert the RF signals into digital and back signals again. The 5G base station will have beamforming massive multiple-input, multiple-output (MIMO) antennas – an array of antennas that can focus and steer multiple beams simultaneously to different targets on the ground, such as a cell phone, using the millimeter wave spectrum. Antenna radomes are protective enclosures, protecting the antenna from winds, rain or ice. The key performance targets for radome materials are low permittivity and low loss tangent at GHz frequencies, while also satisfying cost reduction and processing requirements. Plastics used for making radome shells typically include polypropylene (PP), polycarbonate (PC), polyvinyl chloride (PVC) and sheet molding compound (SMC). Their better processability and lower dielectric constant (Dk) make them more suitable for manufacturing radome shells. Antenna radomes also require critical UV resistance, low-temperature mechanical strength and stable dielectric constant (D_k)/dissipation factor (D_f) values (Gopal and Desai 2020).

1.3.2 Drivers to 5G Hardware System Integration

The complexity of 5G systems is a direct result of the addition of features to established hardware, which requires tighter integration, interoperability, and shielding of all components and devices in a module. Similarly, signals at higher frequencies experience higher losses in free space due to higher attenuation. A solution to mitigate this attenuation is to use highly focused, narrow beams (beamforming and MIMO) to communicate with the target nodes in an efficient manner. This is accomplished by using higher-order antenna arrays that are driven by multiple active components. MIMO beamforming techniques are primarily classified into three categories: digital, analog, and hybrid. In analog beamforming, amplitude and phase variation are controlled as modulated signals at the transmit end, while the received signals from each antenna are summed before the analog-to-digital (ADC) conversion. In digital beamforming, the amplitude/phase variation is controlled as demodulated signals at the transmit end while received analog signals are processed after the ADC. The hybrid beamforming technique takes advantage of both analog and digital beamforming types, which include high flexibility, channel configuration, and low cost and power consumption. The base station can thus serve multiple users in a timeslot (Watanabe et al. 2021).

Analog beamforming is preferred in mobile systems because a single RF chain can address the antenna elements, while beam control is performed with passive elements. Digital beamforming consumes higher power with complex hardware but can have full beam control with several simultaneous beams and multiple apertures for future ultrawide-band communication and software-defined radio features for smart and dynamic spectrum hopping and band-selection. In such an innovative digital beamforming approach, the antenna outputs are encoded and multiplexed and digitized with a single ADC, while demultiplexing and beamforming are performed in the digital domain. A novel onsite coding approach that combines frequency and code multiplexing is developed to reduce the power requirements of the digitizer by 10–32 times. To attain in-band duplex operation to effectively utilize the current-congested spectrum allocation, systems employ time (TDD) or frequency division duplexing (FDD), where the interference between transmit and receive signals is suppressed by differentiating them in time or frequency. However, both require doubling the time or frequency resources. This limitation is addressed with simultaneous transmit and receive (STAR) systems to better utilize the 5G spectrum. In-band full-duplex operation is dependent upon the cancellation of interference between the strong transmitter (Tx) and receiver (Rx) using techniques such as cross-polarization array elements, filtering in the analog processing to reduce Tx harmonics, and digital signal processing to filter out multipath domains (Venkatakrishnan et al. 2018).

The need for low losses demands close integration between antennas with high-power PA (power amplifier) and low-noise-figure LNA. As the antenna element size and pitch are comparable to the wavelength, antenna-in-package (AiP) solutions make this approach feasible with mm-wave communications. Integrating an antenna into a packaging substrate reduces the overall interconnect length between the RFIC and antennas and, therefore, mitigates the feed-line loss and enhances the antenna efficiency. In conjunction with the prior popularity of AiP in the 60-GHz band designed for radar, the trend of AiP for 28-GHz or 39-GHz has been more prominent for consumer electronics (Gu et al. 2019).

The choice and integration of transceiver (TRx) technologies vary with the number of antenna elements and antenna gain to attain the required signal power. The radio-frequency front-end (RFFE), control and calibration circuits are integrated into the CMOS, SiGe BiCMOS, or GaAs transceiver dies and are connected to the antennas, passive components, and power circuitries using routing layers in the package, microbumps, and through package vias (TPVs). A lower number of elements utilizes GaN, while higher antenna elements are driven by SiGe BiCMOS and much higher antenna elements with bulk CMOS. Highly integrated wireless transceivers further benefit from deep submicron to 28 nm technology to achieve single-chip CMOS solutions. CMOS and SiGe PAs are optimized for 25 dBm power and 30% power efficiency, while GaAs achieves 30–35 dBm and GaN provides capabilities beyond 45 dBm. A 64-element array needs up to 5 W per PA to reach 240 W. Each PA must operate in the linear regime to reduce the error vector magnitude, which is more challenging for 64-array antennas. As the efficiency of PA modules with the required 8-dB back-off is estimated as 40%, the increased power dissipation from the combined analog and digital processing requires integrated thermal management.

1.3.3 Materials and Electronic Components for 5G Packaging Technology

The electronic packaging of 5G systems requires the integration of RF, analog and digital functions along with passives and other system components in a single module. These systems ideally exemplify the heterogeneous integration trend. This becomes more important for 5G for several reasons (Watanabe et al. 2021): (a) integration of antennas with transceiver ICS and associated passive and RF power divider networks; (b) addition of sub6 GHz (FR1) in the short term with advances in packaging technologies; (c) new mm-wave bands (FR2) drive the integration of new filters and diplexers along with broadband power amplifiers and switches; and (d) the add-on modules to the existing RFFE put additional emphasis on miniaturization and component integration. The proximity of the transceiver and front-end module is also important to reduce the size and losses. This is achieved by integration of antennas with the RF module as well as simultaneous modeling of a heat dissipation solution to keep active components in acceptable thermal conditions. The integration of power amplifiers with antenna arrays needs to address the challenges with size, cost, and performance. These challenges translate to multilayer fabrication with fine-line features and precise layer-to-layer registration, advanced low-loss materials to reduce conductive losses and cosimulation of circuit, device, package, and thermal solutions. The emerging 3D package integration solutions also underscore the need for isolation between the various circuit blocks. Because of the deployment of such high-power amplifiers and a large antenna array in millions of base stations, cost needs to be addressed for high-volume manufacturing.

1.3.3.1 Packaging Requirements for 5G Systems

System-level packaging in mm-wave technology is partitioned into baseband modules and antenna-integrated transceiver modules. In such packages, interconnections between ICs and other elements, such as antennas, passive components, and PCBs, must satisfy several requirements. One of the most critical requirements is impedance control, especially in the analog domain. In mm-wave antenna-in-package (AiP) solutions, interconnections between transceiver ICs and antennas should result in low insertion loss and acceptable return loss over the frequency range of interest. The other key requirement is the form factor related to interconnect techniques. As shown in Fig. 1.9, two conventional interconnect techniques are wire-bonding and flip-chip interconnections, whereas the emerging technique is referred to as IC embedding or fan-out packaging. Although flip-chip and fan-out interconnections were originally developed for high-performance computing (HPC) or mobile processor applications, their interconnection attributes, such as fine pitch and low electrical parasitics, are becoming critical in RF/mm-wave packages, such as baseband modules and antenna-integrated modules (Watanabe et al. 2021).

Despite the maturity and cost effectiveness in the packaging industry, interconnection using the wire-bonding technique has been identified as one of the key

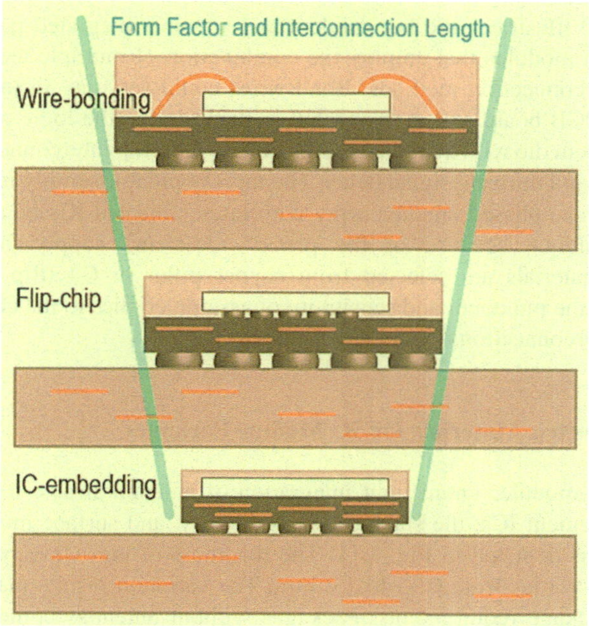

Fig. 1.9 Trends of interconnection assembly methods in RF/mm-wave technology for miniaturization. Wire-bonding as the most traditional low-I/O interconnection, flip-chip interconnection down to 25 μm, and IC-embedding or fan-out interconnection. (Adapted from Watanabe et al. 2021 with permission from IEEE)

challenges in system-on-package (SoP) because of the significant signal loss and impedance discontinuity caused by the bond wire, which degrades the performance of the RF/mm-wave system chain. Interconnection using the flip-chip technique offers better performance than wire bonding, as the bump height is smaller than the length of the bond wires. The flip-chip technique is also preferable for a small form factor, which provides more than 800 inputs/outputs (I/O). The flip-chip technique started from solder bumps with a diameter range of 75–200 μm, while copper-pillar interconnections with solder caps currently reach diameters less than 40 μm. The copper-pillar technique not only provides high-density I/O but also offers lower conductive loss. The evolution of IC-embedding or fan-out packaging has been imperative. Embedded wafer-level ball grid array (eWLB) and fan-out wafer-level packaging (FOWLP) are drawing attention as mm-wave packages. This embedding technology eliminates the use of wire bonding, which not only introduces significant high-frequency loss and parasitics but also increases the footprint for high-pin count dies. Transceiver ICs are embedded in a reconfigured molded wafer with a compression molding process. Multiple redistribution layers (RDLs) are formed to fan-out the baseband signals, and through-mold vias are utilized for vertical interconnections. In conjunction with molding-compound-based waver-level IC-embedding technology, laminate-based and glass-based panel-level embedding technology is also emerging as an alternative candidate (Ravichandran et al. 2019; Che et al. 2020).

Figure 1.10 illustrates the key modem and antenna-integrated packages. Most cases are such modules that employ the combination of multiple techniques from the three interconnection methods. Bond wires could serve as an interconnection between the PCB board and memory that is stacked atop the logic or modem die, while the modem die with a high pin count entails flip-chip interconnects to provide a clean signal and mitigate signal delay. The most popular assembly method to integrate a mm-wave phased antenna array with interconnected ICs is, as of now, the flip-chip technique because of the process cost and supply-chain maturity. Conductive materials are selected from copper pillar or C4 (flip chip) bumps, depending on the pin count and sensitivity of assembled dies to the conductive loss caused by interconnections (Watanabe et al. 2021).

1.3.3.2 Dielectric Materials for 5G Module Packages

5G mm-wave modules entail tight integration of antenna arrays, transceiver ICs, power-management ICs, the stack of logic-memory, and surface mounted passive components, as depicted in Fig. 1.11. The thickness of mm-wave packaging substrates usually varies from 0.15 to 1.2 mm. The variation results mainly from the antenna-in-package requirements. Packages without antennas could employ thin substrates below 300 μm. However, antenna-integrated modules require higher thickness, as antenna arrays offer higher bandwidth with a thicker substrate due to more separation of the ground plane from antenna patches. The thickness of all modules (packaging-substrate thickness plus mold height) also varies from 0.5 to

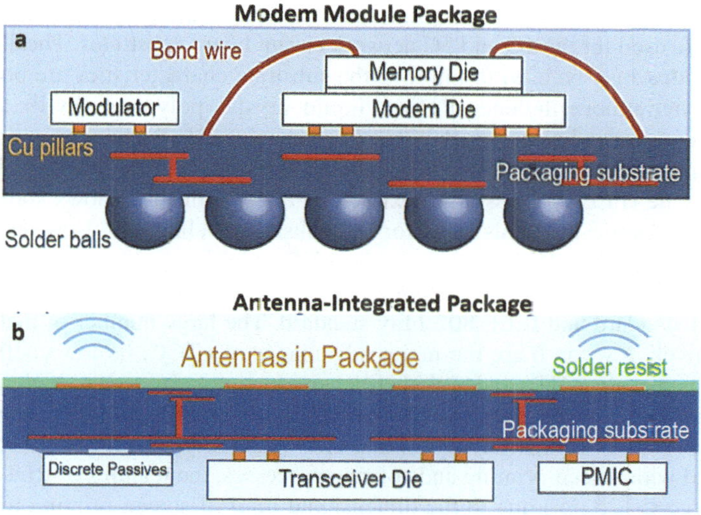

Fig. 1.10 Examples of system-integrated mm-wave packages (Adapted from Watanabe et al. 2021 with permission from IEEE): (**a**) modem SoP; (**b**) antenna-integrated package

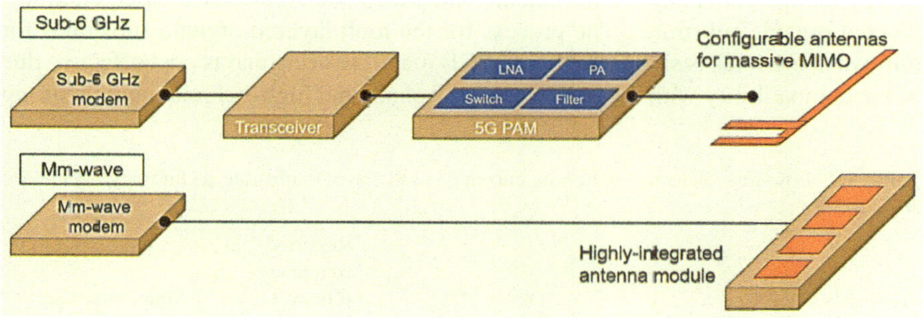

Fig. 1.11 An example of tight integration in RFFE architectures for 5G electronics. (Adapted from Watanabe et al. 2021 with permission from IEEE)

2 mm. In addition to the thickness requirement for the antenna, the system components and interconnects (Fig. 1.11) determine the total thickness and area of the entire package (Watanabe et al. 2021).

The selection of dielectric materials is critical to obtain the desired electrical performance of ICs and antenna arrays integrated in 5G module packages. Well-designed dielectric materials provide an improved link budget, low signal dissipation, high signal or power density, desired frequency response of passive components, small footprint of elements, lower module thickness, high antenna efficiency and EIRP, beam width, angular coverage, energy consumption, and miniaturization of antennas. The substrate materials should have very low dielectric constant (D_k), dissipation factor (D_f) and moisture absorption values at low material and manufacturing costs. Alkali-free glass fiber base cloth and polytetrafluoroethylene (PTFE) are used for the 4G and 5G network circuit board substrates. The alternative materials that may be used to improve the substrate characteristics are polyimides (PI), polytetrafluoroethylene (PTFE), liquid crystal polymers (LCP) and low-temperature co-fired ceramics (LTCC). Further materials that can be used in substrates are modified epoxy, phenolic resins, Polyphenol Ether (PPE), Polyether Ether Ketone (PEEK), glass and flexible ceramics. Table 1.2 shows some typical low-loss dielectric materials used for 5G substrate technology. Low-temperature cofired ceramic (LTCC) has been employed for high-frequency antenna modules such as WiGig, alternatively known as 60 GHz WiFi, which includes the IEEE 802.11ad standard and IEEE 802.11ay standard. The large number of metal layers and relatively low tan δ are the major advantages of LTCC. Its low coefficient of thermal expansion (CTE) and high thermal conductivity also lead to high reliability and robustness to temperature variations. The major challenges are the low density of signal routings in internal layers and size scalability. As LTCC substrates are fabricated with screen-printing and cofiring processes, the features of RDL are generally large (>125 μm) due to the dimensional limit of screen-printing masks and alignment accuracy between layers. This large feature size results in a low density of signal routing and entails a high number of metal layers, resulting in modules (Gopal and Desai 2020; Watanabe et al. 2021).

The mainstream platform for antenna integrated packages is still based on multi-layered organic substrates. The process for the multilayered organic substrates for mm-wave modules is similar to that of PCB manufacturing and is cost-effective due to the compatibility with the existing supply chain and high-volume manufacturing

Table 1.2 Low-loss dielectric materials currently used for 5G substrate technology (Watanabe et al. 2021)

Materials		Dk	Df	Reported frequency (GHz)	Major suppliers
Organic	Bismaleimide Triazine (BT)	3.4	0.004–0.005	10	Mitsubishi
	Polyphenylethers (PPE)	3.25–3.4	0.002–0.005	1–50	Panasonic, Risho Kogyo
	Liquid-crystal Polymer (LCP)	2.9	0.0025	10	Rogers, Murata
	Polytetrafluoro-ethylene (PTFE)	2.2	0.0009	10	Rogers, DuPont
Inorganic	Low-temperature cofired ceramic (LTCC)	6	0.0018	60	Hitachi Metals, Kyocera, TDK
	Borosilicate glass	5.4	0.005–0.009	10–60	AGC, Corning, Schott, 3DGS, NSG
	Fused silica	3.8	0.0003–0.0004	10–60	AGC, Corning, Schott, 3DGS, NSG

for consumer electronics. The organic materials are designed to show lower tan δ than the traditional FR4, where tan δ is higher than 0.02 in the mm-wave frequency bands. The mainstream copper-clad laminates (CCL) and prepregs used for the formation of multilayered organic substrates comprise four classes of polymers (Watanabe et al. 2021): (1) bismaleimide triazine (BT); (2) polyphenylene ether (PPE); (3) liquid-crystal polymer (LCP); and (4) polytetrafluoroethylene (PTFE), as listed in Table 1.2. Unlike glass-cloth epoxy resin (e.g., FR4), glass-cloth PPE substrates feature a high glass-transition temperature (Tg), low water absorption, and low dielectric constant (both Dk and Df). PPE-based substrates are typified by MEGTRON6 (core and prepreg) from Panasonic and CS-3376C from Risho Kogyo. Similar to PPE resin, the lamination of the BT-based core with build-up dielectrics forms multilayered organic substrates. BT-based laminates provide low CTE, low shrinkage, and high peel strength with copper. LCP and PTFE are gaining attention because of their low tan δ values in the mm-wave frequency range. Unlike PPE and BT, these materials require high-temperature and high-pressure processes for lamination and thermal compression.

Halogenated substrates are not usually preferred for handset applications due to environmental and safety concerns. The mechanical and process limitations of organic laminates due to their low modulus and high CTE are addressed with emerging inorganic substrates, as listed in Table 1.2.

Glass substrates offer a wide range of D_k (3.7–8) and D_f (0.0003 for fused silica to 0.006 for alkaline-free borosilicate), smooth surface, good dimensional stability (<2 μm for 20-mm substrates), large-panel scalability, ability to form fine pitch through vias, stability to temperature and humidity, and tailorable CTE depending on the packaged components. However, glass-based packaging still poses challenges associated with process immaturity and higher cost resulting from lack of supply chain readiness and the nature of glass, such as brittleness or robustness and difficulty in handling (Hayashi et al. 2019).

In conjunction with the substrate materials, the formation of RDL and microvias becomes more critical to form high-density interconnects (<10 μm) and to enable tight integration of components. Dry-film dielectric materials show good compatibility with panel-level packages with double-side RDLs and offer a wide thickness variation (>5 μm). Low-loss or low tan δ materials are developed in the industry for mm-wave packages, as shown in Table 1.3. Low-loss dry films, which are typically epoxy-based, usually contain silica or ceramic fillers to lower the dissipation factor (tan δ). The fillers increase Young's modulus, decrease CTE, and lead to technical challenges such as the narrow process window, poor adhesion to substrate materials, and reliability in harsh environments. The other method to form dielectrics is the use of liquid-based dielectrics, such as polyimides. As opposed to dry films, liquid-based dielectric materials are more compatible with wafer processes or build-up layers on one side of the substrate. Nonfiller photosensitive or photoimageable low-loss liquid-based dielectric materials designed for 5G mm-wave applications are emerging to form fine features below 5 μm. Low-loss liquid-based dielectric materials from major suppliers are listed in Table 1.4. These materials are, however, not

Table 1.3 Low-loss build-up dry films for high-density signal routings used in high-frequency circuits (Watanabe et al. 2021)

Material supplier	Dk	Df	Applied frequency	CTE (ppm/K)	Tg (°C)
Ajinomoto	3.3	0.0044	5.8 GHz	20	153
DOW	2.57	0.0032	1 MHz	63	250
Hitachi Chemical	3.3	0.0034	5 GHz	17	233
Sekisui	3.3	0.0037	5.8 GHz	27	183
Taiyo Ink	3.3	0.0025–0.003	5–60 GHz	20	160

Table 1.4 Low-loss photosensitive (Photo-imagable) dielectrics for build-up dry films for high-density signal routings used in high-frequency circuits (Watanabe et al. 2021)

Material supplier	Dk	Df	Applied frequency	Minimum line width/ space (L/S)	Elongation (%)
DOW	2.65	0.0008	<20 GHz	18 μm	8
Hitachi Chemical	2.4	0.0018	10 GHz	–	–
Sekisui	2.6	0.0048	<40 GHz	8 μm	>50
Taiyo Ink	2.9	0.0030	1 GHz	30 μm	–

still in the high volume manufacturing phase. The other challenge is that these materials are spin-coated or slit-coated and have difficulty in forming films with thicknesses greater than 40 μm (Tomikawa et al. 2018).

1.3.3.3 Microwave Circuit Design and Materials

Ideally, devices, components, and interconnects should maintain stable performance under various signals, powers, and thermomechanical loads. The packaging structure affects not only the signal and power distribution but also the electrical performance of each component or device. This demands accurate and reliable multiphysics modeling and analyses from electrical, materials, chemical and mechanical standpoints (Geyik et al. 2019; Sivapurapu et al. 2019).

5G mm-wave technology entails the codesign of phased antenna arrays with transceiver or beamforming ICs to meet the power distribution and signal integrity requirements and passive components to support all functions with minimal parasitics and interference. The codesign should also consider electromagnetic compatibility and thermal design (Gu et al. 2019).

For instance, the integration of multiple components into a package requires more complicated signal routing and power delivery networks in a limited space to attain the required signal and power integrity and the control of electromagnetic interference (EMI). The dielectric thickness for signal routing and power distribution can be 15 μm and lower. As in lower frequency applications in the RF/microwave bands, control of electromagnetic interference (EMI) is also critical in mmWave bands. Traditional shielding techniques are comparatively less effective due to their smaller wavelengths. EMI control in the mmWave range is critical and hence the need for absorbers. Absorbers are available in various forms, including polyurethane foams, silicone or urethane elastomers, and a range of thermoset forms. Polyurethane foam may be chosen based on cost or low weight. An elastomer may be chosen for material toughness or other environmental properties as desired. A thermoset material may be chosen for a complex shaped part, which is best made using injection molding techniques (Gopal and Desai 2020; Watanabe et al. 2021).

Microwave circuits such as oscillator modules and dielectric resonator modules are made up of ultra/superconducting materials and ceramic materials that offer improved frequency, stability, supply current and voltage, operating temperature, etc. Electronic components in 5G communications require improved signal transmission and device miniaturization. Ferrite materials and advanced spin-ferrite materials (based on ferric oxide, bismuth tetroxide, yttrium oxide, cupric oxide, calcium carbonate, gadolinium (III) oxide, indium (III) oxide, vanadium (V) oxide and manganese (III) oxide) are used to improve signal transmission and device miniaturization. Wireless charging in 5G communication can be achieved using iron-based micro/nano magnetic powder material (combination of iron, silicon, boron niobium, molybdenum, nickel and copper). Isolators used in 5G communication are made up of magnetic materials. For 3–6 GHz, Triplate or Stripline (below the resonance values of yttrium iron garnet-based materials) and for mmWaves, Microstrip or SIW are used (High Magnetization Ni-Zn Ferrite and Li-based Spinels). Dielectric substrates used in filters are based on materials such as liquid crystal material, ceramic materials and Roger Corp's Rogers5880. These are used in manufacturing filters such as cascade high-pass filters, low-pass filters, and bandpass filters (Gopal and Desai 2020).

Chip material usually comprises gallium nitride wafers. For instance, power amplifiers for 5G communications are made up of semiconductor gallium nitride and polybutylene PBT-package power amplifier for 28 GHz. Other materials, such as silicon-germanium (SiGe) power amplifiers, also operate at higher frequencies, albeit at lower power rates than gallium-based devices. Indium phosphide (InP) offers an alternative to gallium arsenide (GaAs) in providing increased performance at higher frequencies while consuming less power. Among the circuit materials that provide desirable characteristics for 5G amplifiers, Rogers' RO4385™ circuit laminates are low-cost circuit materials that maintain consistent performance across wide temperature ranges. The materials have a dielectric constant (Dk) of 3.48 in the z-axis at 10 GHz, tightly controlled within ±0.05. They are ideal for competitive applications and can be fabricated with standard epoxy/glass (FR-4) processes. RO3003™ laminates consist of PTFE with ceramic filler. They have a dielectric constant (Dk) of 3.0 in the z-axis at 10 GHz tightly controlled within ±0.04. They feature extremely low loss at higher frequencies that helps obtain the maximum gain from the active devices in an amplifier circuit, even at the various millimeter-wave bands expected to serve the many backhaul links of future 5G wireless networks (Deng et al. 2019; Gopal and Desai 2020).

1.3.3.4 Thermal Conductors and Thermal Management for 5G

In conjunction with the traditional material requirements for dielectrics in IC packaging, such as compatible CTE and Young's modulus, the new class of requirements, especially for AiP, leads to the development of low-loss dielectric materials. Low-loss materials are translated to materials with a low dissipation factor or loss tangent (tan δ). Low tan δ mitigates dielectric loss in in-package interconnects, feedlines, and antennas and thus increases the antenna efficiency. The high losses in the power amplifiers make thermal management more challenging. In addition, there are studies on advanced RF conductors other than conventional copper, such as graphene, graphene composite, copper paste, artificial magnetic conductors, metasurface, and metaconductors. These materials provide lower resistivity, superior impedance control, or lower manufacturing cost than traditional copper patterning (Hwangbo et al. 2018).

The reduced distance and the increased number of component densities call for integrated heat-spreading structures from the thermal-management standpoint. 5G mm-wave technology is expected to consume more power than the previous generations of wireless technologies. Efforts to address thermal management challenges have been made mostly with electrical and thermal codesign and analysis (Gu et al. 2019; Watanabe et al. 2021).

1.3.3.5 Integration of Passive Components

Passive components play a key role in wireless system implementations, as there is a need to provide matching impedances for components such as PAs and LNAs, filtering, tuning and biasing. Passive components also make up functions such as couplers, baluns, power combiners and dividers, filters, phase shifters, circulators and isolators. Duplexers are usually paired with RF ICs for their nominal operation on the system level. It is typically estimated that passive components account for 90% of the component count, 80% of the size and 70% of the cost.

1.3.3.5.1 Discrete Lumped Circuits for sub6 GHz 5G Bands

The evolution of passive components started with low-temperature cofired ceramic (LTCC), which became very popular as surface-mount technology (SMT) components for resistors, inductors and capacitors. A comparison of five different substrate technologies for passive components in terms of performance, thickness, size and some other parameters is given in Fig. 1.12a. Moreover, a comparison of filter performance (Q-factor) vs. footprint is given in Fig. 1.12b. Acoustic wave technologies have surpassed the typical lumped and distributed LC networks in the sub6 GHz range in realizing high-performance filters, resonators, oscillators and delay lines and have found many applications in 4G and LTE networks.

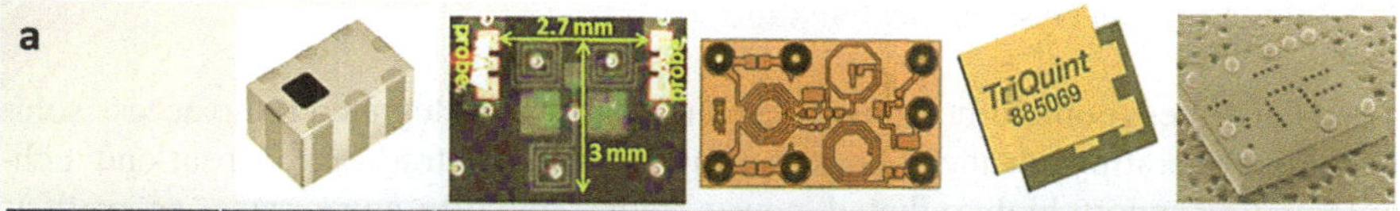

	LTCC	Laminate	Silicon	Accustic Wave	3D Glass
Performance	Low loss	Substrate loss	Substrate loss	High selectivity	Low loss
Thickness	Thick	Thin	Thin	Thin	Thin
Size	Small footprint	Moderate footprint	Moderate footprint	Small footprint	Small footprint
Density	Low density	Moderate density	Moderate density	High density	High density
Frequency	Low - high	Low - moderate	Low - moderate	Low	Low - high
Cost	Mass production	Less in demand	High	High	Not in market

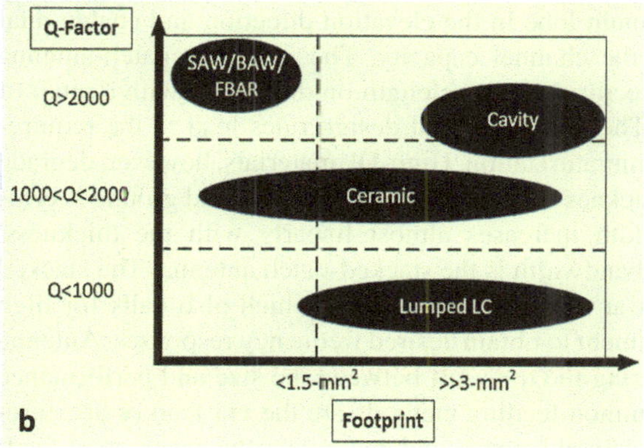

Fig. 1.12 (**a**) Evolution of passive component technologies; and (**b**) comparison of various filter technologies at 1 GHz in terms of performance (Q-factor) and footprint). (Adapted from Watanabe et al. 2021 with permission from IEEE)

1.3.3.5.2 Distributed Components for mm-Wave

Various planar passive components, such as filters, power dividers, couplers and baluns, have been designed and fabricated with their evolution in terms of reduction in size while improving performance metrics. Some of the transmission lines of choice are the microstrip, conductor-backed coplanar waveguide (CBCPW) and substrate-integrated waveguide (SIW). The technologies include LTCC, organic laminates, state-of-the-art integrated fan-out wafer-level packaging (InFO WLP) and ultrathin laminated glass for passive component realization (Matthews 2019).

1.3.3.6 Antenna Systems in Package

The 5G wireless communication system entails highly integrated radio access solutions, incorporating advanced phased-array antenna and transceiver front-end technology to support high radiated power, large signal-to-noise ratio, as well as beamforming, scanning in elevation and azimuth directions within a wide range. As the antenna element size and pitch are correlated with the wavelength, antenna-in-package (AiP) solutions become more feasible at mm-wave frequencies, unlike discrete antennas for 4G/LTE. Major package-integrated antenna structures are patch antennas and dipole antennas. A greater number of antenna elements (N) results in additional gain (10 log N). A 2×2 antenna array in Qualcomm QTM052 is reported to have a gain of 20 dBi, which comprises 5 dBi per element, 6 dBi from summation, 6 dBi from array formation and an additional 3 dBi gain from single polarization.

Patch antennas feature the main lobe in the elevation direction and enable dual polarization, which increases the channel capacity. The size of a patch-antenna array is approximately $\lambda/2$ (effective half wavelength on dielectric) with a pitch of $\lambda_0/2$ (half-wavelength in air). These fundamental design rules lead to the requirement of high-Dk materials for miniaturization. High-Dk materials, however, degrade the bandwidth and gain. The thickness or separation of the patch and ground is critical since the antenna bandwidth increases almost linearly with the thickness. Another technique to increase bandwidth is the stacked-patch antenna. The stacked patches create two resonances at different frequencies, which also calls for high accuracy of layer-to-layer alignment to obtain desired frequency responses. Antenna designs are performed considering the trade-off between the size and performance (i.e., bandwidth and gain). Common feeding methods are the via-feed or aperture-coupling feed. Via-feeding is relatively simple and guarantees the accuracy of feeding points into the antenna patch. However, the challenge is the limited bandwidth because of the inductive reactance caused by the via. Conformal by shielding is employed to emulate the coaxial feed to reduce the undesired inductance and enable better impedance matching. In contrast, the aperture-coupled feed leads to a higher bandwidth than the via feed, while the layer-to-layer alignment is extremely critical; a few-micron shift between layers causes the frequency shift and may result in not covering the targeted frequency bands.

Dipole antennas are employed to cover the azimuth direction with a single polarization. Since the length of the dipole is nearly $\lambda_0/2$, the design process is relatively simple. Dipole antennas generally offer wide bandwidths compared to patch antennas. The gain is controlled by placing an additional number of directors in a modified antenna topology referred to as the Yagi-Uda antenna. The dipole antenna and Yagi-Uda antenna are fed by two differential transmission lines to reverse the phases of the provided signals. Coplanar-waveguide and stripline signal routing are more common in multilayered packages because of their inherent shielding features and minimal cross-talk with other nearby transmission lines, components, and antennas.

There are several options to implement mm-wave antenna arrays, as shown in Fig. 1.13. The first option refers to implementing antennas directly on the PCB board (Fig. 1.13a). The most notable advantage is the lower cost than the other

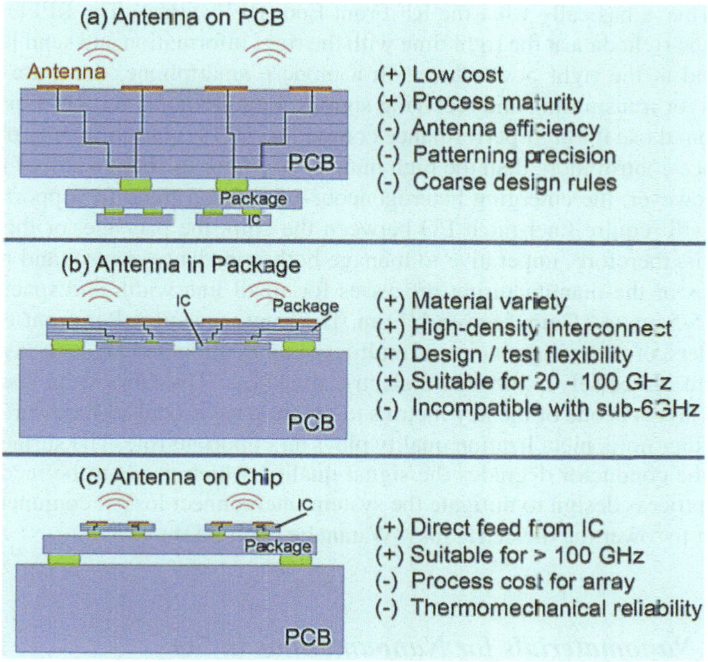

Fig. 1.13 Three approaches for mm-wave antenna implementation (Adapted from Watanabe et al. 2021 with permission from IEEE): (**a**) Antennas on PCB; (**b**) antennas in package; and (**c**) antennas on chip or wafer

options because of the supply chain maturity. The major challenges of antennas on PCBs are the manufacturing process and tolerances. The relatively coarse design rules do not allow designers to layout fine structures in transmission-line widths, spaces, via diameters and pitch, and accurate layer-to-layer alignment. Figure 1.13b illustrates the most viable option for 5G mm-wave applications, the implementation of AiP. In contrast, Fig. 1.13c shows the direct antenna implementation onto the IC wafer. Although the antenna-on-chip approach offers the lowest feedline loss and parasitics and provides direct integration with other front-end circuitry, the challenges include antenna efficiency, process scalability for large arrays, yield and cost, thermomechanical reliability issues, and design flexibility.

1.3.3.7 High-Precision Patterning in Heterogeneous Package Integration for 5G

In current mobile devices, there is a sophisticated circuitry that is responsible for converting information from the near-zero frequency baseband signals used to convey information and data to radio-signals that can be received or transmitted over the air. This is basically what the RF Front End (RFFE) does. The RFFE needs to process the right data at the right time with the right information and send it with the right band at the right power level. In a modern smartphone, there are multiple antennas for transmitting and receiving signals. Requirements for RFFE packaging differ from those for high-performance computing. RFFE conventionally prioritizes impedance control more than the high number of I/O or the bandwidth of interconnects. However, the emerging heterogeneous-integration trend to support massive MIMO will require finer-pitch I/O between the chip, the package, or the antenna array. It is, therefore, imperative to manage both impedance control and minimize tolerances of the manufacturing processes for small line width and spacing (L/S) down to 5/5 μm to 2/2 μm or even 1/1 μm. The emergence of AiP in front-end packages underscores the importance of multilayer fabrication and layer-to-layer alignment with a precisely controlled dielectric thickness. The fabrication accuracy of multiple layers is one of the key metrics to obtain good model-to-hardware correlation. Furthermore, metallization quality plays an important role. The surface roughness of the conductor degrades the signal quality, which needs to be incorporated into the process design to mitigate the system interconnect loss in conjunction with the effort to lower the dielectric loss (Watanabe et al. 2021).

1.3.4 Nanomaterials for Nanoantennas in 5G

5G wireless communications require antennas with a greater capacity, wider wireless spectrum utilization, high gain, and steer ability due to the cramped spectrum utilization in the previous generation. Conventional antennas are unable to serve the new high frequency due to limitations in fabrication and installation, mainly in smaller sizes. Metallic nanoparticles are frequently used to conduct inks used for antennas. These particles, due to their high surface area, can interact with atmospheric water or oxygen, causing the antenna to oxidize and degrade more rapidly than bulk metals. The use of carbon-related nanomaterials such as graphene and CNTs has promising antennas with smaller sizes and thinner dimensions that are capable of emitting high frequencies because the available bandwidth is inversely proportional to the antenna size. Nanoantennas almost two orders of magnitude below the dimensions of current on-chip antennas are appropriate for 5G (Scidà et al. 2018; Hao et al. 2020).

As the total consecutive available bandwidth for millimeter waves is not enough to support the terabit-per-second data rate, terahertz band communications are envisioned as a key wireless technology to satisfy the future demands within 5G and beyond. As shown in Fig. 1.14, graphene, CNTs, metallic nanomaterials, and metamaterials can be used for millimeter wave and even terahertz band frequencies. The electromagnetic properties, namely, the permittivity and permeability of graphene, CNTs, metallic nanomaterials, and metamaterials, are tunable and highly dependent on their structures. The properties of nanoantennas include radiation efficiency indicating the efficiency of antennas in performing the conversion from electric signals

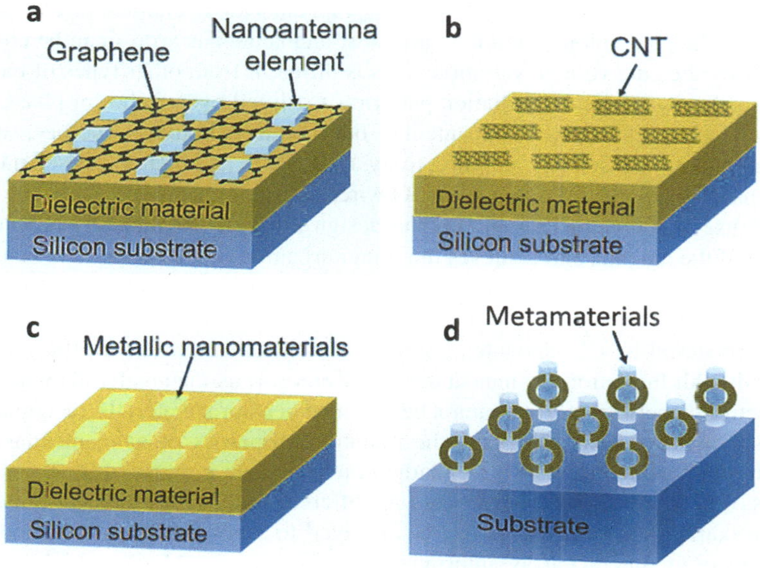

Fig. 1.14 Schematic diagram of (**a**) graphene-based nanoantennas, (**b**) CNT-based nanoantennas, (**c**) metallic nanomaterial-based nanoantennas, and (**d**) metamaterial-based nanoantennas. (Modified from Hao et al. 2020 with permission from Walter de Gruyter and Company)

to electromagnetic waves, radiation directivity indicating the concentration of radiation in the direction of maximum radiation, and radiation pattern representing the antenna radiation as a function of direction. By comparison, nanomaterial-based nanoantennas have the following characteristics (Hao et al. 2020):

1. As the conductivity of graphene can be tuned via chemical potential by means of chemical doping and electrostatic bias voltage, the radiation properties of graphene-based nanoantennas can be tuned via chemical potential. Compared to graphene, CNT-based nanoantennas have less tunability with more plasmonic loss due to the curvature effect. However, the radiation properties of metallic nanoantennas cannot be easily tuned, and SPP waves on such nanoantennas exhibit large ohmic losses.
2. The graphene-based nanoantennas have higher radiation efficiency than CNT-based and copper-based nanoantennas. Moreover, the radiation efficiency of a single CNT as a nanoantenna is lower than that of copper-based nanoantennas. The low radiation efficiency of copper-based nanoantennas is correlated with the low conductivity of copper, and the low radiation efficiency of a single CNT as nanoantennas is due to the large reactance resulting from classical and quantum effects of the nanometer radius.
3. The graphene-based nanoantennas have higher directivity than CNT-based nanoantennas, and the directivity of CNT dipole antennas is higher than that of copper nanoantennas in the low terahertz band frequency range.

4. The gain of nanoantennas patterned on graphene can be enabled or disabled by tuning the gate voltage, and the pattern of nanoantennas arrays can be changed only by the gate voltage variation. This is different from other types of nanoantenna arrays, where the radiation pattern is modified by switches or phase shifters. The graphene-based nanoantennas have better radiation properties, such as high radiation efficiency and directivity, than CNTs and metallic-based nanoantennas, showing great potential in 5G wireless communications.
5. The use of metamaterials in antenna design not only dramatically reduces the size of the antenna and achieves the miniaturization of antenna size but can also improve nanoantenna performance, such as enhancing bandwidth, increasing gain, and generating multiband frequencies of antenna operation. The metamaterial-based nanoantennas can overcome the restrictive efficiency and bandwidth limitation for nanoantennas. Moreover, as metamaterials with novel electromagnetic properties cannot be obtained in natural materials, metamaterial-based nanoantennas can make the radiation properties of nanoantennas more controllable and promising. Depending on the design purpose of the nanoantennas, the metamaterials can be used as different functions of the nanoantennas. For example, metamaterials can be arranged to surround the nanoantenna elements of an antenna array, improving the antenna gain. Metamaterials can also be used as a superstrate placed above the radiation surface, increasing the obtained bandwidth of the nanoantennas.

In addition, it is important to control the energy consumption of nanomaterials in nanoantennas. This is because the heat caused by the electromagnetic loss of nanoantennas can result in an environment with an elevated temperature degrading the performance of nanoantennas. For example, the imaginary part of the permittivity that determines the dielectric loss part of the electromagnetic loss is not constant and highly dependent on the temperature; the imaginary part of the permittivity increases with an increase in the temperature, resulting in an increase in the dielectric loss. The high-temperature absorption nanocomposites can be used in nanoantennas to reduce energy consumption. For example, CNTs and graphene can be incorporated with ceramics such as SiC and SiO_2 because of their high strength, low oxidation, high thermal stability, and thermal conductivities at higher temperatures. Metallic and ceramic nanoparticles can be deposited on the surface of CNTs and graphene (Hao et al. 2020).

1.4 Challenges in 5G and Beyond – 6G

6G is expected to utilize even higher frequencies and channel bandwidths than 5G, primarily by operating in the terahertz (THz) gap frequency range, resulting in massive data rates of 100 Gbps – 1 Tbps, compared to 10–20 Gbps of 5G. If 5G is known for its revolutionary flexibility, 6G will likely be known for using artificial intelligence (AI) by capitalizing on the backbone of flexibility offered by 5G. The other use cases are also expected to significantly advance with the advent of 6G: massive URLLC to provide true microsecond latency and mMTC to connect hundreds of billion devices. A capability factor of 10–100 is expected for 6G compared to 5G. The key technologies to enable 6G are AI, advanced RF, optical and network technologies. As a result, pervasive AI and machine learning will dominantly create smart, self-configurable mesh networks to support scalability with minimal human interaction. With THz radios, smart sensors and IoT driven by requirements such as HD artificial/virtual reality (AR/VR), extended reality (XR), local processing (wireless AI) and others, it would be imperative to identify potential new technologies to realize it, support the evolution of use cases and identify new ones to provide a ubiquitous, ultralow latency, high-fidelity network regardless of the nature of the environment (Saad et al. 2019).

Many of the traditional approaches used for RF packages are not applicable to sub-THz and THz radios. Sub-THz or 6G communications focus on heterogeneous package integration, as illustrated in Fig. 1.15, by incrementally advancing the system components such as precision antenna arrays, low-loss interconnects and waveguides and active devices. As the frequency increases into the THz region, it will not workable to employ the techniques used for low frequencies because of multiple challenges associated with signal loss, dimensions, and materials. Current interconnection techniques, such as wire bonding or solder-based bumps, are too bulky; therefore, excitation by multimodes, radiation, and reflection adversely affects the electrical performance. As the skin depth is approximately 120 nm at 300 GHz, higher frequency packaging will require shorter interconnects and smooth surfaces to mitigate conductor loss. Radiation loss must be minimized as well with good impedance matching and high-precision manufacturing with tolerance below 1 μm. Dielectric loss is more predominant in sub-THz or THz bands, which translates to the need for accurate characterization of potential materials at those high frequencies. Another key metric in THz applications is antenna implementation. While the

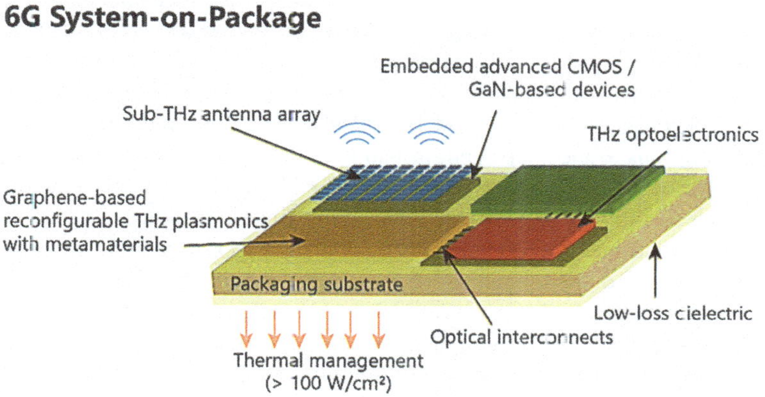

Fig. 1.15 Conceptual diagram of heterogeneously-integrated quasi-optical THz package for 6G communications. (Adapted from Watanabe et al. 2021 with permission from IEEE)

AiP approach is viable, antennas or radiators on ICs will play a significant role. The integrated antenna may include graphene-based radiation sources and detectors, THz antenna arrays (e.g., 1024 elements in 1 mm^2 at 1 THz), lenses, and intelligent metasurfaces with space-time-frequency coding and low-loss interconnects. Figure 1.15 shows a conceptual diagram of a heterogeneously integrated quasi-optical THz 6G package for a device with an antenna, where sub-THz or THz waves will be radiated into the air from the reconfigurable AiP or the on-chip antenna. Technical challenges such as thermal management and EMI shielding must also be addressed for such high-frequency systems (Song 2017; Watanabe et al. 2021).

1.5 Outlook and Future Perspectives

Millimeter wave and THz-enabled subsystems will dominate future communication networks for broadband wireless mobile connectivity, latency-sensitive applications such as vehicle-to-vehicle and vehicle-to-network, and massive connectivity needs in the IoT. They will also create mm-wave imaging, sensing, and other applications that simultaneously demand high bandwidth, reliability and zero perceived latency. The associated product segments are classified into base stations, customer premise equipment and user equipment. Despite the wide range of power (<1–30 W), communication range (few meters to hundreds of meters) and bandwidth requirements, underlying technologies have several similarities in package architectures, materials and processes. The key challenges with mm-wave communications, such as large path losses, interconnect losses, power-hungry devices, and other technology limits, have been systematically addressed with advances in III-V devices, IC-package codesign, innovative beamforming architectures, and test methodologies, making 5G systems a reality today (Bennis et al. 2018).

Unlike with 4G, antenna arrays need to be integrated into 3D packages with smaller sizes, making antenna integration the mainstream 5G package architecture. Multilayer organic (MLO) packages with transceiver dies flip-chip-attached to the backside of the package and antenna arrays on the top side, known as the AiP phased-array system, are widely adapted in today's UE and base station products. Ultralow transmission losses, below 0.1 dB/mm, fine lines and spaces with 1–5% process variations to meet stringent in-band loss and out-of-band rejection specifications have been the key packaging focus for 5G modules. Various passive elements, such as filters, power dividers and other functions, are embedded into the packages as distributed components. Organic packages, however, are limited in line-width control due to dimensional instability, requiring more layers with low process precision and high via-transition losses, resulting in thicker packages. This has been paving the way for inorganic substrates such as glass as an alternative to meet the density and impedance continuities across the packages (Westberg et al. 2019).

THz or 6G communications are emerging as the next frontier with 100–1000X increase in network speed compared to 4G and LTE, leading to new applications based on virtual reality and intense high-resolution video communications and connectivity of a much larger number of devices, such as sensors for IoT. Compact and high-output power sources and high-responsivity low-noise detectors that can operate in the THz band are required to surmount the high path loss at these frequencies. Key innovations, such as the integration of graphene-based radiation sources and detectors, THz antenna arrays, lenses and intelligent metasurfaces with space-time-frequency coding and low-loss interconnects, will create the next set of research and development challenges for academia and industry (Watanabe et al. 2021).

References

Asbeck PM, Rostomyan N, Özen M, Rabet B, Jayamon JA (2019) Power amplifiers for mm-Wave 5G applications: technology comparisons and CMOS-SOI demonstration circuits. IEEE Trans Microw Theory Techn 67(7):3099–3199

Bennis M, Debbah M, Poor HV (2018) Ultrareliable and low-latency wireless communication: tail, risk, and scale. Proc IEEE 106(10):1834–1853

Che FX, Yamamoto K, Rao VS, Sekhar VN (2020) Panel warpage of fan-out panel-level packaging using RDL-first technology. IEEE Trans Compon Packag Manuf Technol 10:304–313

Deng R, Chen B, Li H, Zhang K, Zhang T, Yu Y, Song L (2019) MXene/Co3O4 composite material: stable synthesis and its enhanced broadband microwave absorption. Appl Surf Sci 488:921–930

Dong R, She C, Hardjawana W, Li Y, Vucetic B (2019) Deep learning for hybrid 5G services in mobile edge computing systems: learn from a digital twin. IEEE Trans Wirel Commun 18(10):4692–4707

Geyik CS, Mekonnen YS, Zhang Z, Aygn K (2019) Impact of use conditions on dielectric and conductor material models for high-speed package interconnects. IEEE Trans Compon Packag Manuf Technol 9(10):1942–1951

Gopal V, Desai F (2020) Materials used in 5G communication. https://www.patent-art.com/knowledge-center/materials-used-in-5g-communication-27/#:~:text=Materials%20generally%20used%20for%20manufacturing,techniques%20to%20impart%20desired%20performance

Gu X, Liu D, Baks C, Tageman O, Sadhu B, Hallin J, Rexberg L, Parida P, Kwark Y, Valdes-Garcia A (2019) Development, implementation, and characterization of a 64-element dual-polarized phased-array antenna module for 28-GHz high-speed data communications. IEEE Trans Microw Theory Techn 67(7):2975–2984

Hao H, Hui D, Lau D (2020) Material advancement in technological development for the 5G wireless communications. Nanotechnol Rev 9:683–699

Hayashi K, Kidera N, Sato Y (2019) Low-loss glass substrates formulated with a variety of dielectric characteristics for millimeter-wave applications. In: 2019 IEEE 69th electronic components and technology conference (ECTC), pp 712–717

Holma H (2019) Technology components. In: Holma H, Toskala A, Nakamura T (eds) 5 G technology: 3 GPP new radio. Wiley, New Jersey, pp 27–47

Holma H, Toskala A, Nakamura T, Uitto T (2019) Introduction. In: Holma H, Toskala A, Nakamura T (eds) 5 G technology: 3 GPP new radio. Wiley, New Jersey, pp 1–12

Hu F (2016) Opportunities in 5G networks: research and development perspectivea. CRC Press, Boca Raton

Hwangbo S, Rahimi A, Yoon Y (2018) Cu/co multilayer-based high signal integrity and low rf loss conductors for 5 g/millimeter wave applications. IEEE Trans Microw Theory Techn 66(8):3773–3780

Jang J, Chung M, Hwang SC, Lim Y-G, Yoon H-J, Oh TK et al (2016) Smart small cell with hybrid beamforming for 5G: theoretical feasibility and prototype results. IEEE Wirel Commun 23:124–131

Jiang L, Edmondson J, Ghaffarzadeh K (2020) 5G technology, market and forecasts 2020–2030: 5G technology and materials innovations, infrastructure, user equipment, vertical applications and NB-IoT. https://www.idtechex.com/ko/research-report/5g-technology-market-and-forecasts-2020-2030/753

Larsson EG, Edfors O, Tufvesson F, Marzetta TL (2014) Massive MIMO for next generation wireless systems. IEEE Commun Mag 52:186–195

Matthews P (2019) Approaching the 5G mmWave filter challenge – key specifications for mmWave filtering and available options. Microwave J. https://www.microwavejournal.com/articles/32228-approaching-the-5g-mmwave-filter-challenge

Nakamura T (2020) 5G evolution and 6G. In: 2020 international symposium on VLSI design, automation and test (VLSI-DAT), Hsinchu, Taiwan, pp 1–1. https://doi.org/10.1109/VLSI-DAT49148.2020.9196309

Ravichandran S, Yamada S, Ogawa T, Shi T, Liu F, Smet V, Sundaram V, Tummala R (2019) Design and demonstration of glass panel embedding for 3D system packages for heterogeneous integration applications. Journal of Microelectronics and Electronic Packaging 16(3):124–135

Saad W, Bennis M, Chen M (2019) A vision of 6G wireless systems: applications, trends, technologies, and open research problems. ArXiv abs/1902.10265

Scidà A, Haque S, Treossi E, Robinson A, Smerzi S, Ravesi S, et al. (2018) Application of graphene-based flexible antennas in consumer electronic devices. Mater Today 21:223–230

Sivapurapu S, Chen R, Mehta C, Zhou Y, Bellaredj MLF, Jia X, Kohl PA, Huang T, Sitaraman SK, Swaminathan M (2019) Multi-physics modeling and characterization of components on flexible substrates. IEEE Trans Compon Packag Manuf Technol 9(9):1730–1740

Song H (2017) Packages for terahertz electronics. Proc IEEE 105(6):1121–1138

Tomikawa M, Araki H, Kiuchi Y, Shimada A (2018) Low dispersion loss polyimides for high frequency applications. In: 2018 IEEE CPMT symposium Japan (ICSJ), pp 167–170

Venkatakrishnan SB, Alwan EA, Volakis JL (2018) Wideband RF self-interference cancellation circuit for phased array simultaneous transmit and receive systems. IEEE Access 6:3425–3432

Watanabe AO, Ali M, Sayeed SYB, Tummala RR, Pulugurtha MR (2021) A review of 5G front-end systems package integration. IEEE Trans Compon Packag Manuf Technol 11(1):118–133. https://doi.org/10.1109/TCPMT.2020.3041412

Westberg E, Staudinger J, Annes J, Shilimkar V (2019) 5G infrastructure RF solutions: challenges and opportunities. IEEE Microw Mag 20(12):51–58. https://doi.org/10.1109/MMM.2019.2941631

Chapter 2
Semiconductor Solutions for 5G

Abstract RF hardware for 5G user equipment will be increasingly integrated, with filter banks, high density switches, antenna tuning, LNAs and PAs integrated into RFFEs with system on chip (SoC) technologies. Silicon semiconductors are one of the most prominent technologies for implementing mmWave circuits in 5G. III-V semiconductors have the potential to revolutionize the electronics industry and enable a wide range of innovations, from 5G to IoT and smart vehicles. Compound semiconductors have diverse properties that will benefit these applications. Future RFFEs may use RF silicon on insulator (SOI), SiGe BiCMOS or RF CMOS SoCs, GaN or GaN-on-silicon or GaN-on-SiC, which integrate the PA, LNA, switches and control functions to operate mmWave phased array beamforming antenna systems. It is possible that future RF silicon technologies can be further integrated or combined with other technologies to include filtering and the digital hardware required to enable hybrid beamforming modules. Future variations of RF SOI or RF CMOS may even be integrated with more advanced digital hardware, such as FPGAs, memory and processors. Baseband processing and accessory DSP functions could be implemented in the package for compact 5G mmWave solutions. This chapter will give a brief review of various semiconductor materials and their technologies for 5G applications.

2.1 Evolution of 5G Semiconductor Technologies

A number of semiconductor technologies capable of operating at frequencies up to 90 GHz are currently available. Each has particular strengths and weaknesses with respect to various applications identified for millimeter wave transmission systems. First-generation products based on single function GaAs pHEMT MMIC technology facilitated the demonstration of high-performance communication links up to 86 GHz. Subsequent generations have seen the introduction of highly integrated silicon-based chips (SIGe, SIGEBICMOS, CMOS and in future FD-SOI) and the parallel development of multifunctional GaAs, GaN and other compound

C. Tong, *Advanced Materials and Components for 5G and Beyond*, Springer Series in Materials Science 327, https //doi.org/10.1007/978-3-031-17207-6_2

semiconductor chipsets offering similar levels of baseband to RF functionality. Future requirements for fronthaul and for backhaul of 5G will require capacities achievable only by using very large amounts of contiguous spectrum. This spectrum exists at frequencies above 90 GHz; therefore, the development of semiconductor technologies operating above 90 GHz and up to 300 GHz is becoming increasingly important (Rüddenklau et al. 2018).

Figure 2.1 illustrates the general semiconductor system in 5G. The RF hardware for 5G user equipment will be increasingly integrated, with filter banks, high density switches, antenna tuning, LNAs and PAs integrated into RFFEs with systems on chip (SoC) technologies. Foundry processes that are available and in production for RF analog components are III-V (primarily GaAs) and silicon (primarily SiGe/BiCMOS and CMOS, used in WiGig 60 GHz thus far). New processes under development include GaN, InP and GaAs mHEMT; however, none of these is currently available for commercial volume production. Future RFFEs may use RF silicon on insulator (SOI), SiGe BiCMOS or RF CMOS SoCs, GaN or GaN-on-silicon or GaN-on-SiC, which integrate the PA, LNA, switches and control functions to operate mmWave phased array beamforming antenna systems. It is possible that future RF silicon technologies can be further integrated or combined with other technologies to include filtering and the digital hardware required to enable hybrid beamforming modules. Future variations of RF SOI or RF CMOS may even be integrated with more advanced digital hardware, such as FPGAs, memory and processors. Baseband processing and accessory DSP functions could be implemented in the package for compact 5G mmWave solutions (Asbeck et al. 2019).

The millimeter-wave (mmWave) application domain has historically been dominated by III-V MMIC semiconductor technologies (primarily GaAs). These are ideally suited for the RF front ends of mmWave systems, such as power amplifiers and low noise amplifiers, as well as enabling oscillators with excellent phase noise characteristics (Fig. 2.1). Now, SiGe:C HBT-based technologies can be used for mmWave transceiver front end integration in small-cell applications with limited output power (usually intended to use V-band frequencies) but rather the quality factor of the on-chip passive devices, such as inductors, capacitors and transmission lines for matching and tuning and their accurate characterization in the mmWave frequency domain. E-band applications for macrocells usually require high-order QAM modulation (>128), which can be achieved with GaAs components or a combination of SiGe/BiCMOS transmitter/receiver and GaAs PAs, LNAs and VCOs.

CMOS implementation promises higher levels of integration at reduced cost if volumes scale to several million parts per year due to the higher speed of scaled technology. Some developments, especially chipsets available and targeting WiGig 60 GHz, have shown that CMOS can be used to operate at mmWave frequencies, as the CMOS transistor transition frequency (f_T) approaches 400 GHz. However, the performance for point-to-point links is worse compared to SiGe or GaAs components in terms of phase noise and noise figure for the same distances (e.g., >100 m). In parallel with the usual CMOS planar bulk technologies, fully completed SOI (FDSOI) technology is known to be very promising for providing high speed at low voltage. It is fully compliant with designs already available for bulk CMOS technologies guaranteeing successful IP porting. SiGe-BiCMOS combines

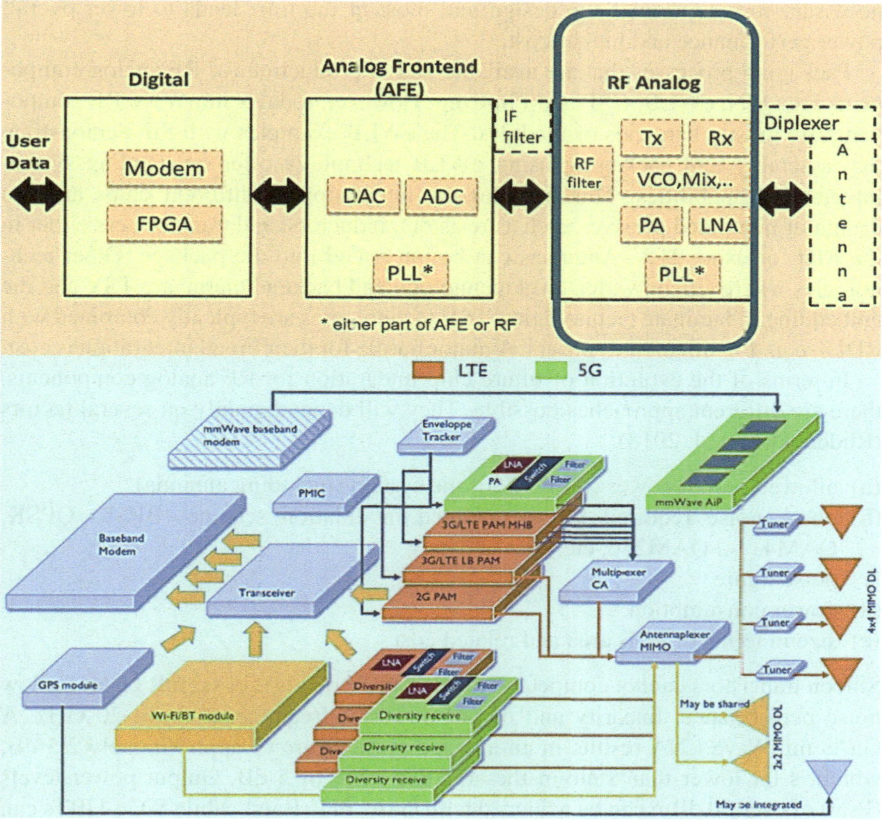

Fig. 2.1 Semiconductor system in 5G. (Modified from Rüddenklau et al. 2018 with open access (Hindawi), and Yole 2019, credit: https://www.3dincites.com/2019/03/the-5g-revolution-is-pushing-innovations-for-rf-front-end-sip/)

high-performance bipolar transistors and CMOS technology in a single chip. Bipolar transistors provide the high speed and gain that are critical for high-frequency analog sections, while CMOS technology is excellent for building simple, low-power logic gates. Solid know-how in design, architecture and process integration enables leading-edge SiGe BiCMOS technology that makes it possible to integrate the RF, analog and digital parts on a single chip. This drastically reduces the number of external components while optimizing power consumption. BiCMOS process technology today offers a level of performance that was previously only attainable with more expensive technologies such as gallium arsenide (GaAs). Compared to bulk CMOS, BiCMOS with silicon germanium (SiGe) heterojunction bipolar transistors (HBT) allows a much higher cutoff frequency at a given technology node together with a higher voltage capability. To reach similar frequencies, bulk CMOS designs have to use much smaller process nodes. This makes it necessary to compromise the design and most of the time leads to lower overall power performance and higher cost.

Packaging processes that are available and in production for RF analog components are QFN, eWLB, CSP and FlipChip. However, today's mmWave RF components are mostly bare dies or modules. The eWLB examples with SiP demonstrate that assembly and packaging using eWLB technology offer outstanding system integration capabilities. This includes the integration of different chips and the design of integrated passives such as resistors, inductors, and transformers either in the RDL or using TEV. Antennas can be integrated into the package. Other technologies where silicon wafer-level technology and backend merge are TSV and die embedding in laminate technologies. TSV technologies are typically combined with RDLs, e.g., for silicon interposers. A major hurdle for their broad integration is cost.

In terms of the evolution of future chip integration for RF analog components, there are different approaches possible. They will depend mainly on several factors (Rüddenklau et al. 2018):

(a) allowed output power and EIRP of the system (including antenna)
(b) phase noise required for the defined modulation scheme (BPSK, QPSK, QAM4, …, QAM256, etc.)
(c) noise figure
(d) power consumption
(e) size in terms of PCB area and related cost

Silicon transistors cannot compete with III-V compounds (GaAs, InP, GaN) for low noise performance, linearity and output power at frequencies above 20 GHz. A GaAs mmWave LNA results in an average noise figure of approximately 2.5 dB, which is far lower than state-of-the-art SiGe LNA of 5 dB. Output power levels (Psat) of over 30 dBm can be achieved with GaAs in E-Band, while SiGe-HBTs can reach 19 dBm (Psat). Silicon RF ICs do, however, allow the integration of multiple application-specific functionalities on a single silicon chip (RF ASIC) with excellent yield and uniformity plus the possibility of integrating the different calibration schemes required to take into account RF impairments (not possible or much more complex to implement in GaAs). The level of integration is a factor to be considered. A high level of integration makes the chip very specific and could increase the development time in a first design but reduces the production test and simplifies module assembly. A good compromise for high-end applications (e.g., E-band high power, QAM256) is to use compound semiconductors for the front ends (LNA of the receiver input and power amplifier of the transmit output) and silicon semiconductors for the lower frequency mixed signal functions and control/digital elements.

III-V compound devices can realize systems to expand the use of the electromagnetic spectrum above 90 GHz. However, improvements in the high-frequency capability of CMOS/BiCMOS technology have made it possible to consider it as a low-cost, low-performance alternative to III-V compound devices. A major obstacle to further investigation is the state of regulation for the use of frequencies >90 GHz for commercial backhaul/fronthaul applications (Rüddenklau et al. 2018).

2.2 Effect of CMOS Technology Scaling on Millimeter Wave Operations

The scaling theory developed by Mead and Dennard allows a "photocopy reduction" approach to feature size reduction in CMOS technology, and while the dimensions shrink, scaling theory causes the field strengths in the MOS transistor to remain the same across different process generations (Mead 1972; Dennard et al. 1974).

Transistor scaling is the primary factor in achieving high-performance microprocessors and memories. Each 30% reduction in CMOS IC technology node scaling has (1) reduced the gate delay by 30%, allowing an increase in maximum clock frequency of 43%; (2) doubled the device density; (3) reduced the parasitic capacitance by 30%; and (4) reduced energy and active power per transition by 65% and 50%, respectively (Borkar 1999).

High-frequency RF and mmWave communication applications are also drawing benefits of CMOS scaling because the transition frequency (f_T) of the MOSFET significantly increases with scaling. With a 28 nm CMOS process, f_T has been improved to 250 GHz. Having a large value of f_T is an important design requirement for building high frequency mmWave circuits. Another large advantage of scaling is the reduction in the noise figure. Almost all RF receiver front end circuits in communication systems employ a low noise amplifier (LNA) as a first block, and achieving a low noise figure is one of the most important design requirements of an LNA. It is evident that scaling of CMOS technology has made it possible to implement mmW circuits on silicon; however, there are numerous challenges associated with scaling (Juneja et al. 2021):

1. The operating frequency requirements of mobile applications have risen to >20 times, for example, 2.6 GHz in 4G to 28 GHz in 5G. However, the transition frequency of the MOSFET only increases by approximately 5 times, from 50 GHz in the 180 nm process to 250 GHz in the 28 nm process. Since the MOSFET speed does not increase very fast in comparison to the operating speed requirements, this presents several circuit design challenges.
2. The intrinsic gain of the MOSFET and its supply voltage decrease with scaling, resulting in a lower voltage gain in the amplifier circuit. The intrinsic gain, which is given by gm/gds, is 25 in the 180 nm process and 5 in the 45 nm process, where gds is the channel conductance. The effect of a lower supply voltage is significant in circuits such as LNAs that employ cascode topology to minimize reverse isolation.
3. The relatively slower transistor speed and lower supply voltage have put forward a need to use inductive elements in RF circuits to employ resonance. However, these passive inductive elements, either lumped inductors or distributed transmission lines, can occupy up to 50% of the chip area on silicon, which is a large pain area for designers. Scaling down the on-chip inductor size has been very slow because, as per electromagnetic theory, a large-size inductor is required to

obtain the desired inductance value as well as the desired quality factor. The on-chip inductor in comparison to its off-chip counterpart suffers from huge transmission, radiation, conduction and substrate losses, which leads to a poor quality factor that saturates at approximately 20.

To scale on-chip inductors while meeting the required performance parameters, such as the inductance value and quality factor, the following approaches have been proposed (Pacurar et al. 2012; Juneja et al. 2021):

- Different inductor geometries, including square, octagon and circular geometries, are shown in Fig. 2.2;
- Different layout optimization techniques, such as shielding, stacking, and nesting, are shown in Fig. 2.3a–c;
- Quality factor enhancement techniques;
- Loss reduction techniques;
- Use of a planar integrated waveguide as an alternative to the inductor and transmission line;
- Different materials, such as carbon nanotubes and multilayer graphene, are used to implement on-chip inductors.

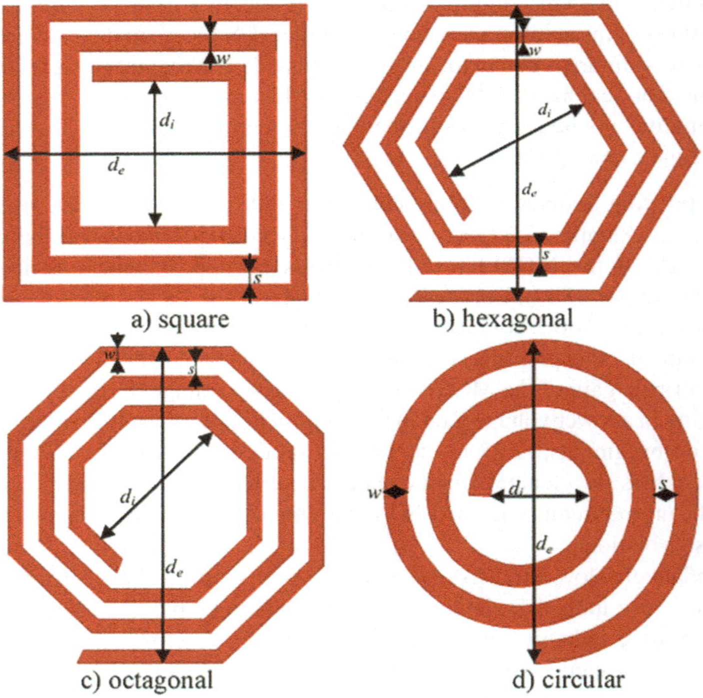

Fig. 2.2 Spiral inductors geometrical shapes used in CMOS. (Adapted from Pacurar et al. 2012 with permission from IEEE): (**a**) Square, (**b**) hexagonal, (**c**) octagon and (**d**) circular inductor

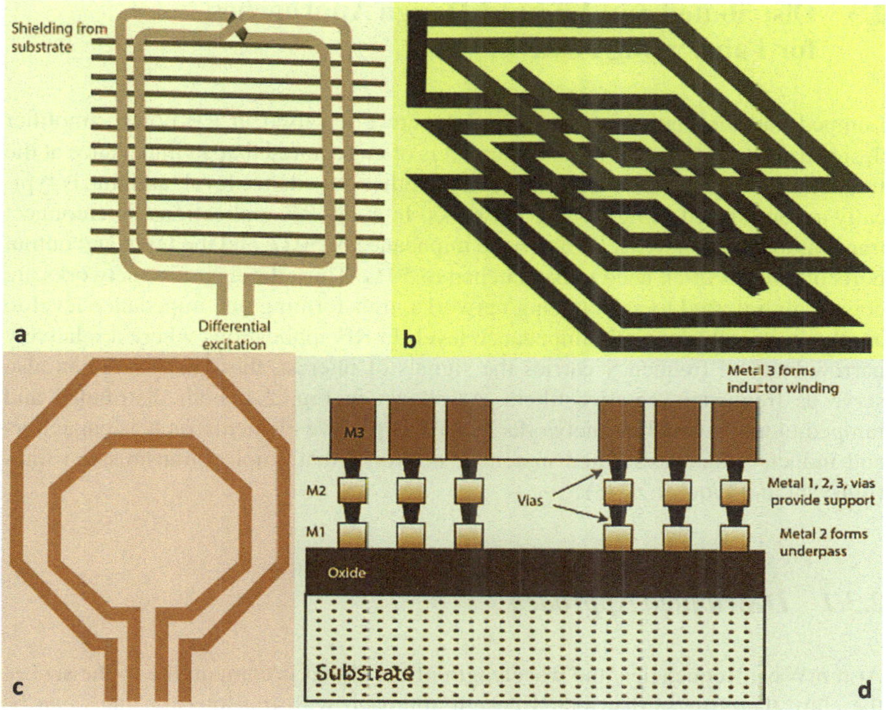

Fig. 2.3 Different layout optimization techniques for inductors. (Adapted from Juneja et al. 2021 with permission from Elsevier): (**a**) Shielding, (**b**) Stacking, (**c**) Nesting, (**d**) Cross sectional view of oxide etched inductor

Figure 2.3d shows the fabrication process of implementing spiral inductors in CMOS, based on the oxide etching technique underneath the inductor substructure to reduce capacitive coupling and to improve the quality factor (Q). A 2.2 nH octagonal spiral inductor was fabricated using this technique, indicating an improvement in Q from 10 to 15 at 3 GHz. The maximum resonance frequency of the inductor shifted to 6.5 GHz with Q = 19. Therefore, this technique of oxide under etching can also be explored for implementing inductors at mmWave frequencies (Büyüktas et al. 2010).

Using these topologies and fabrication techniques, different inductors may be implemented either using a lumped element approach or distributed element approach at mmWave frequencies, as described in the next section.

2.3 Distributed and Lumped Design Approaches for Fabricating Passives

Lumped and distributed passive networks are often used in RF power amplifier design as matching networks and filters. It is often required that an impedance at the frequency of operation be transformed to another impedance level, and this is typically accomplished using passive networks. In many RF applications, interconnect transmission lines have a characteristic impedance of 50 Ω, and the input and output ports of circuits often need to be matched to 50 Ω. Thus, these passive networks are commonly referred to as matching networks, transforming one impedance level to another to match a specific impedance level. In RF applications where a relatively narrow band of frequency carries the signals of interest, these networks can also serve as frequency-selective filters. As shown in Fig. 2.4, both distributed and lumped-element matching networks consist of passive elements such as capacitors and inductors and distributed matching networks that employ transmission lines (Shirvani and Wooley 2003).

2.3.1 Distributed Approach

At mmWave frequencies, the wavelength of the signal is comparable to the size of the chip; therefore, a distributed design approach with transmission lines can be used to achieve resonance in the circuit. In Fig. 2.4, a series resistor (R) indicates

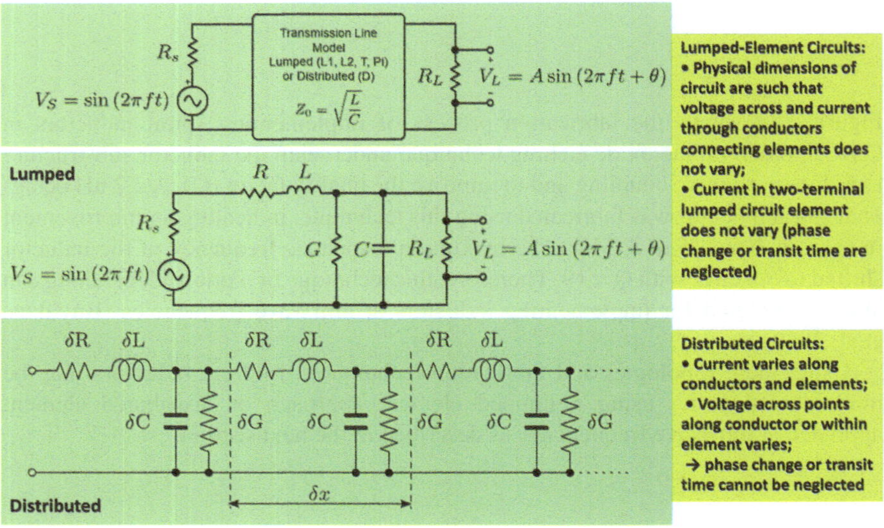

Fig. 2.4 Illustration of transmission line model with lumped or distributed approach

conductive losses due to skin effects, and shunt conductance (G) indicates losses in silicon oxide resulting in a capacitive quality factor (Qc). When a transmission line is implemented on silicon for mmWave frequency operation, the quasi TEM mode of propagation exists, and as a result, precise values of small inductances can be obtained. In addition to the quasi TEM mode, there are two other modes that can exist in planar transmission lines implemented on silicon, which are called the skin depth mode and slow-wave mode, as shown in Fig. 2.5. The existence of propagating modes is a function of the operating frequency and resistivity of the silicon substrate (P_{Si}). A low resistivity silicon substrate will have the properties of a conductor and therefore will lead to a skin depth mode of propagation, which is not desirable. However, the slow-wave mode of propagation in a transmission line can be obtained by incorporating an additional layer into the standard planar transmission line structure (Juneja et al. 2021).

Two types of transmission lines can be designed in CMOS. The first one is the microstrip transmission line, which is implemented using the top metal layer for signal propagation and the bottom metal layer for the ground plane, as shown in Fig. 2.6a. The limitation of the microstrip transmission line is that the ground plane is implemented in close proximity to the signal plane, leading to a high capacitive quality factor (Qc). As a result, an inductor of only small values can be implemented using a microstrip line because the inductive quality factor degrades for higher inductance values (Rao et al. 2016).

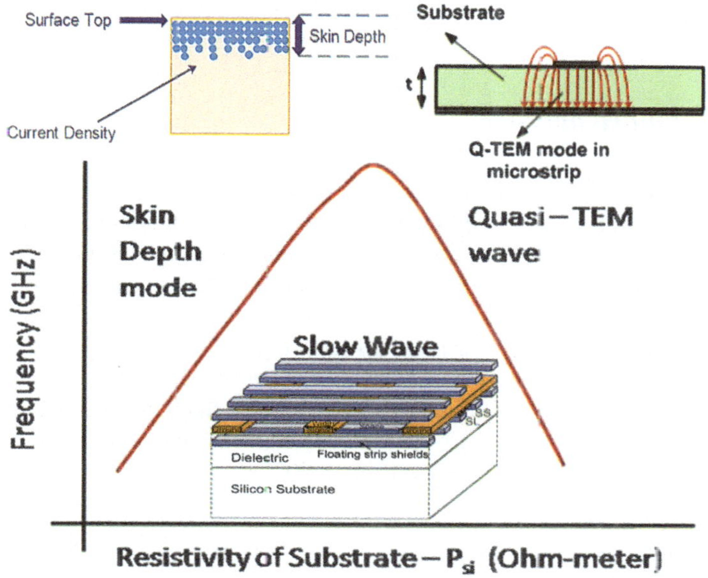

Fig. 2.5 Propagating modes of planer transmission line in Silicon. (Modified from Juneja et al. 2021 with permission from Elsevier)

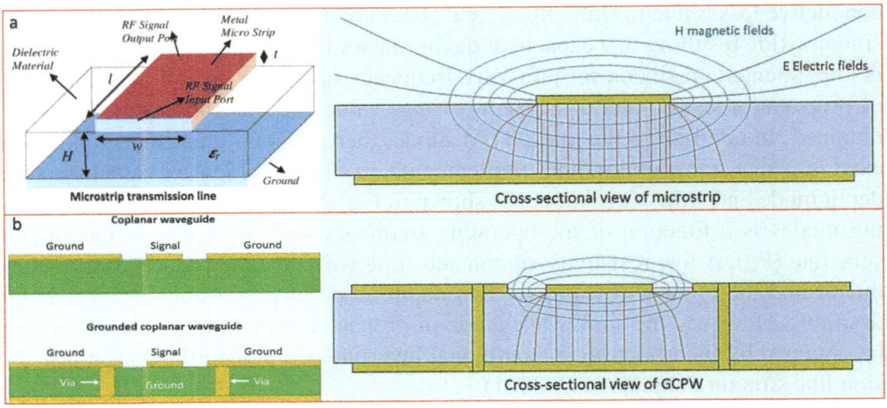

Fig. 2.6 Microstrip transmission line (**a**) and coplanar waveguide/grounded coplanar waveguide (**b**). (Modified from Coonrod 2015, credit: http://www.globalcommhost.com/rogers/acs/techsupporthub/en/docs/2015_IMS_MicroApps_microstrip_vs_coplanar_John_C_final_032315_SE.pdf)

The second type of transmission line is the coplanar waveguide (CPW), which offers a higher Q_L than the microstrip line, but at the same time, it occupies a larger area. As shown in Fig. 2.6b, CPW has two ground planes surrounding the signal line, and it is possible to optimize both Qc and Q_L by varying the separation between the ground planes and signal line. The microstrip line, on the other hand, offers only fixed Q_L and Q_C. To improve the performance of standard CPWs, different design modifications have been proposed, such as GCPWs (Coonrod 2015).

2.3.2 *Lumped approach*

For lumped inductor implementation, square, octagon and circular planer topologies, as shown in Fig. 2.2, could be designed. Each of these topologies has its own advantages and disadvantages. Square implementation has the least number of sides and therefore is easier to design; it gives a higher value of inductance but at the expense of the quality factor. Circular implementation, on the other hand, gives the highest quality factor, but it is most difficult to layout. Hexagonal or octagonal implementation also gives a high quality factor but at the expense of a larger foot print in comparison to a square inductor. To optimize the inductance value, quality factor and resonating frequency of the lumped element, one or more of the following parameters can be modified (Juneja et al. 2021):

- track width of the inductor,
- inner radius,
- number of turns and
- track spacing.

A small foot print of the inductor results in lower substrate losses and a higher quality factor, but there is a trade-off between the substrate loss and series resistance of a lumped element. To minimize foot prints on silicon, passive components are fabricated close to each other in circuits, which leads to coupling effects between them. Coupling lowers the quality factor as well as the self-resonating frequency of the lumped element and therefore must be addressed seriously for mmWave applications. By using the lumped element approach, different inductor topologies and on-chip passive circuits have been proposed for implementation at mmWave frequencies, such as lumped inductor/resonator using SiGe technology, bandpass filter (BPF) design incorporated with metal–insulator-metal (MIM) capacitors using the BiCMOS process, compact bandpass filter design with lumped inductor and MIM capacitor designed on silicon based Integrated Passive Device (IPD) technology, active tunable capacitor implemented using CMOS process technology, and decoupling capacitors with a combination of MIM, metal oxide metal (MOM) and metal oxide semiconductor (MOS) capacitors. In addition to passive components implemented with different silicon technologies, there are other process technologies, especially III-V semiconductors, that could be preferred for implementing both active and passive mmWave circuits (Chakraborty et al. 2016; Natsu et al. 2019).

2.4 Comparison of Silicon and III-V Semiconductors

For high frequency mmWave applications, III-V semiconductor technology such as the usage of GaAs, InP, GaN, etc., offers better performance in comparison to CMOS technology and is therefore a natural choice for the implementation of mmWave circuits. GaAs/InP has higher electron mobility, higher power gain, lower noise and body effects due to the high resistivity of the substrate, and hence, it is possible to implement high-Q passives with III-V technology. Comparably, CMOS has many inherent disadvantages, such as low resistivity of the substrate (10 Ω-cm), which leads to signal loss, high sheet resistance of the polysilicon gate (approximately 10 Ω/square), which leads to higher noise in the circuit, and lower power gain. However, the thermal conductivity of GaAs and InP is lower than that of CMOS, and therefore, the device density of GaAs and InP is lower than that of CMOS. GaN has thermal conductivity comparable to silicon, and it also has a very high breakdown voltage. Wideband gap GaN HEMT is therefore an ideal choice for implementing power amplifiers at mmWave frequencies due to its higher power added efficiency at these frequencies. Table 2.1 shows the different material properties of GaAs, InP, GaN and silicon (Cathelin and John 2008; Juneja et al. 2021).

Despite the many benefits of semiconductor III-V technology, CMOS technology for high frequency mmWave applications has been extensively explored. One of the prime reasons for the same is that with the scaling of CMOS technology, it is possible to obtain a transition frequency (f_T) that is comparable to f_T of III-V semiconductors, and f_T is one of the key measures of transistor performance while working at such high RF frequencies. The other important reason is economics because

Table 2.1 Typical material properties of GaAs, InP, GaN and Silicon (Juneja et al. 2021)

Materials	f_T (GHz)	ρ (Ohm-cm)	μ (cm²/V-s)	Bg (eV)	K (W/m-K)	V_B (V/cm)
Silicon	250 GHz for 28 nm CMOS process	10	1.5×10^3	1.12	150	3×10^5
GaAs	>500 GHz	10^7 to 10^9	8×10^3	1.42	50	4×10^5
InP	>700 GHz for InP HEMT	10^7	5×10^3	1.35	67	5×10^5
GaN	>500 GHz	10^6 to 10^9	1.2×10^3	3.4	130	30×10^5

f_T: Transition frequency, ρ: Resistivity, μ: Electron mobility, Bg: Bandgap, K: Thermal conductivity, V_B: Breakdown voltage.

for the large scale commercial success of 5G using mmWave frequency band cost is going to be the biggest factor. CMOS technology, owing to its well-established processes, offers many benefits over semiconductor III-V technology (Juneja et al. 2021):

- It is cost efficient
- It has a higher manufacturing capacity
- It offers higher integration
- It has a higher manufacturing yield

Figure 2.7 summarizes the basics of CMOS technology and its limitations and benefits over III-V technology. The nonlinear circuit element in CMOS is the MOSFET or metal oxide semiconductor field effect transistor. CMOS technology is associated with very large-scale integrated circuits (VLSIs), where a few million or even billions of MOSFET transistors are integrated into a single chip or die. The dominant use of CMOS technology in the fabrication of VLSI chips is due to its reliability, low power consumption, considerably low cost and, most importantly, scalability. Some CMOS technology-based low noise amplifier or other component designs have been proposed for 5G applications. To make the best use of the benefits offered by both semiconductor III-V technology and CMOS technology, a heterogeneous integration technique has been proposed: InP HEMT and InP HBT chiplets are integrated on CMOS wafers and used for circuits working close to 1 Hz applications (Radisic et al. 2015).

In addition to CMOS, there are two other silicon technologies that are being used for implementing mmWave circuits, namely, SiGe BiCMOS and silicon-on-insulator (SOI). BiCMOS technology is extensively used in mmWave applications and has a transition frequency up to 300 GHz (for 55 nm process technology). In addition to a higher transition frequency, BiCMOS technology offers better RF performance in comparison to bulk CMOS technology and is ideal for implementing mmWave circuits for 5G applications. However, in comparison to bulk CMOS, BiCMOS has higher power consumption, low integration density and high cost of volume production (Sarkar and Floyd 2017).

Silicon-on-insulator (SOI) is another silicon-based process technology that is being used for implementing mmWave circuits, and its performance is comparable to BiCMOS. The 28 nm SOI technology has a transition frequency >300 GHz,

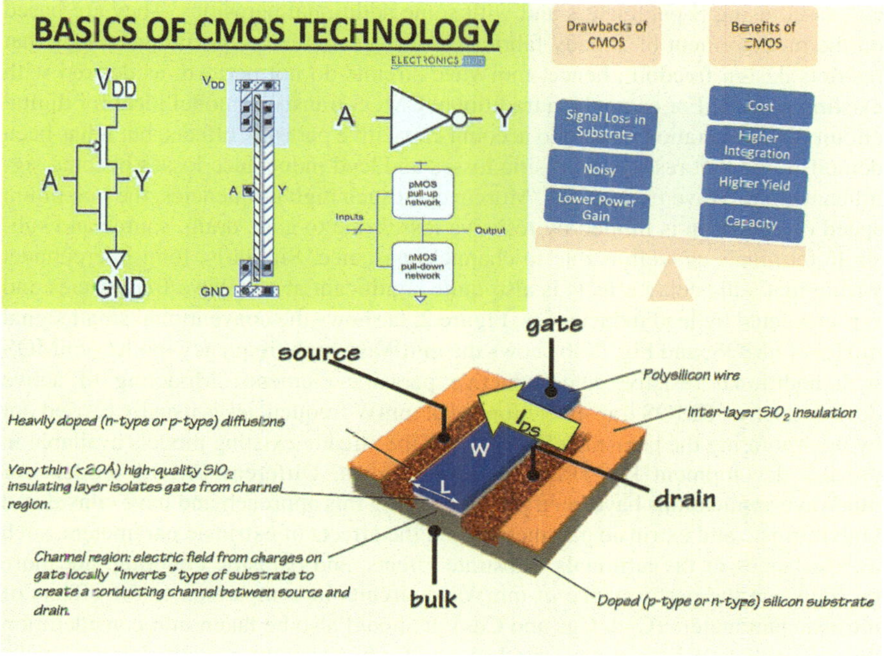

Fig. 2.7 Basics of CMOS technology and its benefits and drawbacks. (Modified from Ward 2017, credit: M.I.T. Department of Electrical Engineering and Computer Science)

which is comparable to that of BiCMOS process technology. SOI technology overcomes two important limitations of the bulk CMOS process: the low resistivity of the substrate and the high sheet resistance of the polysilicon gate (Kodak and Rebeiz 2019).

It is now evident that different silicon technologies, including bulk CMOS, BiCMOS and SOI technology, have been prominently used for implementing mmWave circuits for 5G applications. Even in the presence of better performing semiconductor III-V technologies, silicon technologies will remain at the forefront for designing different mmWave circuits for 5G applications. However, the existing silicon device models that are available in EDA tools for low-frequency operations cannot be directly used for building mmWave circuits (Juneja et al. 2021).

2.5 Transistor Model Design Challenge in CMOS Technology

The greatest challenge with respect to designing mmWave circuits using any EDA tool is that the available CMOS transistor models in these tools are not optimized for such high-frequency operations. These existing models are black box models expressed using S-parameters and with some additional parasitics. They are based on the measurement of already fabricated devices and have specific geometry that restricts design freedom; hence, mmWave circuits do not perform as desired with existing models. For example, a traditional CMOS transistor model ideal for digital circuit implementation takes into account capacitive parasitic effects, but it has been demonstrated that resistive parasitic losses and lead inductance losses become significant at mmWave frequencies. Moreover, at such high frequencies, the maximum speed of operation is limited by resistive losses due to gate, drain, source and substrate terminals as well as due to channel resistance. Similarly, long interconnect wiring that adds delay effects is also quite significant at mmWave frequencies and is represented by lead inductances. Figure 2.8a shows the conventional small signal model of nMOS, and Fig. 2.8b shows the mmWave high frequency model of nMOS with additional resistive and inductive parasitic elements. Modeling of active devices such as CMOS transistors for use at mmW frequencies is usually carried out by incorporating the parasitic elements into the already existing models available in process development kits (PDKs) of EDA tools. Different CMOS models for mmWave applications have been proposed using this approach and have considered both intrinsic and extrinsic parameters, and the effects of extrinsic parameters, such as resistances of the terminals, substrate effects, and parasitic coupling, are more profound on the performance of mmWave circuits in comparison to the effects of intrinsic parameters (Cgd, Cgs and Cds). It should also be taken into consideration that available transistor models are designed for fixed layout. Specific layout details, such as connections to various terminals, also affect parasitic elements and must be taken into account (Doan et al. 2005; Sialm et al. 2006).

Figure 2.9 shows some examples of layout design approaches for nMOS at mmWave frequency. By having multiple gate connections on both sides of the transistor and by using a low resistivity Metal 1 ring for gate connections, for example, the effect of parasitic gate resistance can be minimized to a great extent. Similarly, using multiple short gate fingers in design is also a desirable layout technique at mmWave frequency (Heydari et al. 2007; Pornpromlikit et al. 2011).

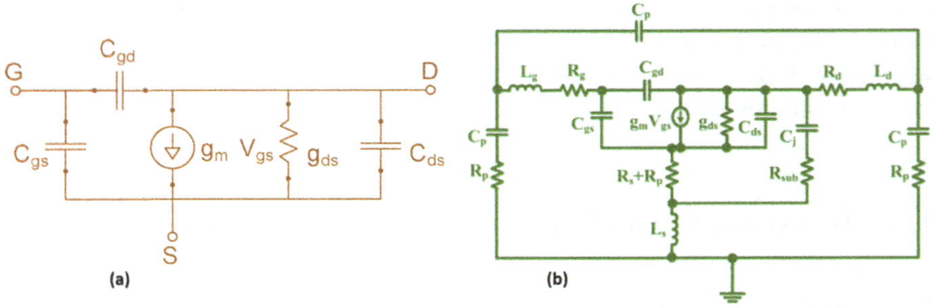

Fig. 2.8 Comparison of (**a**) conventional small signal model of nMOS and (**b**) mmWave high frequency model of nMOS with additional resistive and inductive parasitic elements. (Adapted from Doan et al. 2005; Sialm et al. 2006 with permission from IEEE)

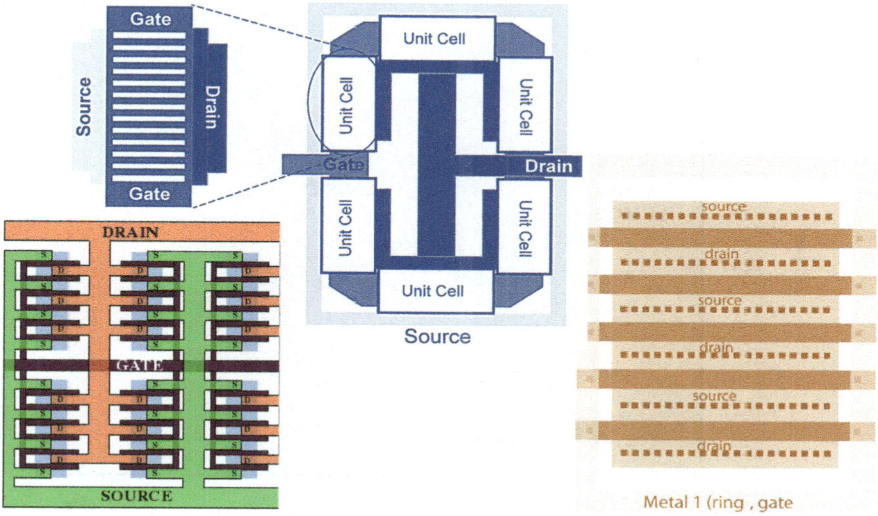

Fig. 2.9 Example layout design approaches for nMOS at mmWave frequency. (Modified from Juneja et al. 2021 with permission from Elsevier)

2.6 GaN and GaN-on-SiC Wide Bandgap Semiconductors for 5G Applications

5G technologies and infrastructure design need to deliver connectivity that meets new requirements for bandwidth, latency and data speed. It will not only require densification on the macro level – meaning more base stations – but also densification of power on the device level. Today's telecommunications infrastructure design requires technologies that best match a number of criteria for the application, including heat, speed, power, efficiency, size and cost. There are several types of semiconductors available for 5G RF design, as shown in Fig. 2.10. Silicon germanium (SiGe) technology uses a relatively low operating voltage of 2 V to 3 V but is very attractive for its integration benefits. GaAs has been used widely for power amplifiers in microwave frequencies and has operating voltages of 5 V to 7 V. Silicon LDMOS (laterally diffused metal oxide semiconductor) technology operating at 28 V has been used in telecommunications, but it is mainly useful below 4 GHz, so it is not as widely used in broadband applications. GaN technology operating at 28 V to 50 V on a substrate with low loss and high thermal conductivity, such as silicon carbide (SiC), has opened up a range of new possibilities. Today, GaN on silicon technology is limited to operation below 6 GHz. The RF losses associated with the silicon substrate and its lower thermal conductivity compared to SiC compromise the gain, efficiency, and power as the frequency increases (Benson 2017).

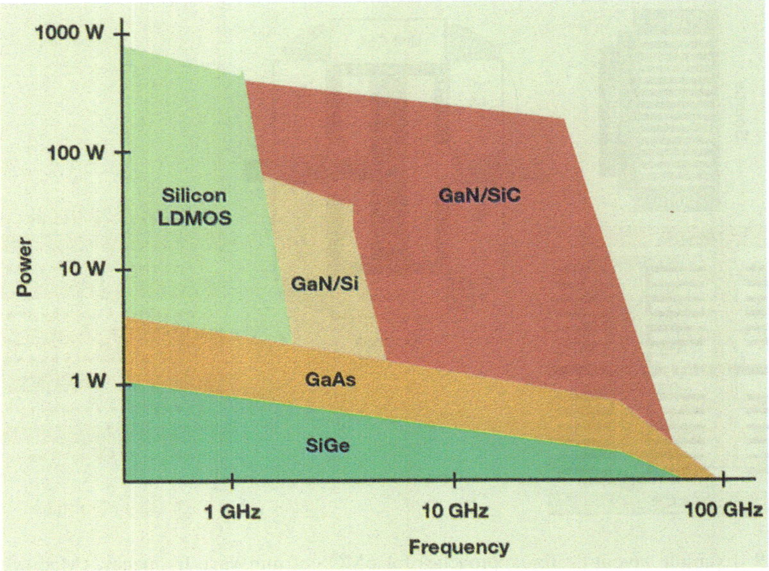

Fig. 2.10 Comparing power and frequency of different semiconductor materials in the microwave range, which includes mmWaves. (Adapted from Benson 2017, credit: https://www.analog.com/en/analog-dialogue/articles/rf-power-amplifiers-go-wide-and-high.html)

2.6.1 Characteristics of GaN Devices Applied in 5G Technology

GaN is an extremely stable compound and a direct bandgap semiconductor with a very wide bandgap (3.4 eV) compared with other materials, such as Si and GaAs. It is also a hard material with a high melting point of approximately 1700 °C. GaN crystals usually have hexagonal Wurtzite structures under atmospheric pressure. These tremendous characteristics have created good conditions for its application in the field of 5G technology (Wang and Sheng 2020):

(a) High pressure resistance and radiation resistance

GaN crystals possess stronger chemical bonds; thus, they are capable of withstanding electric fields numerous times higher than those of silicon devices without collapse. Therefore, it possesses the characteristic of high pressure resistance; the bandgap width of GaN is large, electrons are not easily excited to the conduction band, and the interference signal has little effect on the device; thus, it is resistant to radiation. These characteristics permit the device to resist higher voltage to emit 5G signals with higher power. In addition, its radiation resistance allows the signals away from other disturbing signals, keeping the 5G signals in a relatively steady circumstance, capable of transmitting the information more accurately.

(b) High frequency

The gate charge of GaN is relatively low, and the gate charge must be supplemented in each switching cycle. Therefore, the lower the gate charge is, the more likelihood the device can achieve the state of high frequency. GaN is capable of working at a frequency as high as 1 MHz without loss of efficiency, but this excellent performance cannot be kept in regard to other materials; for example, silicon has difficulty reaching more than 100 kHz. In addition, GaN has a strong chemical bond and high loading capacity, so the distance between each electrical terminal of the transistor is shortened many times; thus, the conversion time of electrons is shorter. Moreover, the bottom of the conduction band of GaN is at the Γ point, and the energy difference between it and the other energy valleys of the conduction band is large; thus, it is not easy to produce intervalley scattering, and the electron drift rate is not easily saturated. Coupled with the heterojunction formed by GaN and semiconductor materials such as AlGaN, a two-dimensional electron gas will be formed, which has high mobility. Therefore, its electronic devices possess faster switching characteristics. In addition, such a characteristic will play a significant role in 5G application for a fast calculation speed.

(c) High operating temperature

Since GaN has a vastly wide bandgap, the intrinsic excitation of GaN is weaker than that of other narrow bandgap semiconductors at the same temperature, and thus, its device has a higher signal-to-noise ratio for signal transmission. Therefore, it has a higher operating temperature, and this property is of great significance for the circuit to operate under higher power with a higher temperature condition, such as the 5G base station. In addition, GaN has high thermal conductivity and excellent heat dissipation performance, especially when integrated with a SiC substrate, which are conducive to its operation under high-temperature conditions.

(d) Low energy loss

GaN has a high pressure-bearing energy level; thus, the distance between each terminal of the transistor can be designed to be shorter, which makes it possible to achieve lower resistance loss. In addition, the high mobility and high carrier concentration also decrease the resistivity; therefore, the electronic devices obtain low conduction resistance and further reduce the resistance energy loss. This characteristic can reduce the energy consumed on 5G signals transmitting, making it possible to realize the idea that the lower the energy it consumes, the higher the power the signal it emits.

2.6.2 GaN Power Integration for MMIC in 5G Technology

The 5G architecture employs multiple-input multiple-output (MIMO) and beamforming technology to direct signal power for increased overthe-air data rates. Massive MIMO beamforming requires a multiplicity of RF circuitry for each antenna element in the phased-array transceiver system. Therefore, size, cost and power density are crucial figures of merit for both the base station and handset architectures. While other semiconductor technologies are better suited for handsets due to cost, battery voltage and RF power requirements, GaN is a natural candidate for base-station deployment. Continual efforts to customize GaN for lower operating voltages and higher operating frequencies enable the development of switches, LNAs and frequency conversion circuitry. Eventually, it will be possible to integrate the multiplicity of RF chains into a single or several GaN MMICs (Yuk et al. 2017).

2.6.2.1 GaN Power Integration for MMICS

The GaN power integration processes for MMICs (Monolithic microwave integrated circuits) feature the monolithic integration of normally ON Schottky gate AlGaN/GaN HEMTs, Schottky barrier diodes, thin-film resistors (TFRs), metal-insulator-metal (MIM) capacitors and inductors. Figure 2.11 demonstrates the schematic cross-sections of a typical 100 nm GaN-on-Si MMIC process (a) and a high breakdown-voltage millimeter-wave GaN HEMT integrated circuit (b). GaAs devices in MMICs offer high performance wideband-gap (WBG) solutions up to 500 GHz for mobile devices, communications infrastructure, and aerospace applications. High frequency (operating frequency >50 GHz) power amplifiers (PAs) are the major category of GaN MMICs, while other categories include high linearity low noise amplifiers (LNAs), voltage-controlled oscillators (VCOs), transmitter/receiver and modulator components (Fujitsu 2009; Rocchi 2016).

For instance, diodes are essential devices to provide a high-voltage single-direction blocking state in power ICs and can also be used as voltage level shifters, which take advantage of their forward voltage drop. The most common is the Schottky barrier diode (SBD), which is formed by depositing Ni-based metal on an AlGaN or GaN surface, as shown in Fig. 2.12a. Moreover, lateral field effect rectifiers (L-FERs) use a Schottky-gate-controlled 2D EG channel between the anode and cathode. The lateral device structure and process compatibility with AlGaN/GaN HEMT make L-FER a well-qualified candidate in smart power IC platforms, as shown in Fig. 2.12b. In Fig. 2.12c, the tri-gate Schottky diodes are based on the tri-anode structure with Schottky contact to the multichannels through the fin sidewalls to obtain a lower turn-on voltage. It also exhibits great potential in working with normally off AlGaN/GaN Fin-FETs (Ma et al. 2019).

GaN integration can convert the power to where silicon power cannot go. More technologies in Si electronics have been applied to GaN integration. The dedicated Si CMOS gate drivers can be packaged in the same case or directly bonded with

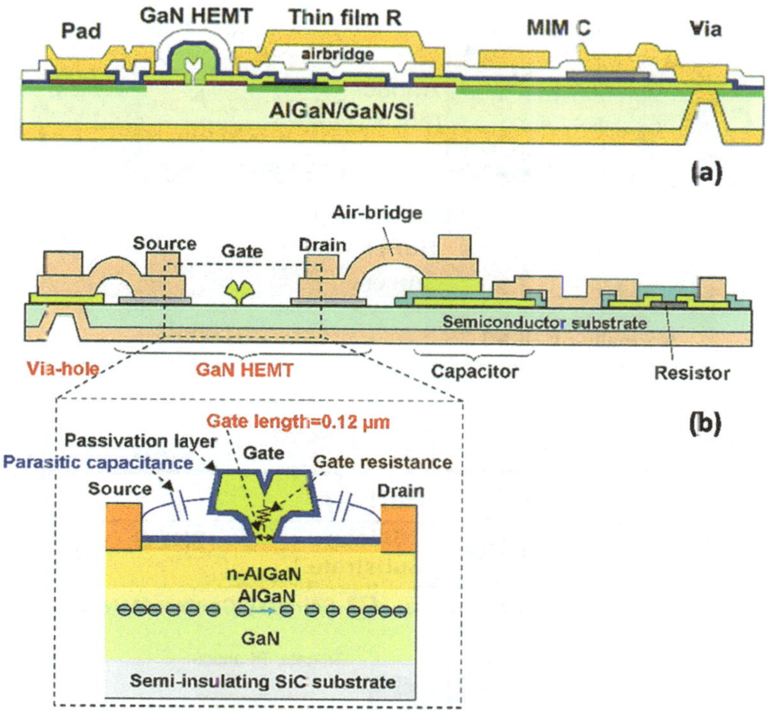

Fig. 2.11 Schematic cross-sections of 100nm GaN-on-Si process for MMICs (**a**) and high breakdown-voltage millimeter-wave GaN HEMT integrated circuit (**b**). (Modified from Fujitsu 2009; credit: Fujitsu Limited; and Rocchi 2016 with permission from IEEE)

GaN in 3 dimensions. More approaches are being proposed. However, GaN integration should be established on a stable platform with well-calibrated and modeled active and passive components. Therefore, the performance of the GaN power ICs can be reliably predicted. The technology of discrete GaN devices is improving rapidly, and GaN power integration should be updated simultaneously. To exploit most of the theoretical GaN performance, the GaN power integration road follows the path of Si power IC, but on a fast lane (Sun et al. 2020).

2.6.2.2 GaN Base Station PAs

In MIMO, each antenna is driven by its own PA; therefore, it is important to meet the power and linearity requirements while minimizing variation across cells. The development of 5G GaN-based small-cell base station PAs is important for compactness, reduced weight, and low cost while retaining high power and efficiency for ease of deployment. A thorough understanding of how the unique attributes of GaN, such as breakdown voltage, self-heating, trapping, field plate design and transconductance shape, impact the operating frequency, power, efficiency (PAE),

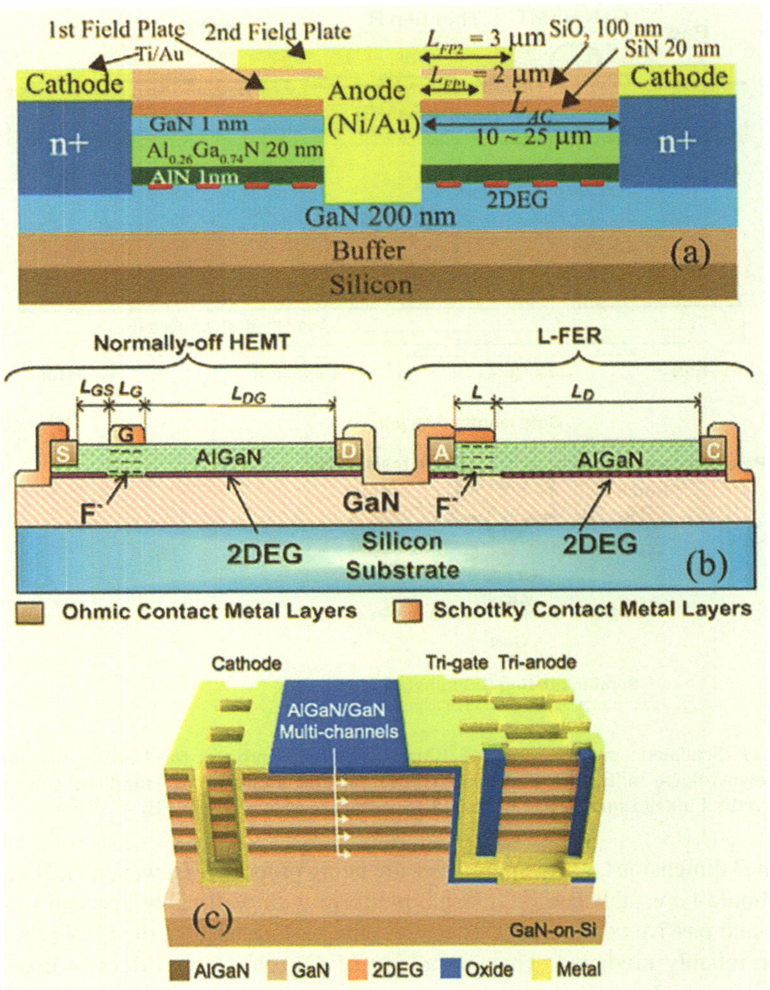

Fig. 2.12 Schematic structure of (**a**) a typical Schottky barrier diode. (Adapted from Zhu et al. 2015 with permission from IEEE), (**b**) a L-FER with a normally-OFF HEMT. (Chen et al. 2008 with permission from AIP Publishing) and (**c**) a tri-gated Schottky barrier diode. (Ma et al. 2019 with permission from IEEE)

linearity (harmonics, EVM, ACPR, IIP3, AM-AM, AM-PM), ruggedness, and transient behavior is critical. The desire for high density power will be satisfied by GaN in ways that existing GaAs FET and Si LDMOS solutions cannot. This will require extensive efforts in modeling high-power devices using dynamic DC and RF techniques (Micovic et al. 2016).

A 28 GHz Class AB PA MMIC measuring 1.8 mm × 1.7 mm using a single 0.20 µm GaN 8 × 100 µm FET device was designed. The circuit layout is shown in Fig. 2.13a, and the simulated performance is shown in Fig. 2.13b. This demonstration PA produces a small-signal gain = 10.03 dB, and at input P1 dB = 27 dBm produces Pout = 36.01 dBm with PAE = 51.1%, as shown in Fig. 2.13b. As a PA output stage of a single MIMO transmitter, the design indicates that presently available GaN technology is capable of first-generation 5G systems. For base station deployment, this PA can serve as a building block for the Doherty architecture shown in Fig. 2.14a (Yuk et al. 2017).

2.6.2.3 GaN Frequency Synthesis

The number of antenna elements in 5G MIMO systems increases tremendously. With the increased number of transmitters, new challenges in accurately generating and distributing coherent local oscillator (LO) power will arise. One direct way of addressing these issues is to amplify the LO power using a PA. However, since 5G carrier signals will initially start in the sub66 GHz range, compatibility with lower frequency cellular bands ranging from GSM850/900 to DCS/PCS to LTE frequencies is necessary. Therefore, the MIMO signal distribution and compatibility problem may be solved using high power frequency multiplication to provide adequate power at the desired frequencies. Eventually, the same techniques might be applied for generating a mm-wave 5G LO signal. Generation of a high-frequency, high-power LO signal from a lower-frequency reference can be achieved using high-power GaN frequency multipliers. The resulting output can then be precisely distributed to each massive MIMO chain using a passive network, as shown in Fig. 2.14b. Frequency multiplication allows GaN devices to provide power at above f_T using harmonic enhancement techniques. The development of GaN technology for high harmonic generation without breaking down is another possible area of technology development (Yuk et al. 2017).

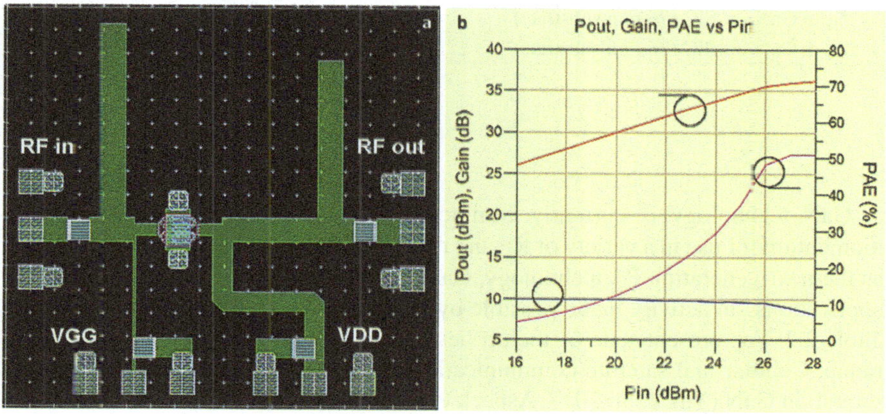

Fig. 2.13 5G PA using a 0.20 um GaN 8 × 100 um FET (**a**) MMIC and (**b**) Pout, Gain and PAE performance. (Adapted from Yuk et al. 2017 with permission from IEEE)

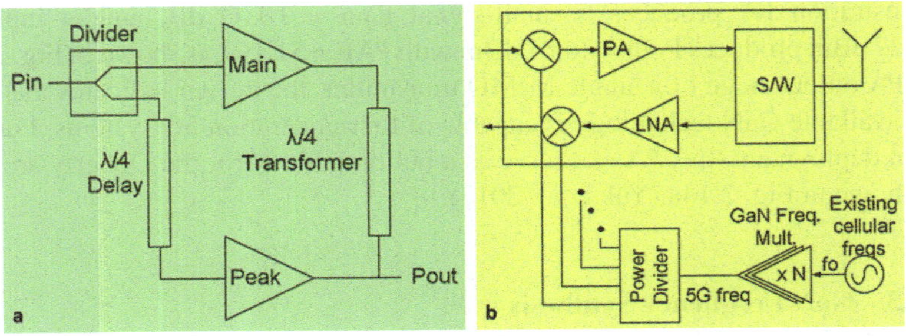

Fig. 2.14 GaN Doherty PA for base stations (**a**) and Frequency conversion and distribution using high power GaN frequency multipliers (**b**). (Adapted from Yuk et al. 2017 with permission from IEEE)

Table 2.2 Commercially available GaN foundry services (Yuk et al. 2017)

Foundry service			
Process	Bias	Frequency	Discretes
0.25 μm GaN-on-SiC	28–40 V	18 GHz, 30 GHz	Y
0.40 μm GaN-on-SiC	28 V, 50 V	8 GHz	
0.25 μm GaN-on-Si	N/A	N/A	N
0.50 μm GaN-on-Si	N/A	N/A	
0.50 μm GaN-on-SiC	N/A	N/A	
0.20 μm GaN (100 mm substrate)	N/A	60 GHz	Y
0.10 μm GaN	N/A	>70 GHz	N
0.25 μm GaN-on-SiC (100 mm substrate)	40 V, 48 V	10 GHz, 18 GHz	Y
0.15 μm GaN-on-SiC (100 mm substrate)	28 V	40 GHz	
0.50 μm GaN-on-SiC (100 mm substrate)	65 V	10 GHz	
0.50 μm GaN-on-SiC (76 mm substrate E-mode)	N/A	N/A	N
0.15 μm GaN-on-SiC (76 mm substrate)	N/A	Ka-band	
0.50 μm GaN-on-SiC (76 mm substrate)	40 V	X-band	

GaN technology is currently commercially available and continues to gain momentum for use in a variety of RF and microwave industries. Primarily cultivated as the next-generation PA technology, GaN is being developed for different circuit applications, an activity made possible by the range of foundry offerings shown in Table 2.2. The present state-of-the-art lies within the 0.10–0.15 μm channel length range. Cellular and satellite communications are two vital areas that will fuel the growth in GaN (Yuk et al. 2017, Asbeck et al. 2019).

References

Asbeck PM, Rostomyan N, Özen M, Rabet B, Jayamon JA (2019) Power amplifiers for mm-wave 5G applications: technology comparisons and CMOS-SOI demonstration circuits. IEEE Trans Microw Theory Techniq 67(7):3099-3109

Benson K (2017) GaN breaks barriers – RF power amplifiers go wide and high. Analog Dialog 51-09, September 2017. https://www.analog.com/media/en/analog-dialogue/volume-51/number-3/articles/rf-power-amplifiers-go-wide-and-high.pdf

Borkar S (1999) Design challenges of technology scaling. IEEE Micro 19(4):23–29

Büyüktas K, Koller K, Müller K, Geiselbrechtinger A (2010) A new process for on-chip inductors with high Q-factor performance. Int J Microw Sci Technol 2010:517187

Cathelin A, John P (2008) Silicon technologies to address mm-wave solutions. In: mm-Wave Silicon Technology, Springer, Boston. pp 25–57

Chakraborty S et al (2016) A broadside-coupled meander-line resonator in 0.13-um SiGe technology for millimeter-wave application. IEEE Electron Device Lett 37(3):329–332

Chen W, Wong K-Y, Huang W, Chen KJ (2008) High-performance AlGaN/GaN lateral field-effect rectifiers compatible with high electron mobility transistors. Appl Phys Lett 92(25):Art. no. 253501

Coonrod J (2015) Microwave PCB structure considerations: microstrip vs. grounded coplanar waveguide. In: IEEE 2015 international microwave symposium, 17–22 May, 2015, Phoenix, AZ. http://www.globalcommhost.com/rogers/acs/techsupporthub/en/docs/2015_IMS_MicroApps_microstrip_vs_coplanar_John_C_final_032315_SE pdf

Dennard RH, Gaensslen FH, Yu H-N, Rideout VI, Bassous E, LeBlanc AR (1974) Design of ion-implanted MOSFETs with very small physical dimensions. IEEE J Solid State Circuits SC-9:256–268

Doan C, Emami S, Niknejad AM, Brodersen RW (2005) Millimeter-wave CMOS design. IEEE J Solid State Circuits 40(1):144–155

Fujitsu (2009) Fujitsu develops world's first millimeter-Wave Gallium-Nitride transceiver amplifier chipset. https://www.fujitsu.com/global/about/resources/news/press-releases/2009/0930-02.html

Heydari B, Bohsali M, Adabi E, Niknejad AM (2007) Millimeter wave devices and circuit blocks up to 104 GHz in 90 nm CMOS in solid-state circuits. IEEE J 42(12):2893–2903

Juneja S, Pratap R, Sharma R (2021) Semiconductor technologies for 5G implementation at millimeter wave frequencies – design challenges and current state of work. Eng Sci Technol 24(1):205–217

Kodak U, Rebeiz GM (2019) A 5G 28-GHz common-leg T/R front-end in 45-nm CMOS SOI with 3.7-dB NF and -30-dBc EVM with 64-QAM/500-MBaud modulation. IEEE Trans Microw Theory Technol 67(1):318–331

Ma J, Kampitsis G, Xiang P, Cheng K, Matioli E (2019) Multichannel tri-gate GaN power Schottky diodes with low ON-resistance. IEEE Electron Device Lett 40(2):275–278

Mead C (1972) Fundamental limitations in microelectronics – I. MOS technology. Solid State Electron 15:819–829

Micovic M, Brown DF, Regan D, Wong J, Tang Y, Herrault F, Santos D, Burnham SD, Tai J, Prophet E, Khalaf I, McGuire C, Bracamontes H, Fung H, Kurdoghlian AK, Schmitz A (2016) High frequency GaN HEMTs for RF MMIC applications. In: 2016 IEEE International Electron Devices Meeting (IEDM), pp 3.3.1-3.3.4

Natsu Y, Takano K, Umeda Y (2019) Comparison of millimeter-wave 0–Ω transmission lines in 0.18 lm. CMOS technology. In: 12th IEEE global symposium on millimeter waves (GSMM), Sendai, Japan, pp 38–40

Pacurar C, Topa V, Racasan A, Munteanu C (2012) Inductance calculation and layout optimization for planar spiral inductors. In: 2012 13th international conference on optimization of electrical and electronic equipment (OPTIM), Brasov, pp 225–232. https://doi.org/10.1109/OPTIM.2012.6231846

Pornpromlikit S, Dabag HT, Hanafi B, Kim J, Larson LE, Buckwalter JF (2011) A Q-band amplifier implemented with stacked 45-nm CMOS FETs. In: 2011 IEEE compound semiconductor integrated circuit symposium (CSICS), pp 1–4

Radisic V et al (2015) Submillimeter wave InP technologies and integration techniques. In: 2015 IEEE MTT-S international microwave symposium, Phoenix, AZ, pp 1–4

Rao KS, Prakash MD, Thalluri LN (2016) Cantilever and circular disc structure based capacitive shunt RF MEMS switches. 2016 international conference on electrical, electronics, communication, computer and optimization techniques (ICEECCOT), Mysuru, pp 336–338. https://doi.org/10.1109/ICEECCOT.2016.7955241

Rocchi M (2016) Advanced III/V MMIC process roadmaps for terahertz applications. In: Proceedings of the IEEE MTT-S international microwave workshop series on advanced materials and processes RF THz applications (IMWS-AMP), pp 1–2

Rüddenklau U, Michael G, Andrea P, Barrett M, Wambacq P, Sellars M, Rao RM (2018) ETSI White Paper No. 15 – mmWave semiconductor industry technologies: status and evolution. https://www.etsi.org/images/files/ETSIWhitePapers/etsi_wp15ed2_mmWave-Semiconductor_Technologies_FINAL.pdf

Sarkar A, Floyd BA (2017) A 28-GHz harmonic-tuned power amplifier in 130-nm SiGe BiCMOS. IEEE Trans Microw Theory Technol 65(2):522–535

Shirvani A, Wooley BA (2003) Design and control of RF power amplifiers. Springer, Cham, pp 71–85

Sialm G, Kromer C, Ellinger F et al (2006) Design of low-power fast VCSEL drivers for high-density links in 90-nm SOI CMOS, microwave theory and techniques. IEEE Trans 54(1):65–73

Sun R, Lai J, Zhang B (2020) GaN power integration for high frequency and high efficiency power applications: a review. ID: 210971995. https://doi.org/10.1109/ACCESS.2020.2967027Corpus. https://ieeexplore.ieee.org/stamp/stamp.jsp?tp=&arnumber=8962057

Wang K, Sheng C (2020) Application of GaN in 5G technology. J Phys Conf Ser 1699:012004. https://iopscience.iop.org/article/10.1088/1742-6596/1699/1/012004/pdf

Ward S (2017) Computation structures. M.I.T. Department of Electrical Engineering and Computer Science. https://computationstructures.org/notes/cmos/notes.html

Yole Développement (2019) The 5G revolution is pushing innovations for RF front-end SiP. http://www.yole.fr/iso_upload/News/2019/PR_RF_SIP_Market_5Gimpact_SYSTEMPLUSCONSULTING_YOLE_March2019.pdf

Yuk K, Branner GR, Cui C (2017) Future directions for GaN in 5G and satellite communications. In: 2017 IEEE 60th international Midwest symposium on circuits and systems (MWSCAS), Boston, MA, 6–9 August 2017

Zhu M, Song B, Qi M, Hu Z, Nomoto K, Yan X, Cao Y, Johnson W, Kohn E, Jena D, Xing HG (2015) 1.9-kV AlGaN/GaN lateral Schottky barrier diodes on silicon. IEEE Electron Device Lett 36(4):375–377

Chapter 3
Design and Performance Enhancement for 5G Antennas

Abstract 5G technology supports smartphones and different IoT devices to provide various services, such as smart buildings, smart cities, and many others, which require 5G antennas and beamforming approaches with low latency, low path loss, and stable radiation patterns. 5G antenna architecture can be classified into two major categories. SISO (single input single output) and MIMO (multiple input multiple output) are based on input output ports. Both are further classified as wideband and multiband based on their frequency response. The SISO antennas can also be classified into single element and multielement antennas, which are suitable for integration with 5G devices to support IoT. The MIMO antennas can be categorized into a multielement antenna with and without a metal rim for both wideband and multiband. MIMO antennas are the best candidate for smartphones, while massive MIMO antennas can be used at base stations. In MIMO metal rim antenna design, the use of carrier aggregation reinforces the transmission rate. Additionally, design features such as orthogonal polarization boost isolation, thereby enhancing the overall efficiency. Additionally, antennas can be classified based on their types, which are elaborated in detail with their performance enhancement techniques. These enhancement methods have a profound effect on the electrical and physical properties of an antenna, which in turn enhances the overall performance of an antenna. This chapter provides a brief review of different 5G antenna designs with their performance enhancement techniques, as well as different antenna beamforming approaches for 5G technology.

3.1 5G Antenna Classification

The antenna is one of the pivotal parts of the 5G device, which is required to work at an enhanced gain, bandwidth, and lesser radiation losses. Therefore, antenna design for 5G devices becomes crucial with different performance enhancement techniques regarding the type of antenna. The 5G antenna architecture can be classified into two major categories. SISO and MIMO based on input output ports. Both are further classified as wideband and multiband based on their frequency response.

C. Tong, *Advanced Materials and Components for 5G and Beyond*, Springer Series in Materials Science 327, https://doi.org/10.1007/978-3-031-17207-6_3

The SISO antennas can be classified into single element and multielement antennas, which are suitable for integration with 5G devices that support IoT. The MIMO antennas can be categorized into a multielement antenna with and without a metal rim for both wideband and multiband. MIMO antennas are the best candidate for smartphones, while massive MIMO antennas can be used at base stations. In MIMO metal rim antenna design, the use of carrier aggregation reinforces the transmission rate. Additionally, design features such as orthogonal polarization boost isolation, thereby enhancing the overall efficiency. Additionally, antennas can be classified based on their types. All these antenna types are elaborated in detail with its performance enhancement technique. These enhancement methods have a profound effect on the electrical and physical properties of an antenna, which in turn enhances the overall performance of an antenna (Kumar et al. 2020):

3.1.1 Classification Based on Input and Output Ports

Based on input and output ports, as shown in Fig. 3.1, the antenna can be broadly classified as (Kumar et al. 2020):

(a) Single Input Single Output (SISO) antenna

The SISO antenna is either a single or multielement antenna for 5G applications and is easy to design and implement. To achieve a high gain, the size of a

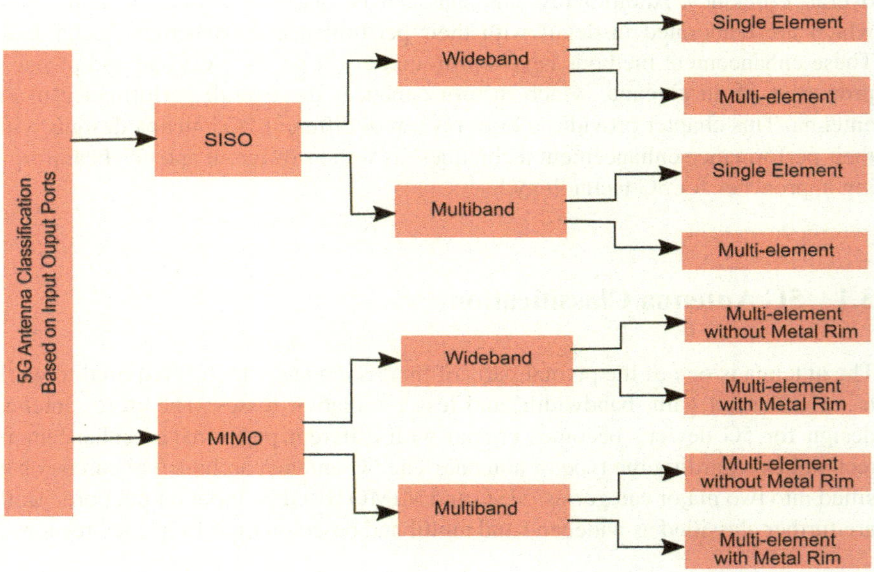

Fig. 3.1 5G Antenna classification based on input output ports. (Adapted from Kumar et al. 2020 with open access (IEEE))

single element antenna is large. At frequencies above 6 GHz, the signals suffer from higher propagation losses, and the quality of service degrades. Therefore, it is required to replace a single element antenna with a multielement antenna to achieve uniform and good performance. In fact, a multielement antenna is mainly used to enhance the gain of an antenna at the cost of increased size and design complexity.

(b) Multiple-Input Multiple-Output (MIMO) antenna

Wireless communication is prone to interference, multipath fading, and radiation losses. Additionally, it becomes severe at higher frequencies. To overcome these issues, the utilization of multiple-input multiple-output (MIMO) antennas becomes very important as it enhances the transmission range without increasing the signal power. Thus, MIMO design can be used in 5G to achieve low latency, maximum throughput, and large efficiency. In MIMO, more signals can be launched intelligently by using multiple antennas, thus significantly enhancing the channel capacity.

The method used to reduce the number of antennas in MIMO is to use multiband antennas that provide coverage of different wireless applications. Furthermore, MIMO antennas can be classified depending upon their frequency band as wideband and multiband antennas. The wideband and multiband antennas can be further classified into multiple elements with a metal rim and multiple elements without metal rim antennas. The metal rim antenna provides excellent mechanical strength as well as aesthetic appearance to mobile phones. Additionally, in compact devices for achieving a higher transmission rate, the MIMO antenna with improved isolation is preferred. Different types of enhancement techniques are used in various antenna structures to increase the gain, improve isolation (mutual coupling) among antennas, bandwidth, envelope correlation coefficient (ECC), and efficiency. The electromagnetic interaction between antenna elements in MIMO is called mutual coupling (MC). In this process, energy is absorbed by the receiver of one antenna when another antenna is radiating energy. Hence, it is essential in MIMO to reduce mutual coupling between antenna elements. It can be calculated mathematically as follows (Nadeem and Choi 2019):

$$MC_{mn} = 1 - \frac{1}{N}\sum_{m}\sum_{m \neq n} \exp\left(-\frac{2x_{mn}}{\lambda}(\alpha + n\pi)\right) \tag{3.1}$$

where MC_{mn} and x_{mn} are the mutual coupling and the distance between mth and nth antenna elements, respectively. The parameter α controls the coupling level, and N is the number of MIMO elements. It is generally calculated in the form of scattering parameters and measured in dB.

3.1.2 Classification Based on Antenna Types

Another classification method can be based on antenna types, as shown in Fig. 3.2. Different types of antennas suitable for 5G applications can be classified as follows (Kumar et al. 2020):

(a) Monopole Antenna

It consists of a straight microstrip line of λ/4 length, where λ is the wavelength of the resonant operating frequency of an antenna. Its basic structure can be changed into different shapes, such as conical, spiral and others modified for different applications with different requirements.

(b) Dipole antenna

It consists of two straight microstrip lines each of λ/4 length, and feeding is provided between two microstrip lines. Thus, the total length of the dipole antenna is λ/2.

(c) Magneto-electric (ME) dipole antenna

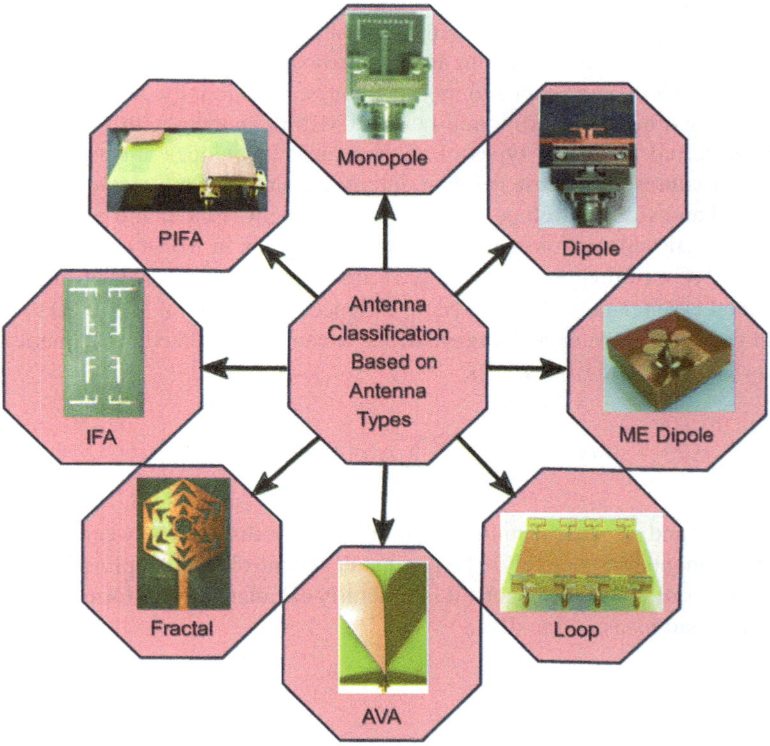

Fig. 3.2 5G Antenna classification based on antenna types. (Adapted from Kumar et al. 2020 with open access (IEEE))

It consists of a planar electric dipole and a vertically short planar magnetic dipole. The feeding is provided to the magnetic dipole from the bottom side of the substrate.

(d) Loop Antenna

It consists of a circular, rectangular, square or any other shape of a ring. The radius of the loop antenna is smaller than the wavelength.

(e) Antipodal vivaldi antenna (AVA)

It consists of two conductors on both sides of the substrate, and they are mirror images of each other. The upper conductor acts as a radiator, whereas the bottom conductor acts as a ground.

(f) Fractal Antenna

It consists of a repetition of the same structure multiple times. It is designed by using an iterative mathematical rule. The fractal antenna can be of different shapes, such as rectangle, circle, star, triangle, and leaf.

(g) Inverted F Antenna (IFA)

It consists of a microstrip line with one bend, and feeding is given to the straight part of the microstrip line. The feed point is near the bent part, and hence, the overall look of an antenna is of inverted F type.

(h) Planar inverted F antenna (PIFA)

It consists of the patch antenna and ground plane, which are connected by using a shorting pin, and feeding is provided from the bottom side of the substrate. As it resonates at quarter wavelength, it requires less space.

The advantages and disadvantages of different antennas are presented in Table 3.1 (Kumar et al. 2020).

3.2 Performance Enhancement Techniques for 5G Antenna Design

Various performance enhancement techniques have been employed in antenna designs that aim to target one or more parameter enhancements, such as bandwidth, gain, efficiency, reduction in the mutual coupling, and compact size. Figure 3.3 shows different antenna performance enhancement and decoupling techniques and their targeting parameters for SISO and MIMO antennas (Yue et al. 2019; Kumar et al. 2020).

3.2.1 General Antenna Performance Enhancement Techniques

Figure 3.3 shows the important antenna performance enhancement techniques that can be employed for 5G antennas, while their advantages and disadvantages are listed in Table 3.2. These techniques include the following (Dixit and Kumar 2020a; Kumar et al. 2020):

Table 3.1 Advantages and disadvantages of different antenna types (Kumar et al. 2020)

Antenna type	Advantages	Disadvantages
Monopole	Simple to design and fabricate In multi-element monopole antenna design, it can be easily rotated in any direction	Less gain Requires large area of ground Gives poor response in bad weather condition
Dipole	Simple to design and fabricate Receives balanced signal	Less gain Cannot be used for long range communication Low bandwidth
Magneto-Electric (ME) Dipole	High front to back ratio Low side lobe and back lobe level Wide bandwidth Low cross polarization	Design and fabrication is complex Costly
Loop	Easy to design Provides good channel capacity	As single element loop antenna cannot meet the 5G requirements, multi-element loop antenna is required Low gain
Antipodal Vivaldi Antenna (AVA)	Enhances the gain Provides wider bandwidth Gives stable radiation pattern	Requires more space Low gain at lower frequencies
Fractal	It helps to miniaturize antenna size Provides wider bandwidth Good impedance matching Provides consistent antenna performance over the operating range	Design is complex Limitation on repetition of fractal design
Inverted F Antenna (IFA)	Smaller in size Good impedance matching due to intermediate feeding	Narrow bandwidth Low gain
Planar Inverted F Antenna (PIFA)	Low profile Good impedance matching Enhances front to back ratio	Narrow bandwidth Low gain

(a) Substrate Choice

The main requirement of an antenna implementation is the appropriate selection of a substrate. Various substrates with different permittivities and loss tangents are available for antenna fabrication. To increase gain and reduce power loss, a substrate with less relative permittivity and low loss tangent must be selected.

(b) Corrugation

The corrugation means removal of a metal part (rectangular, sine, triangular, or square shape) from the edge of a radiator. It helps to improve bandwidth and the front-to-back ratio.

(c) Multielement

Further gain of an antenna can be increased by the multielement antenna. It also enhances the antenna bandwidth and efficiency. In applications where a

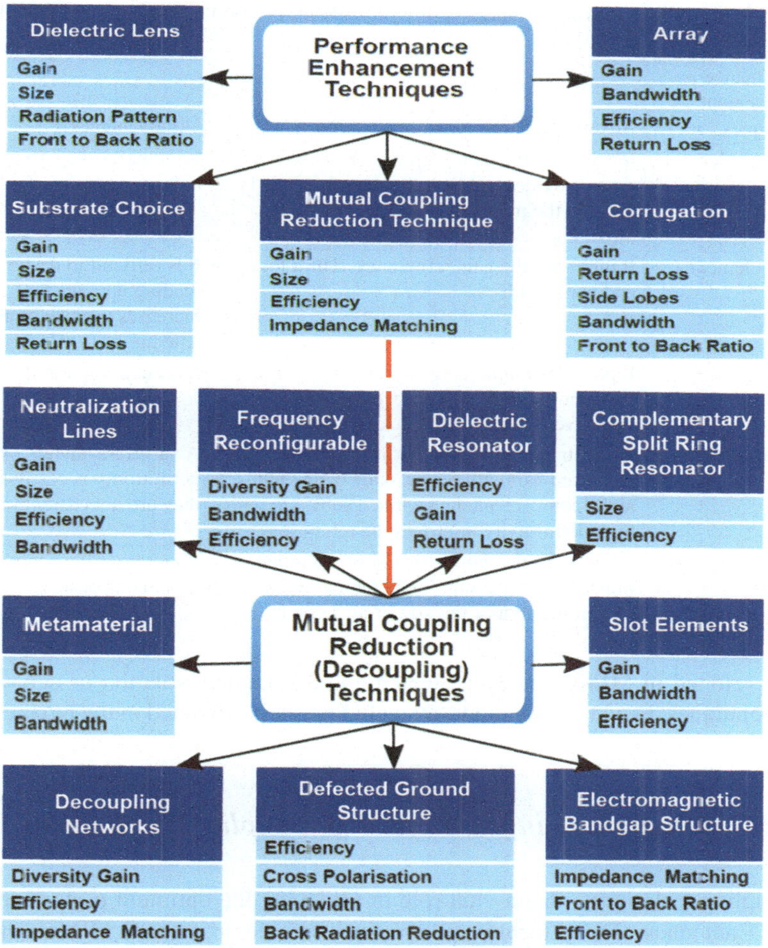

Fig. 3.3 Various performance enhancement techniques for 5G antenna. (Modified from Kumar et al. 2020 with open access (IEEE))

single element antenna cannot fulfil the requirements, such as high gain and wide bandwidth, a multielement is more effective.

(d) Dielectric Lens

Electrostatic radiation is transmitted in one direction by the dielectric lens, which leads to an increase in the gain and directivity of an antenna. There are different shapes of a dielectric lens, and it is designed by using the same or different substrate material with the same or different substrate.

(e) Mutual coupling reduction techniques

In multielement antenna design, the antenna elements affect the performance of each other. To reduce this, different mutual coupling techniques have been

Table 3.2 Advantages and disadvantage of different antenna performance enhancement techniques (Kumar et al. 2020)

Performance enhancement techniques	Advantages	Disadvantages
Dielectric lens	It enhances the gain, improves front to back ratio, provides stable radiation pattern, and radiates the maximum energy in the front direction	It increases the size of an antenna
Multi-element	It improves the gain, efficiency, return loss, and bandwidth	It is difficult to design the feeding network and increases the size of an antenna
Corrugation	It provides improved gain, return loss, and bandwidth. Also, as it reduces side and back lobe levels, the front to back ratio increases	It reduces input impedance
Substrate choice	A substrate having low permittivity gives enhanced gain, efficiency, wide bandwidth, and a compact antenna while a substrate with high permittivity improves the return loss	A substrate having low permittivity is costly and they are not easily available
Mutual coupling reduction	It enhances the gain, efficiency, and input impedance matching. Some mutual reduction techniques reduce the size of an antenna	It increases the complexity of antenna design

developed in MIMO antennas, which are also named isolation or decoupling techniques. Few of these important techniques are explained in the next section.

3.2.2 Mutual Coupling Reduction (Decoupling) Techniques

Decoupling techniques play a vital role in achieving the optimum performance of MIMO antennas. These techniques are an unavoidable part of MIMO antenna design. Their advantages and disadvantages are presented in Table 3.3. These mutual coupling reduction techniques mainly include the following (Nadeem and Choi 2019; Kumar et al. 2020):

(a) Neutralization Lines

Using a metallic slit or lumped element, neutralization lines pass electromagnetic waves between antenna elements to reduce mutual coupling. It reduces the antenna area and improves bandwidth when connected between ground planes. With the change in the location of a point on the neutralization lines, impedance changes, thereby changing the effective bandwidth.

(b) Decoupling Network

In a decoupling network, cross admittance is transformed to a purely imaginary value by adding discrete components or transmission lines. This technique employs a plane decoupling network that acts as a resonator to reduce mutual coupling. The decoupling network includes pattern diversity for multielement,

Table 3.3 Advantages and disadvantage of mutual coupling reduction (decoupling) techniques (Kumar et al. 2020)

Mutual coupling reduction (decoupling) techniques	Advantages	Disadvantages
Defected ground structure	It is easy to implement, to enhance the bandwidth, to improve the front to back ratio, and to increase the efficiency	Its analysis is the challenging issue
Dielectric resonator antenna	It enhances efficiency, bandwidth, and gain	Its structure is complex
Complementary split ring Resonator	It improves diversity gain and reduces antenna size	It provides low bandwidth
Neutralization lines	It is a compact antenna. It gives wider bandwidth and enhanced efficiency	Its structure is complex
Slot or parasitic element	It enhances diversity gain, bandwidth, and efficiency	It is difficult to design and to decide the position of slot or parasitic element
Frequency reconfigurable	It provides compact size and supports multiple wireless standards. Also, it improves diversity gain, bandwidth, and efficiency	It required external components
Electromagnetic bandgap structure	It provides good front to back ratio and impedance matching	Its structure is complex
Metamaterial	It enhances the diversity gain, bandwidth, and ECC. Also, it is compatible for integration with another components	It is difficult to design and decide the position of metamaterial unit cells
Decoupling network	It improves diversity gain and impedance matching	It's gain is low and the design is complex

dummy load, and coupled resonator techniques. It is a cost-effective solution to improve isolation.

(c) Electromagnetic Bandgap (EBG) Structure

It acts as a medium for the transmission of electromagnetic waves. The EBG structure is made up of dielectric or metallic material and has a periodic arrangement. Because of this periodicity-independent resonance, it can produce more than one bandgap. The EBF structure provides low mutual coupling and high efficiency.

(d) Dielectric Resonator

An antenna that contains a dielectric resonator is called a dielectric resonator antenna (DRA). DRA provides high gain, high radiation efficiency, and low loss. DRA can also provide high isolation with dual-band properties.

(e) Defected ground structure (DGS)

It is the structure where the slots or defects are consolidated on the ground plane of the antenna. DGS can be used to provide maximum efficiency, low mutual coupling, and wide bandwidth.

(f) Metamaterials

Different types of metamaterials include single negative, electromagnetic, electromagnetic bandgap, double negative, anisotropic, isotropic, terahertz, chiral, tunable, photonic, frequency selective surface based, and nonlinear metamaterials. Metamaterials are designed by using two or more materials. Using a metamaterial, it is possible to have an antenna with low mutual coupling, high gain, bandwidth, and compact size.

(g) Slot Elements

It is used to enhance impedance bandwidth using the coupling method in the ground plane or the radiation patch. The slot antenna is used to provide wide bandwidth, high gain, high efficiency, and high mutual coupling value.

(h) Complementary Split Ring Resonators (CSRR)

CSRR is used for isolation improvement, to perform filtering function, and to provide lower mutual coupling. CSRRs are also used to provide high efficiency while miniaturizing the size of the antenna. The CSRR is made up of two concentric ring structures with slots opposite each other.

(i) Frequency Reconfigurable

It is based on switching techniques. In the reconfigurable antenna, to increase the frequency range and envelope correlation coefficient, varactor diodes, MEMS switches, and p-i-n are used. The reconfigurable antenna structure can provide lower mutual coupling, a high value of diversity gain, and efficiency.

3.3 Structural Design and Building Materials of 5G Antennas

The antenna can be broadly classified as SISO and MIMO based on input output ports. The MIMO antennas are further classified based on their design, which is multielement without a metal rim and multielement with a metal rim for both wideband and multiband applications. Structural design and building material selection are crucial for each kind of antenna to enhance its performance.

3.3.1 SISO Wideband Antennas

SISO antennas for 5G applications can be categorized into single element and multielement antennas.

3.3.1.1 Single Element Antenna

The single element antenna is easy to design, implement, and fabricate. To reduce the size of an antenna, some SISO 5G antennas with a single substrate layer and multilayer structures have been designed, as shown in Table 3.4. Compared with a single substrate layer antenna, the multilayer design can enhance the gain of a compact antenna (Kumar et al. 2020).

A multilayer and circularly polarized antenna design is shown in Fig. 3.4. Figure 3.4a depicts the design of an antenna that contains two substrates of RT/Duroid 5880 and three layers of copper. Out of these three layers, the top layer is the metasurface layer, which is designed by 4 × 4 square rings. The middle copper layer is the radiator patch, and the bottom copper layer is a ground. The square ring of the metasurface is a series combination of capacitor and inductor, which affects the resonance frequency and hence the bandwidth of an antenna. As depicted in Fig. 3.4b, the bandwidth of the antenna without a metasurface is 24.6–28.7 GHz, whereas the bandwidth of the antenna with a metasurface is 24–34.1 GHz. Furthermore, the metasurface also improves the gain (9.5–11 dBic) and axial ratio bandwidth (24.1–29.5 GHz), as shown in Fig. 3.4c. This gain enhancement of a compact wideband antenna is achieved by practically using multilayer and metasurface enhancement technologies (Hussain et al. 2020).

3.3.1.2 Multielement Antennas

The most important requirements of 5G antennas are high gain, stable radiation pattern, and wider frequency band. These requirements cannot be satisfied by a single element antenna and hence different multielement antennas designed for 5G applications. The comparison of the wideband 5G multielement antennas is given in

Table 3.4 Comparison of single element antennas (SISO wideband) (Kumar et al. 2020)

Antenna type	Substrate	Size (mm^3)	Number of substrate layers	Gain (dBi)	Frequency band (GHz)
Dipole	FR4	40 × 10 × 1	1	2–2.5	3.08–5.15
Antipodal Vivaldi antenna	FR4	40 × 24 × 1.6	1	5–9.53	25–33.4
Circular slot	Nelco NY9220	20 × 16 × 0.508	1	8–9	20–28
Microstrip patch	RO4003C, Taconic TLX-9	90 × 96 × 2.878	2	8.59–10.43	3.24–3.8
ME dipole	Arlon 25N	40 × 40 × 10.516	2	6–8	4.98–6.31
Dielectric resonator antenna	Teflon, ceramic, Rogers 5880	75 × 75 × 15.428	3	6–9.2	3.1–5.1
Microstrip patch	RT/Duroid 5880	12 × 12 × 1.02	2	9.5–11	24–34.1

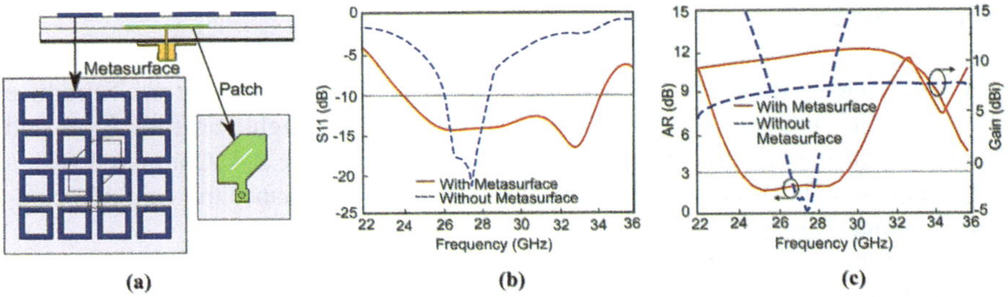

Fig. 3.4 Multilayer antenna design. (Adapted from Hussain et al. 2020 with open access (IEEE)): (**a**) Antenna design; (**b**) Return loss; (**c**) Gain and axial ratio

Table 3.5 Comparison of multi-element antennas (SISO wideband) (Kumar et al. 2020)

Antenna type	Substrate	Size (mm^3)	Number of substrate layer	Gain (dBi)	Frequency band (GHz)
Antipodal Vivaldi antenna	RT/Duroid 5880	28.8 × 24 × 0.254	1	8.2–13.2	24.04–40.85
Antipodal Vivaldi antenna	RT/Duroid 5881	37.6 × 14.3 × 0.254	1	8.5–10.7	23.41–33.92
Fractal	RT/Duroid 5882	32 × 12 × 0.254	1	7.8–10.9	25.28–29.04
Dipole	RT/ Duroid 5883	30 × 35.62 × 4.9	4	10.6–12.61	27.12–29.5
Microstrip patch	Taconic TLY-5	96.1 × 50.5 × 1.016	2	13.83–14.31	26.4–28.92
Microstrip patch	RT/Duroid 5880, Acrylic Polymer	32.1 × 37.45 × 2.124	2	10–12	23–32

Table 3.5. The single substrate layer antennas are generally of moderate size, gain, and bandwidth. To improve the performance of a multielement antenna, a multilayer antenna is designed; however, the overall size of such antennas is large. From Table 3.5, it is observed that a very compact 1 × 4 multielement AVA is designed by incorporating corrugations to achieve a wide bandwidth (Dixit and Kumar 2020b).

The effect of the corrugation technique on the return loss and gain is shown in Fig. 3.5. In Fig. 3.5a, a 1 × 4 multielement AVA design is shown in which corrugations are incorporated at the flat edges of the AVA flares. After incorporating these corrugations, the electric path length of the current at the flat edges increases due to the introduction of inductor (L), resistor (R), and capacitor (C) at the flat edges. This extra RLC circuit changes the resonance frequency of an antenna, as shown in Fig. 3.5b, which proves that the bandwidth of an antenna is increased after incorporating corrugation into it. Furthermore, as the current density increases at the edges, the antenna radiates more energy in the end-fire direction, which in turn enhances

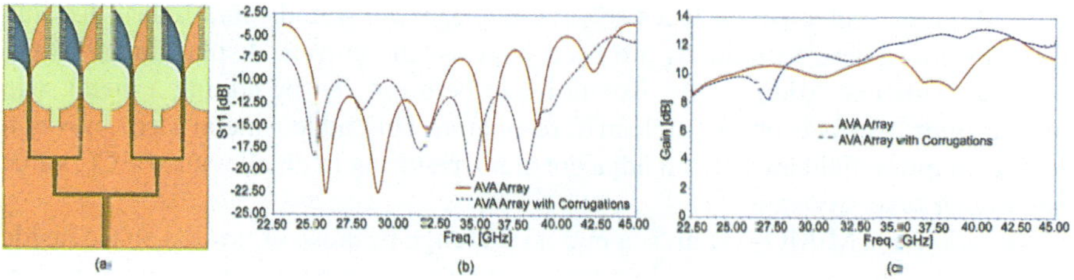

Fig. 3.5 Antenna design with corrugations. (Adapted from Dixit and Kumar 2020b with permission from IEEE): (**a**) Antenna design; (**b**) Simulated return loss; (**c**) Simulated gain

the gain, as depicted in Fig. 3.5c. Thus, the corrugation performance enhancement technique is very useful for bandwidth and gain enhancement (Kumar et al. 2020).

3.3.2 *SISO Multiband Antenna*

SISO multiband antennas can also be classified as single and multielement antennas. Generally, they provide less gain and bandwidth. For example, a dualband antenna operates at 28 GHz and 38 GHz with a low bandwidth of only 3.65 and 2.19 GHz, respectively. A triband antenna is implemented by incorporating a slot technique, but the gain is less. Hence, the SISO multiband is not a good choice for 5G applications (Alibakhshi-Kenari et al. 2016; Deckmyn et al. 2019).

3.3.3 *MIMO Wideband Antennas*

MIMO wideband antennas can be categorized as multielement without a metal rim and multielement with a metal rim.

3.3.3.1 Multielement Without Metal Rim Antennas

The design of MIMO wideband antennas without metal rims mainly focuses on either dual element or multielement antennas.

3.3.3.1.1 Dual Element Antenna Without Metal Rim

In MIMO antennas, dielectric resonator antennas (DRAs) are usually used because of their high efficiency, improved isolation, and enhanced gain. In the MIMO system, the major requirement is to increase the isolation between antenna elements. For instance, the increase in isolation between MIMO DRAs has been achieved with frequency selective surfaces (FSSs), metasurface shields, and hybrid feeding mechanisms. These techniques are used to resist the current displacement within antenna elements. Additionally, isolation can be improved by adding a metal strip on the upper surface of the dielectric resonator. This added metal strip moves a strong coupling field away from adjacent slots, resulting in an alleviated ECC value and higher diversity gain.

Isolation in a MIMO system is a major challenge because of limited space inside the mobile. The neutralization line and decoupling techniques are generally suggested for isolation improvement. In addition, good isolation can be achieved by using a pair of antennas with a self-decoupled structure. This structure is achieved by placing two antenna elements on common ground, which not only improves isolation but also increases the antenna effective length. This self-decoupled antenna structure provides an efficiency of approximately 58%, and the ECC value is less than 0.1. Thus, the self-decoupled structure can provide improved isolation, low ECC, good efficiency, and compact size (Abdalrazik et al. 2017; Zhang et al. 2019).

Table 3.6 compares various MIMO wideband antennas suitable for 5G applications based on various performance parameters, such as antenna size, gain, frequency band, isolation, efficiency, and antenna type used. This shows that the efficiency of the MIMO wideband antenna can be increased by using PIFA and ME dipole antennas, whereas gain enhancement is achieved by using DRA and ME dipole antennas. Additionally, the self-decoupling requires more space compared to the other antennas. It is observed that antennas designed by using DRA provide a higher gain, optimum isolation, much lower ECC value by maintaining a compact size, and high radiation efficiency in the millimeter wave band (Kumar et al. 2020).

Figure 3.6 shows the DRA structure, S parameter, gain, and ECC for MIMO DRA. As shown in Fig. 3.6a, two rectangular-shaped DRAs are mounted on a Rogers 5880 substrate. To enhance the isolation, a metal strip is printed on the top surface of the DRAs. Figure 3.6b depicts the S parameter, which shows that the isolation is improved after adding the metal strip. A large diversity gain and channel capacity are achieved, as shown in Fig. 3.6c. The ECC value achieved is less than 0.013 in the 28 GHz band, and the diversity gain is greater than 9.9 dB (Zhang et al. 2019).

Table 3.6 Comparison of dual element without metal rim antennas (MIMO wideband) (Kumar et al. 2020)

Antenna type	Size (mm^3)	Gain (dBi)	Isolation (dB)	Frequency range (GHz)	Efficiency (%)
DRA	20 × 20 × 2.54	9.9	24	27.25–28.59	–
Monopole	150 × 75 × 0.8	–	17	3.4–3.6	58
PIFA	50 × 100 × 3.00	3	25	2.7–3.6	80–92
ME dipole	60 × 60 × 8	8.2	25	3.3–4.36	89.5

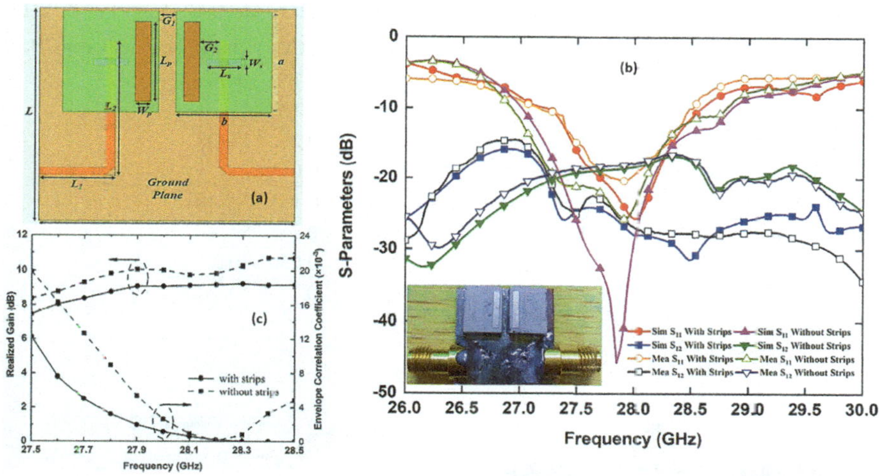

Fig. 3.6 Dielectric resonator antenna design. (Adapted from Zhang et al. 2019 with permission from IEEE): (a) Top view of MIMO DRA; (b) S parameters; (c) Gain and ECC

3.3.3.1.2 Multielement Antenna Without Metal Rim

An MIMO multielement antenna is used to enhance the gain and the transmission range of the signal, which is a prime requirement of any high-speed communication application. These antennas include structured monopoles, slot antennas, and dual and tri-polarized antennas for 5G applications. Along with MIMO, cognitive radio (CR) is also the core technology for 5G communication. A combination of CR and MIMO creates a more efficient system called filtenna, which provides increased spectrum efficiency and data rate, respectively. To enhance data throughput and radiation efficiency, monopolar patch antennas (MPAs) have been designed to generate monopole radiation patterns. MPAs are also combined in a Y-type structure for MIMO operation with an efficiency of 88% and ECC of approximately 0.1. This antenna is suggested for 5G access point applications (Gomez-Villanueva and Jardon-Aguilar 2019).

MIMO advantages can be realized by minimizing mutual coupling. Various mutual coupling reduction techniques have been proposed, such as polarization diversity, orthogonal polarization, DGS, and neutralization lines. The MIMO antenna, which is uniplanar and uses polarization diversity, reduces mutual coupling, which further makes the 5G antenna more resistant to interference and fading (Khan et al. 2015).

A comparison of the MIMO wideband multielement antenna is shown in Table 3.7. This comparison is based on the type of mutual coupling reduction technique, frequency range used, isolation, ECC, and channel capacity. To obtain high isolation and lower ECC values, the polarization technique is usually used in monopole antennas. Enhanced channel capacity of the SIW antenna is obtained with

Table 3.7 Comparison of multi-element without metal rim (MIMO wideband) (Kumar et al. 2020)

Antenna type	Frequency range (GHz)	Isolation (dB)	ECC	Channel capacity (bps/Hz)
Monopole	3.3–4.2	15	0.1	16.5
Filtenna	2.5–4.2	15	<0.5	–
Monopole	3.4–3.6	10	0.2	35–38
Monopole	5.1–5.9	17	0.01	–
Monopole	2.55–2.65	12.5	0.15	38–40
SIW antenna	3.4–3.6	12.5	0.2	57
Slot	3.4–3.6	17.5	0.05	40.8
Inverted F	2.5–7.0	17	0.1	39
Microstrip patch	24.35–31.13	20	–	–

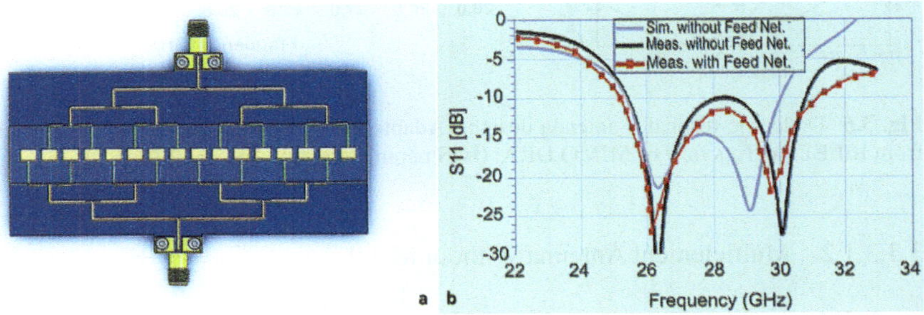

Fig. 3.7 Fabricated multi-element antenna with performance parameters. (Adapted from Khalily et al. 2018 with permission from IEEE): (**a**) Fabricated MSA with parasitic element; (**b**) Simulated S11 with and without feeding network

orthogonal polarization. Enhanced isolation is achieved by using polarization diversity and an F-shaped stub. Out of these antennas, the antenna design with a microstrip patch provides high gain, stable radiation pattern, enhanced isolation, and wider bandwidth (Khan et al. 2015; Khalily et al. 2018).

The fabricated 16 element multielement antenna and its performance parameters are shown in Fig. 3.7. This 16 element antenna is designed by using two multielement antennas in which two substrates of equal size are used. The patch present on the bottom substrate is used as a radiator, whereas the patches present on the top substrate are utilized as parasitic patches for enhancement of bandwidth, as shown in Fig. 3.7a. To decrease mutual coupling between antenna elements, feed lines and radiating patches allowed us to rotate 180 degrees out of phase. In Fig. 3.7b, return loss S_{11} and frequency plot is shown with and without a feeding network. This feeding network operates at a definite phase difference to obtain the E plane and H plane gain. This antenna provides a very high gain of 19.88 dBi and an impedance bandwidth of 5.37 GHz. Hence, this antenna is suitable for 5G applications (Khalily et al. 2018).

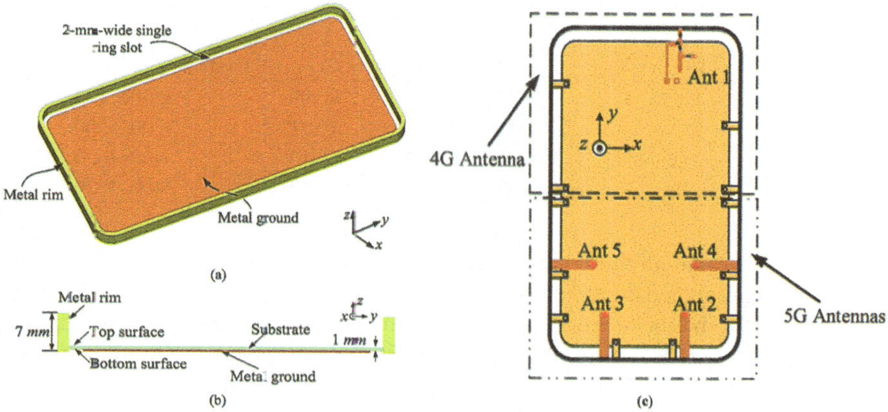

Fig. 3.8 The smartphone metal shell and slot antennas. (Adapted from Chen et al. 2019 with permission from IEEE): (**a**) 3D view; (**b**) Side view; and (**c**) Assembly of single ring slot-based antennas on metal-rimmed 4G/5G smartphones

3.3.3.1.3 Multielement Antenna with Metal Rim

Metal rim-based MIMO antennas are the key elements in the design of smartphone antennas suitable for 5G communication. Loop antennas and slot antennas are widely used to incorporate all the required LTE bands along with 5G bands. Typically, metal rim MIMO antennas can be either fixed or reconfigurable for wideband and multiband applications. Figure 3.8 represents the smartphone metal shell structure. Figure 3.8a represents the 3D view of the smartphone metal shell, Fig. 3.8b represents the corresponding side view, and Fig. 3.8c illustrates the assembly of single ring slot-based antennas on metal-rimmed 4G/5G smartphones. The metal shell consists of a substrate, metal rim, and a 2 mm wide single ring slot. A metal ground present below the substrate is firmly surrounded by an unbroken metal rim. The antenna is designed along the rim of the metal ground (Chen et al. 2019).

Additionally, MIMO metal rim antennas are of two types: reconfigurable and fixed antennas. The important feature of a reconfigurable antenna is that it is a frequency tunable antenna, whereas the frequency band is not variable in a fixed type of antenna. Compared to fixed-type metal rim antennas, reconfigurable (tunable) metal rim antennas increase functionality in communications and radar applications without increasing the size and cost of the system. Additionally, the reconfigurable antenna provides multifrequency operation and polarization diversity. The tunable antennas can be realized using pin diodes and varactor diodes, MEMS devices (switches, phase shifters), and tunable materials such as liquid crystals (Kumar et al. 2020).

Table 3.8 compares some MIMO wideband metal rim antennas suitable for 5G smartphone applications based on various performance parameters, such as antenna type, type of substrate used, efficiency, ECC, and the frequency bands covered.

Table 3.8 Comparison of mutli-element with metal rim antennas (MIMO wideband) (Kumar et al. 2020)

Antenna type	Efficiency(%)	Isolation (dB)	ECC	Frequency band (MHz)
Hybrid IFA	47–70	>11	<0.09	3300–7100
Monopole	57.8–74.7	>12	<0.11	3300–5000
Slot	35.2–64.7	>12.7	<0.13	3400–3600
Slot	>42	>13	<0.15	3400–3600

Taking into account all these parameters, the Slot 1 antenna provides better performance and is designed as a fixed type 4 × 4 MIMO wideband multielement antenna integrated with the metal rim. The dual antenna pair supports the in-phase current mode and slot mode, resulting in a polarization orthogonal mode and thereby enhancing diversity performance. One of the key factors for high isolation includes a symmetrical feeding network (Chang et al. 2019).

3.3.4 MIMO Multiband Antennas

Wireless communication plays an important role in the rapid growth of information technology, where channel capacity is a major issue. The solution for the same is multiband antennas, as it is very difficult to use separate antennas for different applications. Multiple frequencies supported by a single antenna are more suitable for wireless and mobile communication applications because of the compact size requirement.

References

Abdalrazik A, El-Hameed ASA, Abdel-Rahman AB (2017) A three-port MIMO dielectric resonator antenna using decoupled modes. IEEE Antennas Wirel Propag Lett 16:3104–3107

Alibakhshi-Kenari M, Naser-Moghadasi M, Sadeghzadeh RA, Virdee BS, Limiti E (2016) Bandwidth extension of planar antennas using embedded slits for reliable multiband RF communications. AEU-Int J Electron Commun 70(7):910–919

Chang L, Yu Y, Wei K, Wang H (2019) Polarization-orthogonal co-frequency dual antenna pair suitable for 5G MIMO smartphone with metallic bezels. IEEE Trans Antennas Propag 67(8):5212–5220

Chen Q, Lin H, Wang J, Ge L, Li Y, Pei T, Sim C-Y-D (2019) Single ring slot-based antennas for metal-rimmed 4G/5G smartphones. IEEE Trans Antennas Propag 67(3):1476–1487

Deckmyn T, Cauwe M, Ginste DV, Rogier H, Agneessens S (2019) Dual-band (28,38) GHz coupled quarter-mode substrate-integrated waveguide antenna array for next-generation wireless systems. IEEE Trans Antennas Propag 67(4):2405–2412

Dixit AS, Kumar S (2020a) A survey of performance enhancement techniques of antipodal Vivaldi antenna. IEEE Access 8:45774–45796

Dixit AS, Kumar S (2020b) A miniaturized antipodal Vivaldi antenna for 5G communication applications. In: 2020 7th international conference on Signal Processing and Integrated Networks (SPIN), 2020, pp 800–803. https://doi.org/10.1109/SPIN48934.2020.9071075

Gomez-Villanueva R, Jardon-Aguilar H (2019) Compact UWB uniplanar four-port MIMO antenna array with rejecting band. IEEE Antennas Wirel Propag Lett 18(12):2543–2547
Hussain N, Jeong MJ, Abbas A, Kim TJ, Kim N (2020) A metasurface-based low-profile wideband circularly polarized patch antenna for 5G millimeter-wave systems. IEEE Access 8:22127–22135
Khalily M, Tafazolli R, Xiao P, Kishk AA (2018) Broadband mm-wave microstrip array antenna with improved radiation characteristics for different 5G applications. IEEE Trans Antennas Propag 66(9):4641–4647
Khan MS, Capobianco AD, Asif S, Iftikhar A, Ijaz B, Braaten BD (2015) Compact × UWB-MIMO antenna with WLAN band rejected operation. Electron Lett 51(14):1048–1050
Kumar S, Dixit AS, Malekar RR, Raut HD, Shevada LK (2020) Fifth generation antennas: a comprehensive review of design and performance enhancement techniques. IEEE Access 8:163568–163593
Nadeem I, Choi DY (2019) Study on mutual coupling reduction technique for MIMO antennas. IEEE Access 7:563–586
Yue T, Jiang ZH, Werner DH (2019) A compact metasurface-enabled dual-band dual-circularly polarized antenna loaded with complementary split ring resonators. IEEE Trans Antennas Propag 67(2):794–803
Zhang Y, Deng J-Y, Li M-J, Sun D, Guo L-X (2019) A MIMO dielectric resonator antenna with improved isolation for 5G mm-wave applications. IEEE Antennas Wirel Propag Lett 18(4):747–751

Chapter 4
PCB Materials and Design Requirements for 5G Systems

Abstract Printed circuit boards (PCBs) are an important building block in electronic systems. The choice of PCB material for 5G applications depends on the frequency, power level, circuit size and function. Designers have choices for substrate materials from basic epoxy/glass (FR-4), mid-loss materials and, ultimately, high end microwave/millimeter wave materials. The demanding performance requirements of 5G will push the limit of PCB designs from antennas to control functions to amplifier circuits. PCB material parameters such as copper surface roughness, dielectric constant variations, thermal dissipation, passive intermodulation, coefficient of thermal expansion and thickness variations will affect 5G designs more than previous generations that had less stringent performance criteria. PCB designers will have to consider many of these factors in their designs so they need to learn about the various tradeoffs with different types of PCB materials and processes. 5G mmWave applications will have very different material requirements than sub6 GHz applications. At mmWave frequencies, the effects of small variations in material parameters will affect the circuit's performance even more. For higher frequencies into the mmWave range, designers typically need to minimize dielectric permittivity, copper roughness and thickness variations, as they can all negatively affect performance. These considerations must be incorporated into the design using accurate models and simulations to avoid costly iterations. This chapter will provide a brief review of PCB design requirements and the selection of different materials for 5G systems. The effects of various material parameters on loss, thermal dissipation and reliability are also discussed to provide designers with the tools needed to account for these affects.

4.1 The Evolution of Printed Circuit Boards

In electronics, printed circuit boards (PCBs) are used to mechanically support and electrically connect electronic components using conductive pathways or traces etched from copper sheets and laminated onto a nonconductive substrate. After

C. Tong, *Advanced Materials and Components for 5G and Beyond*, Springer Series in Materials Science 327, https://doi.org/10.1007/978-3-031-17207-6_4

populating the board with electronic components, a printed circuit board assembly (PCBA) is formed. Printed circuit boards have changed in various ways since their inception. There were many noticeable developments that occurred over the course of its lifespan, which has made circuit boards so useful in the modern day, where technology has a significant impact on human lives.

4.1.1 History

The inventor of the printed circuit was the Austrian engineer Paul Eisler, who made one in England approximately 1936 as part of a radio set. Approximately 1943, the United States began to use the technology on a large scale to make rugged radios for use in World War II. After the war, in 1948, the U.S. government released the invention for commercial use. Printed circuits did not become commonplace in consumer electronics until the mid-1950s, after the AutoSembly process was developed by the U.S. Army, in which component leads were inserted into a copper foil interconnection pattern and dip soldered. With the development of through-hole technology, this concept evolved into the standard printed circuit board fabrication process in use today. Through-hole technology involves mounting the leads of the component in holes drilled on the board and soldering the leads in place from the underside of the board. Soldering could be done automatically by passing the board over a ripple, or wave, of molten solder in a wave-soldering machine. However, the wires and holes are wasteful since drilling holes are expensive and the protruding wires are merely cut off. Because of the need to drill holes into the board, the available space and routing area were always limited when manufacturing using through-hole technology.

Surface-mount technology (SMT) became a mainstream manufacturing technology for electronics on printed circuit boards in the 1980s, leading to a significant reduction in size, cost, and complexity. SMT allows for more components to be placed in the same space compared to through-hole technology because no holes have to be drilled. Furthermore, components can be placed on both sides of the board. Most importantly, SMT boards can be fabricated in multiple layers, which makes them a great candidate for the implementation of high-speed electronics by providing precision control over the impedance of the traces and electromagnetic interference. Regarding current PCB manufacturing technology, SMT is heavily favored over through-hole. However, a through-hole is still used for simpler boards and is easier to solder by hand due to the larger size as opposed to SMT boards. The vast majority of boards manufactured today employ surface-mount technology, and anyone can utilize CAD software to create a design, send it to a manufacturer, and have their own PCB constructed (Shamkhalichenar et al. 2020).

4.1.2 Materials and Fabrication Process

The most common printed circuit board substrate is known as FR4 (flame retardant 4). The basis of FR4 is composed of woven fiberglass cloth combined with an epoxy resin binder that is flame resistant. FR4 has near-zero water absorption, as well as an excellent strength to weight ratio, and is an excellent insulator regardless of moisture levels in the ambient environment. Typically, the flame-resistant material in FR4 is bromine. In addition, FR4 is easy to manufacture and is usually the cheapest available material. However, other materials are often used depending on the environment in which the board will be placed, the budget, and the required circuit properties.

Polyimide laminates offer performance improvements in every category over FR4, most importantly, higher temperature performance, electrical performance, survivability, and resistance to expansion. The cost to manufacture polyimides, however, is higher than that of FR4. Teflon laminates offer improved electrical properties over both FR4- and polyimide-based substrates; however, they cost significantly more to produce than both and require specialized equipment and a highly skilled workforce. Teflon laminates can be coated onto glass fabric or manufactured as an unsupported mesh, giving it an adaptability factor that neither FR4 nor polyimide possesses.

PCBs are made from substrates with a copper foil or coating bonded to one or both sides. That is, the printed circuits are made of copper, which is either plated or etched away on the surface of the substrate to leave the desired pattern. The copper circuits are coated with a layer of tin or other metals to prevent oxidation. Contact fingers are plated with tin-lead or other metals, then nickel, and finally gold for excellent conductivity.

PCB fabrication is the process that transforms a circuit board design into a physical structure based upon the specifications provided in the design package. This physical manifestation is achieved through the following actions or techniques:

- Imaging the desired layout on copper clad laminates
- Etching or removing excess copper from inner layers to reveal traces and pads
- Creating the PCB layer stack-up by laminating (heating and pressing) board materials at high temperatures
- Drilling holes for mounting holes, through hole pins and vias
- Etching or removing excess copper from the surface layer(s) to reveal traces and pads
- Plating pin holes and via holes
- Adding protective coating to surface or solder masking
- Assembling or mounting electronic components when PCBA is required
- Packing for electronic systems

As PCB technology has progressed over the years, some key features have gradually improved. The most important of these are the increased number of conducting layers (~40 or more), minimum trace spacing (~100 μm or less), and minimum trace width (~100 μm or less).

Many applications require cyclic movement or stretching while still maintaining an electrical connection. To fit this need, flexible circuit boards have become an ever-evolving solution. Additionally, additive manufacturing has come into integrating electronics into parts. There are 2D inkjet styling printing machines that are already used to print circuit boards. 3D printed circuit board technology uses an extruder head that can lay down beads of solder or conductive material on a printing surface in layers. Couple this "wire printing" method with a secondary material head, and manufacturers are gaining the ability to create 3D printed circuit boards with intricate internal wiring. The advantage of this technique is that a given circuit is not constrained by the traditional flat PCB board and can be made to perfectly fit the shape of a given product. Having the ability to design 3D connections in compact spaces is something that is otherwise impossible within the ways that circuit boards are currently produced. With the combination of these two methods, 2D and 3D additive circuit board manufacturing, a drastic change in PCB fabrication is beginning to form.

4.2 RF and High Frequency PCB Technologies

Compared to lower frequency circuits, high-frequency RF/microwave PCBs are sensitive to circuit materials and fabrication processes. Whereas some electrical circuit functions, such as power lines and digital control, may be well supported with low-cost FR-4 substrates and circuit materials, RF/microwave and mmWave circuits require much higher performance circuit materials to minimize signal losses and distortion. Many multilayer mixed-signal PCBs with different electrical functions are a blend of different types of circuit substrate materials, with materials selected according to behavior best suited for the types of circuit functions fabricated on that layer. The vertical transitions from layer to layer can introduce signal integrity issues at high frequencies. This includes impedance mismatch, reflections, electromagnetic interference, bandwidth variation, mode conversion, and insertion loss.

4.2.1 Basic Circuit Configuration of High-Frequency PCBs

High-frequency PCBs are typically based on one of three common circuit configurations – transmission lines (Coonrod 2021): microstrip, stripline, or grounded coplanar waveguide (GCPW) circuits, as shown in Fig. 4.1. The conductors and ground planes are configured differently for each circuit type, with the electric (E) field and current density depicted for each with single-ended transmission lines. The E fields are shown as coupled from the conductors to the ground planes. For

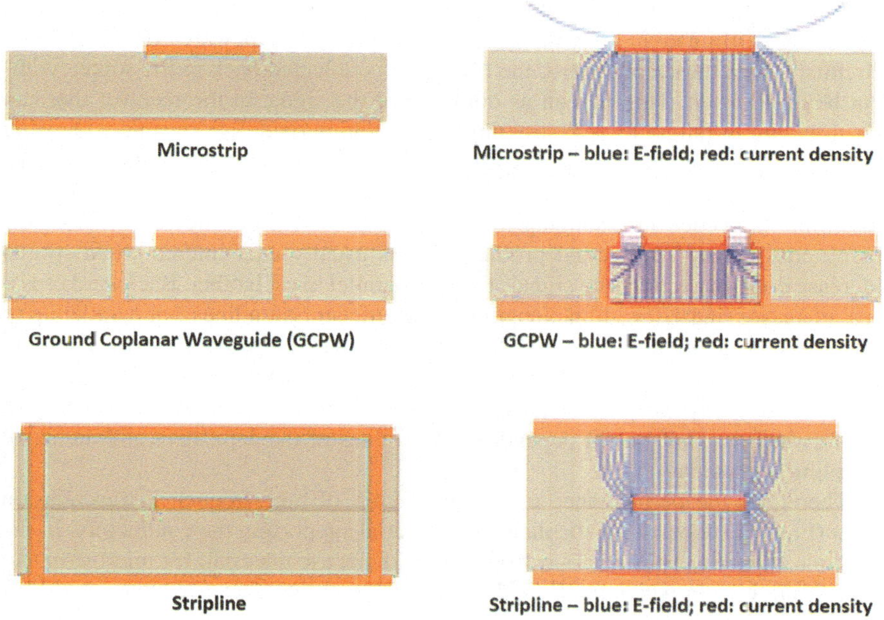

Fig. 4.1 Cross-sectional views (left) for three common high-frequency circuit types, microstrip, grounded coplanar waveguide (GCPW), and stripline circuits, and corresponding electric (E) fields and current density for each circuit type (right). (Adapted from Coonrod 2021, credit: Printed Circuit Design & Fab)

differential circuits, two conductors are used. For a differential microstrip circuit, for example, the E fields would be coupled between two signal conductors on the top layer and the ground plane below. High-frequency circuits based on such transmission-line configurations are sensitive to circuit material parameters and PCB fabrication details, especially at higher frequencies. Moreover, the depictions of the three circuit types on the right side of Fig. 4.1 are rough approximations and will appear differently with changes in frequency. At higher frequencies, the current density will be thinner in a cross-sectional view due to skin effects, while the E fields will be more condensed. Electromagnetic (EM) waves generated by variations in each circuit's E and magnetic (M) fields (not shown) would propagate in a direction perpendicular to the two-dimensional circuit depictions, outward from the page.

From electromagnetic (EM) theory, to achieve highly efficient point-to-point transmission of power and information, the source energy must be guided. Conductors such as wires used to transmit high-frequency communication signals can be regarded as transmission lines that guide the transverse electromagnetic (TEM) waves along them. The electromagnetic energy launched down a transmission line as a signal must be absorbed as it arrives at the end of the transmission line or it will be reflected back to the source, resulting in overshoot that could result in the voltage exceeding the maximum input voltage rating of the input. Since most signal lines terminate in logic inputs at their receive ends, they are effectively open circuits. Under these conditions, the EM wave is reflected back to the source without being inverted. This is seen as doubling of the signal at the receiver that can result in circuit failure by overstressing the insulating oxides or causing a parasitic junction to conduct. Even a clock speed of only 1 Hertz can result in malfunctions. The fastest rise times of the components used in a design determine what level of discipline is needed in managing transmission lines and not the clock frequency being considered. The second problem that is usually encountered when speeds increase is crosstalk between signals that run parallel to each other. Backward crosstalk increases linearly with the length that two transmission lines run parallel up to a point where saturation is reached. Beyond this length, known as the critical length, continuing to run in parallel results in no additional backward crosstalk. The forward crosstalk depends on the rate of change of the input signal, and the polarity of the forward crosstalk signal depends on the values of the capacitive and inductive coupling coefficients.

Therefore, the performance of all three types of high-frequency transmission lines (Fig. 4.1) depends on the dielectric material supporting the conductors. In the stripline case, the conductor is surrounded by dielectric material; for microstrip and GCPW circuits, in which the E fields extend beyond the dielectric material, the air surrounding the circuit contributes to the total dielectric environment of the circuit, so propagating waves in these circuits travel through media with a dielectric constant (D_k) that is a combination of the D_k of the substrate material and air, which is approximately 1. The resulting dielectric environment for microstrip and GCPW circuits is what is known as the effective D_k. The GCPW depicted in Fig. 4.1 is tightly coupled, with small spaces between the ground planes and signal conductor on the top coplanar circuit layer. Larger spaces result in a more loosely coupled GCPW circuit. Tightly coupled GCPWs have a greater percentage of E fields in air than loosely coupled GCPWs, with air contributing to a lower effective D_k for tightly coupled GCPWs than for loosely coupled GCPWs, where a greater percentage of the E fields is in the dielectric substrate material (with a higher D_k than air) (Coonrod 2021).

4.2.2 Transmission Line Parameters Used in RF/High Frequency PCB Design

Advances in integrated circuit (IC) fabrication technology have led to huge increases in both the operating frequencies and edge rates of switching signals in digital circuits. This has driven digital circuit design into frequency domains, including RF, microwave, and/or mm-Wave. As a result, it is important to contemplate the transmission line parameters of the connecting traces used for digital signals in PCBs as much as they have always been considered for RF connections.

1. Characteristic impedance

Electromagnetic waves propagate at the speed of light (3×10^8 m/s) along a lossless conductor surrounded by vacuum and at a slower speed when surrounded by a dielectric material. As an electromagnetic signal propagates along a conductor, the ratio of its voltage amplitude to its current amplitude is called the characteristic impedance, which has units of ohms, denoted with the symbol Z_0. It is a characteristic of transmission lines experienced by propagating electromagnetic waves. The input impedance of an infinite line is equal to the characteristic impedance since the transmitted wave is never reflected back from the end. The general expression for the characteristic impedance of a transmission line is (Le 2019):

$$Z_0 = \sqrt{\frac{R + j\omega L}{G + j\omega C}} \quad (4.1)$$

where j is the imaginary number, R is the characteristic resistance (Ω/m), considering the two conductors to be in series, G is the characteristic conductance (1/(Ω m)) of the surrounding dielectric material, L is the characteristic inductance (Henrys/m), C is the characteristic capacitance (Farads/m), and ω is the angular frequency (radians/s).

For well-functioning transmission lines, R and G are usually too small to be neglected. This assumption is called a lossless transmission line, which is defined as a transmission line that has no line resistance and no dielectric loss. For a lossless transmission line, the characteristic impedance can be simplified as:

$$Z_0 = \sqrt{\frac{L}{C}} \quad (4.2)$$

In particular, Z_0 no longer depends upon the frequency. For a lossless transmission line terminated in Z_0, there is no loss of current across the line, so the voltage remains the same along the line; their magnitudes remain constant along the length of the line and are only rotated by a phase angle. The lossless transmission line model is a useful approximation for many practical cases, such as low-loss transmission lines and transmission lines with high frequency. For both of these cases, R and G are much smaller than ωL and ωC, respectively, and can thus be ignored.

2. Signal speed and propagation delay

The speed of a signal depends on the materials that surround the conductors and how quickly the changing electric and magnetic fields associated with the signal can build up and propagate in the space around the transmission line conductors. That is, the signal is a propagating voltage difference between the signal path and the return path. As the signal propagates, a voltage difference must be created between the two conductors. Accompanying the voltage difference is an electric field between the conductors. In addition to the voltage, a current must be flowing in the signal conductor and in the return conductor to provide the charge that charges up the conductors that generates the voltage difference that creates the electric field. This current loop moving through the conductors will produce a magnetic field. The sudden voltage change creates a sudden electric and magnetic-field change. This kink of field will propagate through the dielectric material surrounding the transmission line at the speed of a changing electric and magnetic field, which is the speed of light or signal in the material. In other words, the propagation and interaction of these fields is described by Maxwell's Equations; if the electric and magnetic fields ever change, the kink they make will propagate outward at a speed that depends on some constants and material properties. Therefore, the speed of the change (kink), or the signal speed, V (m/s), is given by (Bogatin 2018):

$$V = \frac{1}{\sqrt{\varepsilon_0 \varepsilon_r \mu_0 \mu_r}} \tag{4.3}$$

where ε_0 is the permittivity of free space (8.89×10^{-12} F/m), ε_r is the relative dielectric constant of the material, μ_0 is the permeability of free space ($4\pi \times 10^{-7}$ H/m), and μ_r is the relative permeability of the material. In vacuum, the relative dielectric constant and relative permeability are both 1 (in air, they are close to 1). For virtually all interconnect materials, the magnetic permeability of the dielectrics, μ_r, is 1. All polymers that do not contain a ferromagnetic material have a magnetic permeability of 1. Therefore, Eq. (4.3) can be simplified as:

$$V = \frac{c}{\sqrt{\varepsilon_r}} \tag{4.4}$$

where c is the speed of light in vacuum or air for a close approximation (3×10^8 m/s).

In comparison, the relative dielectric constant of materials ε_r is never less than or equal to 1, except in the case of air. In all real interconnect materials, the dielectric constant is greater than 1. This means that the speed of light in interconnects will always be less than 3×10^8 m/s. It is possible in some materials for the dielectric constant to vary with frequency. In other words, the speed of light in a material may be frequency dependent, which is called property dispersion. In general, the dielectric constant decreases with higher frequency. This makes the speed of light in the material increase with a higher frequency. In most applications, dispersion is very small and can be ignored.

The signal propagation delay, which is the inverse of the propagation speed, can be expressed as:

$$t_d = \frac{L}{V} = \frac{L\sqrt{\varepsilon_r}}{c} \tag{4.5}$$

where L is the interconnect length.

Printed circuit boards are usually made up of multiple conductive layers. Conductors on inner layers (stripline) are completely surrounded by PCB insulating material, and conductors on outer layers (microstrip) have PCB insulating material beneath them and air above. This mixed environment for outer conductors causes their propagation delay to be dependent on an effective dielectric constant, which is a combination of that of the PCB material and that of air. The dielectric constant for air is approximately the same as that for vacuum, which is 1. The effective dielectric constant for outer conductors can be estimated as $\varepsilon_{\text{reff}} \approx 0.64\ \varepsilon_r + 0.36$. The standard PCB material is FR4, which has a dielectric constant of approximately 4.0 and an effective dielectric for the outer layers of 2.92. With these values, the propagation delay time is approximately 6.69 ps/mm for the inner layers and approximately 5.51 ps/mm for the outer layers.

The signal wavelength (λ) is equal to the inverse of the propagation delay (t_d) times frequency (f) in Hz (Practical EE 2019):

$$\lambda = \frac{1}{t_d \Delta f} \tag{4.6}$$

When the wavelength is long compared to the conductor length, the system is called a lumped element system and does not need to consider transmission line effects. When the signal wavelength is on the order of the conductor length, the system is a transmission line and needs to take its effects into account. In general, for analog signals, the critical conductor length is 1/4 the wavelength of the highest frequency component – Conductors longer than this are transmission lines; for digital signals, the critical conductor length is 1/6 the physical length occupied by the rising or falling edge, whichever is faster. Conductors longer than this are transmission lines (Practical EE 2019).

3. Impedance mismatch and reflection coefficient

Transmission lines introduce the potential for reflected signals. If a transmission line has no change in characteristic impedance along its length (assuming the transmission line is lossless), a voltage or current signal will be transmitted down the line without attenuation or distortion. If the transmission line is somewhat lossy, then the signal amplitude will be attenuated. If the transmission line has changes in characteristic impedance down the line, then the signal will be distorted due to signal reflections that occur at each place where characteristic impedance changes. When a propagating wave (incident wave) encounters a change in impedance, a wave will be reflected back toward the driver (reflected wave), and a wave will be transmitted beyond the impedance discontinuity (transmitted wave). Figure 4.2 shows transmission line reflection with an arbitrary load. A reflection can occur at either end of the transmission line. The reflection coefficient is defined as the ratio of the reflected voltage to the incident voltage and is a key parameter in RF systems. It can be expressed in terms of characteristic impedance and load impedance (Pozar 2011):

$$\Gamma = \frac{V_0^-}{V_0^+} = \frac{Z_L - Z_0}{Z_L + Z_0} \tag{4.7}$$

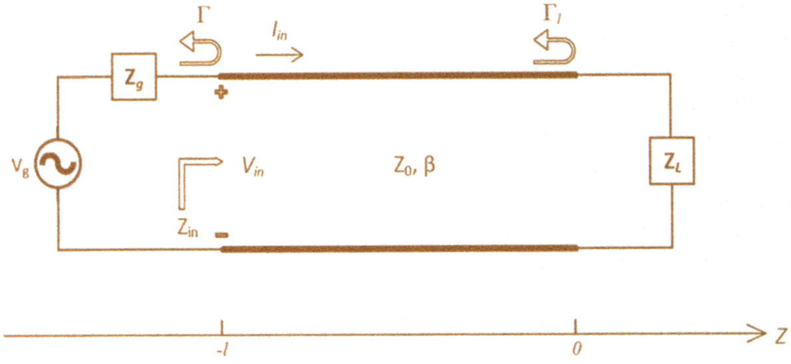

Fig. 4.2 A terminated transmission line (TL) with arbitrary load ZL. (Adapted from Pozar 2011, credit: https://www.3ds.com/products-services/simulia/resources/transmission-line-matching/). The ideal lossless transmission line block is designed for a characteristic impedance of 50 Ω at 3 GHz and an electrical length of 0.5λ: Source voltage Vg = 8 volts; Fixed generator impedance Zg = 20 + j5 Ω; Line impedance of the ideal TL block Z0 = 50 Ω; Frequency of operation 3 GHz

where

$$Z_L = \frac{V(z)}{I(z)} = \frac{V_0^+ + V_0^-}{V_0^+ - V_0^-} Z_0,\ V(z) = V_0^+ e^{-j\beta z} + V_0^- e^{j\beta z}, I(z) = \frac{V_0^+}{Z_0} e^{-j\beta z} + \frac{V_0^-}{Z_0} e^{j\beta z.}$$

This equation indicates that reflections occur when the impedance of a transmission line does not match the impedance of the load. Thus, to achieve maximum power transfer, it is required that the impedance of the generator and load match the characteristic impedance of the transmission line. In a multilayer PCB, the via presents a discontinuity in impedance and can cause reflections. If the load impedance is higher than the characteristic impedance of the transmission line, the reflected wave will have a positive amplitude and constructively interfere with the incident wave. Thus, the reflection coefficient will show an increase once the reflected wave reaches the port. If the impedance is lower than the characteristic impedance of the transmission line, the reflected wave will have a negative amplitude and destructively interfere with the incident wave. This will show a decrease in the reflection coefficient once the reflected wave reaches the port. Based on the time delay and the speed of propagation, the location at which a reflection is induced and whether the junction is higher or lower in impedance can be determined (Le 2019).

4. Driver and receiver

The origin of a signal propagating along a transmission line is called a driver or source, and the destination of the signal is called a receiver or load. The circuitry of the driver and receiver have impedances that are experienced by propagating signals and can cause reflections. For a digital circuit driver, the output impedance is usually low (~20–50 Ω), and the receiver input impedance is high (~1 MΩ). The equivalent resistance of the driver is R_s. The equivalent impedance of the receiver is just R_L, as shown in Fig. 4.3 (Practical EE 2019).

For a digital signal, for example, the driver transitions the logic state of the signal from low to high. The voltage source Vs drives the signal high, and the signal encounters is a voltage divider formed between the source resistance Rs and the characteristic impedance of the transmission line Z_0. The voltage transition wave that actually flows down the transmission line is a divided down version of Vs: Vs * $Z_0/(Rs + Z_0)$. This wave travels down the uniform transmission line unperturbed until it reaches load R_L. If R_L is not equal to Z_0, a reflection is generated. The receiver at the load would receive the transmitted wave, which is the incident wave plus the reflected wave, and the reflected wave would travel back toward the source. Once the reflected wave reaches the source, if Rs is not equal to Z_0, there is an impedance discontinuity, and another reflection is generated, traveling toward the load. The time it takes for the waves to bounce around and reach a steady-state value is called the settling time. When settled, the final value at the load will be the value for a lumped parameter system without transmission line effects. In the circuit shown in Fig. 4.3, the voltage divider value $V_S * R_L/(R_L + R_S)$ involves only the load and source resistances, not the characteristic impedance of the transmission line, because that is a transitory phenomenon (Practical EE 2019).

5. Signal loss in high-frequency PCBs

Maintaining an impedance-matched environment is critical for minimizing the loss of interconnected components in a high-frequency PCB circuit or system, such as the transmission line between the signal generator and load in Fig. 4.3. A passive component, such as a high-frequency transmission line, will exhibit some amount

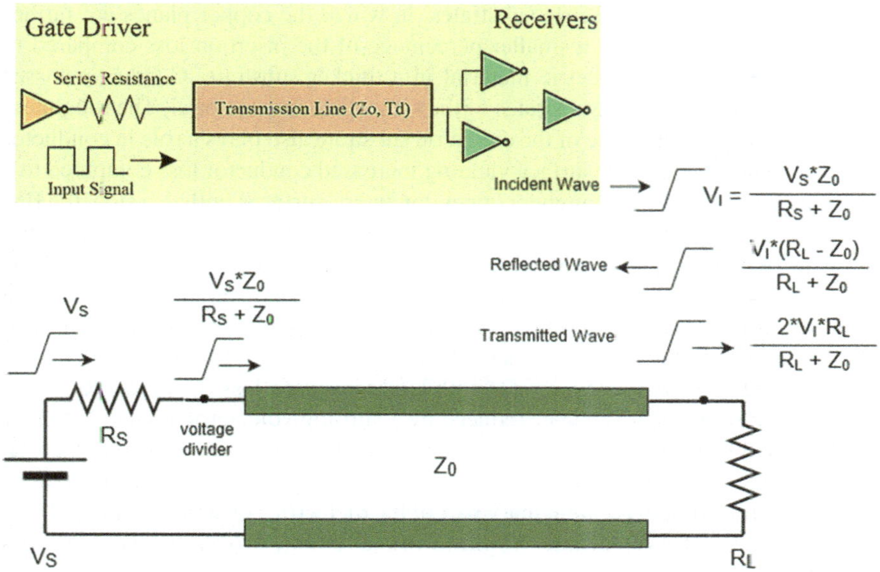

Fig. 4.3 Transmission Line with driver and receiver – Signal driving through transmission line. (Modified from Practical EE 2019, credit: https://practicalee.com/transmission-lines/)

of insertion loss, typically due to the conversion of signal energy to heat, even when impedance matches the generator and load. However, when it is not impedance-matched, signal reflections at mismatched transmission junctions result in return loss, increasing the total loss of the transmission line from a nominal 3 dB in the matched state to 6 dB in the unmatched state. When signal power must be conserved, insertion loss and return loss must be minimized. They can be measured with a vector network analyzer (VNA) by making scattering (S) parameter measurements of S_{21} for insertion loss and S_{11} for return loss (Coonrod 2021).

The insertion loss of a high-frequency PCB circuit is the sum of several different types of loss, including conductor loss (from the metal conductors), dielectric loss (from the substrate materials), radiation loss (from energy radiating away from the circuit), and leakage loss (from energy leaking between the copper planes). A reduction in any one of the four types of losses will result in less insertion loss for a high-frequency circuit (Coonrod 2021).

Leakage losses are typically a concern for substrate materials with relatively low-volume resistivity, such as semiconductor materials, but not for high-frequency circuit materials that typically exhibit high-volume resistivity. Leakage loss can be an issue for high-power circuits but is usually not a concern for the signal power levels used in mmWave circuits. Radiation loss can be a concern at mmWave frequencies, and it can be modeled as part of any investigation of a circuit's total insertion loss, but for now the effects of conductor and dielectric losses are the main concern (Coonrod 2021).

Conductor losses are more dominant for circuits with less dielectric material, i.e., circuits fabricated on thinner substrates where the copper planes are not widely separated. For thicker dielectric substrates, in which the copper planes are farther apart, conductor losses are a smaller percentage of the insertion loss compared to the greater amount of dielectric material in a thicker substrate. Conductor losses increase for thinner circuit substrates of the same dielectric material. The roughness of the copper at the interface of the dielectric substrate also plays a role in conductor loss, with a rougher copper surface yielding increased conductor loss compared to a smoother copper surface. Rougher copper conductor surfaces will also slow the EM wave propagation of high-frequency circuits such as microstrips and increase the phase angle of propagating EM waves as if the D_k (referring to the real part of the dielectric constant) of the substrate material is higher than its nominal value. When a thin substrate is needed for a high-frequency design and low insertion loss is also a key requirement, the type of copper is an important consideration. Smoother copper will yield high-frequency circuits with lower insertion loss (Coonrod 2021).

Similarly, insertion loss can be reduced by controlling dielectric losses. A thinner substrate will result in dielectric loss being a smaller percentage of a circuit's total insertion loss; selecting a substrate material with a lower dissipation factor (D_f, or called tanδ, referring to the imaginary part of the dielectric constant), a material loss parameter, can also contribute to lowering a circuit's overall insertion loss (Coonrod 2021).

6. Precise phase response

   Phase is a critical electrical parameter in many high-frequency circuits, especially in many emerging mmWave applications, such as automotive radars and 5G wireless networks, where phase is the basis for many advanced modulation formats. High-frequency circuits must maintain a consistent phase response, so systems such as radars and wireless communications provide reliable information. A high-frequency circuit's phase response is usually characterized in terms of phase angle or phase velocity. For example, a circuit processing an ideal sine wave would have a 360° phase angle response.

   The phase angle response can be affected by small changes in circuit design and circuit material characteristics. High-frequency circuit design is often based on supporting specific frequency bands and ranges, such as for communications channels, and the physical features of a circuit are very wavelength specific, even involving the use of circuit features with fractional wavelengths, such as quarter- and half-wavelength features. Various circuit material parameters can impact a circuit's phase angle response, such as D_k and copper roughness. For example, the phase response decreases in wavelength with decreasing circuit material D_k value. Compared to smooth, low-profile copper, rougher copper results in slower wave velocity and an increase in the phase angle response. For mmWave circuits based on phase responses to function properly, the phase angle consistency is a key performance parameter. While a 1″ microstrip circuit may exhibit a phase angle response of 360° at 7.6 GHz, a microstrip circuit with that physical length on a substrate material with $D_k = 3.0$ may have a phase angle response of greater than 4000° for a 77 GHz radar, and phase angle variations of as little as ±30° can result in errors in radar detection (such as in automotive collision-avoidance systems). For mmWave circuits at lower frequencies, such as the 26 GHz and 28 GHz frequency bands used in 5G NR systems, phase response is also important for maintaining accuracy in phase-modulated networks, but with the lower frequencies and longer wavelengths, those circuits are less sensitive to variations in phase angle (Coonrod 2021).

4.3 Designing High-Frequency PCBs

The electronic components are developing toward high speed, low consumption, small volume and high anti-interference for mmWaves. PCB design is an important stage of electronic product design. It can realize the connection and function between electronic components and is also an important part of power circuit design. A high-frequency circuit has higher integration and a higher layout density, so it is vital to understand the variables affecting PCB performance and how to make the PCB layout more reasonable and scientific.

4.3.1 Variables Affecting the Performance of High-Frequency PCBs

Many variables affect the performance of a PCB at small wavelengths of mmWave frequencies, starting with how high-frequency signals are launched onto the PCB from a connector interface. An impedance anomaly generally exists at that interface, and a variation in impedance can cause signal reflections, elevated return loss and distortion. The junction of a high-frequency connector to the PCB may be an extremely short distance, and the impedance anomaly may only affect a physical distance of approximately 0.1 "(2.54 mm), but that length can equal a significant fraction of the wavelength at mmWave frequencies and can cause distortion of the wave. For example, at 40 GHz, the wavelength is 0.18" (0.46 mm). Such an impedance anomaly is less of a concern at lower frequencies where the wavelengths are much longer and less affected over that short distance (Coonrod 2021).

Impedance anomalies at fractional wavelengths can impact mmWave circuit performance. An impedance anomaly of one-half wavelength will typically impact wave performance. A one-quarter-wavelength anomaly may also wreak havoc on mmWave circuit performance, but not as much as the one-half-wavelength anomaly. Generally, impedance anomalies of one-eighth wavelength or longer will influence wave behavior, and anomalies should be kept to one-tenth wavelength or shorter to minimize circuit performance problems at mmWave frequencies.

Ensuring that circuit features such as substrate thickness and conductor width are less than one-tenth wavelength at a frequency of interest can prevent performance problems, such as unwanted resonances, at mmWave frequencies. For example, a mmWave circuit fabricated on a circuit substrate with one-half-wavelength thickness at the operating frequency will exhibit resonant conditions between the signal plane and the ground plane due to the thickness of the substrate. A conductor width equal to one-half wavelength at the operating frequency will also cause resonant conditions across the width of the circuit's conductors. By keeping the substrate thickness and conductor widths at one-tenth wavelength or less of the operating frequency, unwanted resonant conditions can be avoided (Coonrod 2021).

4.3.2 High-Frequency PCB Layout Techniques

PCBs can realize the connection and function between electronic components and are also an important part of power circuit design. When the electrical schematic diagram has been designed, several functional plates should be adopted according to the structural requirements and functional division, and the physical size and installation method of each functional plate of the PCB should be determined. The convenience of debugging and maintenance, shielding, heat dissipation and EMI performance should also be considered. It is required to determine the layout wiring plan, such as the key circuits, the signal lines, the details of wiring methods, and the wiring principles to be followed. Several stages of the PCB design process must be examined, analyzed, and modified. Only after the whole layout process is completed and examined through comprehensive rule inspection can it be further processed (PCBWay 2018).

High frequency circuits are often highly integrated with high wiring density. The use of multilayer boards is both a necessary and an effective means to reduce interference. In the PCB layout stage, a reasonable choice of a PCB of a certain size and layers can make full use of an intermediary to set up the block, better nearby grounding, and effectively reduce the parasitic inductance and shorten the length of the transmission of signals but can also greatly reduce the cross interference of signals. All of these methods are good for the reliability of the high-frequency circuit. With the same material, the noise of the four-layer board is 20 dB lower than that of the double-panel board. However, at the same time, there is also a problem; that is, the more layers of PCB boards there are, the more complicated the manufacturing process will be, and the higher the unit cost will be, which requires that in the PCB layout, in addition to selecting appropriate layers of PCB, reasonable component layout planning should be carried out, and proper wiring rules should be adopted to complete the design. Some design guidelines for multilayer PCBs are described below (PCBWay 2018):

1. The fewer alternating leads of the pins between the layers of high frequency circuits, the better.

This means that the less via is used in connection, the better. A via can bring about a distribution capacitance of approximately 0.5 pF, reducing the number of via can significantly increase the speed and reduce the possibility of data error.

2. The shorter the lead between the pins of high-frequency circuits is, the better it is.

The radiation intensity of the signal is directly proportional to the routing length of the signal line. The longer the high-frequency signal lead is, the more easily it can be coupled to its components. Therefore, for high-frequency signal lines such as the signal clock, crystal oscillator, DDR data, LVDS, USB and HDMI, the shorter the routing length needed, the better.

3. The less bend between the pins of high-frequency electronic devices, the better.

High frequency electrical wiring leads had better use of a full straight line. If a breakover is needed, a 45-degree broken line or arc break can be used. This requirement is only used to improve the bonding strength of copper foil in low-frequency circuits, while in high-frequency circuits, meeting this requirement can reduce the external emission and coupling of high-frequency signals.

4. Reducing crosstalk introduced by the signal line parallel to close distance

For high-frequency circuit wiring, attention should be given to the crosstalk introduced by the signal line parallel to a close distance. Crosstalk refers to the coupling phenomenon between signal lines without direct connection. Since high-frequency signals are transmitted along the transmission line in the form of electromagnetic waves, the signal line will act as an antenna, and the energy of the electromagnetic field will be emitted around the transmission line. The unwanted noise signals between signals due to the coupling of electromagnetic fields are called crosstalk. The parameters of the PCB plate layer, the spacing of the signal lines, the electrical characteristics of the driving and receiving terminals and the signal line connection mode all influence the crosstalk. Therefore, to reduce the crosstalk of high-frequency signals, it is required to do the following as much as possible in the layout process:

(a) If there is a serious crosstalk between two lines, a ground line or ground plane can be inserted with, under the condition that the cabling space allows, it can play the role of isolation and reduce crosstalk.
(b) When the space around the signal line itself has time-varying electromagnetic fields, if the parallel distribution cannot be avoided, a large area of "ground" can be arranged on the opposite side of the parallel signal line to greatly reduce the interference.
(c) On the premise of sufficient wiring space, increase the space between adjacent signal lines and reduce the parallel length of signal lines. The clock line should be perpendicular to the key signal lines rather than parallel.
(d) If parallel lines in the same layer are almost unavoidable, they must be perpendicular to each other in the adjacent layers.
(e) In digital circuits, the usual clock signal is that the edge changes rapidly, and the external crosstalk is large. Therefore, in the design, it is advisable for the clock lines to be surrounded by the ground lines and digging more holes for the ground lines to reduce the distributed capacitance and thus reduce the crosstalk.
(f) The low voltage differential clock signal should be used as far as possible for the high frequency signal clock, and the integrity of the perforation should be considered.
(g) The idle input is not to be suspended but to be grounded or connected to a power supply. Because the suspended line may be equivalent to the transmitting antenna, grounding can inhibit transmission.

5. High-frequency digital signal ground line and analog signal ground line for isolation.

The analog ground line, digital ground line and so on connect to the common ground line to use a high frequency choke magnetic bead connection or isolate directly and select the appropriate single point connection. The ground potential of a high-frequency digital signal in a ground wire is not consistent; as a result, a voltage difference is often present. Moreover, the high-frequency digital signal ground wire often contains very rich harmonic components when it connects directly to a digital signal and analog signal ground; the high-frequency signal harmonic will couple to the analog signal interference through the ground. Therefore, in general, the ground line of the high-frequency digital signal and the ground line of the analog signal should be isolated, the single point interconnection mode can be adopted in the appropriate location, or the high-frequency choke magnetic bead interconnection mode can be adopted.

6. Increasing the high frequency decoupling capacitance of the power pins of the IC block.

The power pins of each IC block add a high-frequency demultiplexing capacitance nearby. Increasing the high-frequency decoupling capacitance of the power pins of the IC block can effectively inhibit the interference of high-frequency harmonics on the power supply pin.

7. Avoid loops when wiring.

All kinds of high-frequency signals should not form a loop when wiring. If it is unavoidable, the loop area should be as small as possible.

8. Good signal impedance matching must be ensured.

During the transmission process, when the impedance does not match, the signal will be reflected in the transmission channel, which will make the synthetic signal overshoot, causing the signal to fluctuate near the logical threshold. The basic method to eliminate reflection is to make the impedance of the transmission signal a good match. As the difference between the load impedance and the characteristic impedance of the transmission line is larger, the reflection is larger, so the characteristic impedance of the signal transmission line should be equal to the load impedance as much as possible. At the same time, it should also be noted that the transmission line on the PCB cannot present sudden changes or corners and keeps the impedance of each point of the transmission line continuous as much as possible; otherwise, reflection will occur between each segment of the transmission line.

In addition, high-speed signal routing branches or tree stumps should be avoided when designing PCB signal wiring. A high-frequency signal line easily generates large electromagnetic radiation when walking on the surface. If the high-frequency signal line is wired between the power source and ground line, the radiation generated by absorbing the electromagnetic wave through the power source and the bottom layer will be greatly reduced.

In summary, high-frequency circuits are often highly integrated and have high wiring density. The use of multilayer boards is both a necessary and effective means to reduce interference. In the PCB layout stage, reasonable choice of a certain layer printed circuit board size, the middle tier can be fully used to set the shielding, better implement nearby ground, and effectively reduce the parasitic inductance and shorten the length of the transmission of signals but can also greatly reduce the cross interference of signals. All of these methods are good for the reliability of the high frequency circuit.

4.4 Materials Selection of PCBs for Millimeter Wave Applications

Millimeter-wave frequencies are attractive for 5G communications and other applications for the broad bandwidths available at these frequencies, from 30 to 300 GHz. However, finding cost-effective printed circuit board (PCB) materials that provide high performance levels at these frequencies can be challenging. By understanding the key parameters and characteristics of concern for PCB materials at millimeter-wave frequencies, such as how different circuits behave for different types of PCB materials at millimeter wave frequencies, it is possible to choose properly when selecting PCB materials for use at these higher frequencies. Many of the concerns when designing microwave circuits and selecting PCB materials apply to higher-frequency millimeter-wave circuits, where these concerns can easily become more severe or carry greater impact. These issues include limiting spurious wave mode propagation problems, minimizing conductor and radiation losses, achieving effective signal launch, minimizing unwanted resonances, and controlling dispersion (Coonrod 2013).

Choosing circuit materials for high-frequency printed-circuit boards (PCBs) is a matter of weighing the electrical performance required by the end application against the ease or difficulty of fabricating circuits with those materials. These two factors may not mesh: one material may be well suited for a particular application but may pose challenges in terms of circuit fabrication, and vice versa. By relying on some tangible guidelines designed to evaluate a material in terms of its suitability for circuit fabrication and for meeting the requirements of an application, the process of selecting a PCB for a particular application can be optimized (Coonrod 2012).

4.4.1 High-Frequency PCB Material Selection Guidelines

To minimize wave propagation issues in a high-frequency PCB, a general rule is to use a laminate that is thinner than one-quarter wavelength at an application's highest operating frequency. In practice, better results can be achieved by using a laminate that is thinner than one-eighth the wavelength at the highest operating frequency to eliminate unwanted resonances between different circuit planes in a circuit assembly. Such resonances can interfere with the desired propagation for a circuit and generate surface waves that can disrupt the desired wave propagation. The width of the signal conductors is also related to the thickness of a circuit laminate, since a thinner laminate will use a narrower conductor width. To help eliminate mode issues, the conductor width should be one-eighth the wavelength or less at the highest operating frequency (Coonrod 2013).

These rules for laminate thickness and conductor width apply directly to high-frequency microstrip circuits; other types of circuit configurations may be more forgiving. For grounded coplanar-waveguide (GCPW) circuits, which are also known as conductor-backed coplanar waveguides (CBCPWs), thicker laminates have been shown to exhibit minimal mode issues at millimeter-wave frequencies. As shown in Fig. 4.4a, the microstrip transmission line circuit has a set distance between the signal layer and the ground plane. If that distance (substrate thickness) is one-quarter wavelength, resonance can occur between the copper planes, and this resonance may interfere with the desired wave propagation. Additionally, if the substrate is one-quarter wavelength thick and the conductor width is narrower than

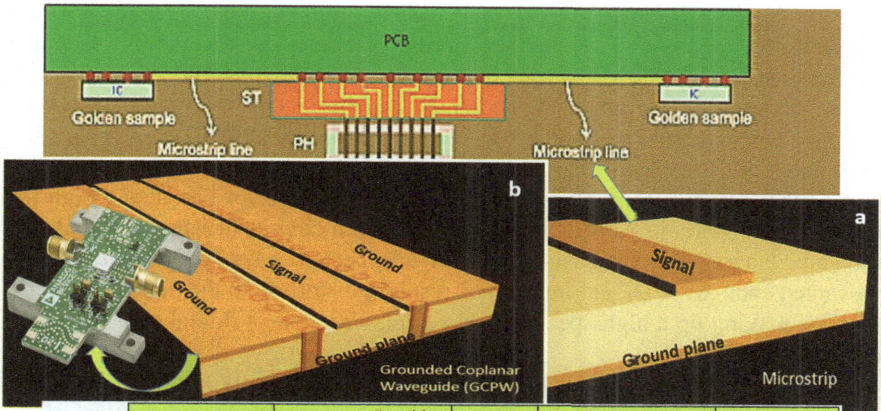

	Dominate Wave Propagation mode	Capable Bandwidth	Feasible Impedance range	Dispersion	Millimeter-Wave Radiation loss	Millimeter-Wave Mode suppression	Performance Sensitive to PCB Fabrication
Microstrip	Quasi-TEM	Moderate	Moderate	High	High	Difficult	Minimal
GCPW	Quasi-TEM	Wide	Wide	Low	Moderate	Moderate	Moderate

Fig. 4.4 Characteristics of microstrip transmission line (**a**) and grounded coplanar waveguide transmission line (**b**). (Modified from Coonrod 2013, credit: https://highfrequencyelectronics.com/Archives/Jul13/1307_HFE_CircuitMaterials.pdf)

one-quarter wavelength, a resonance may not occur or be marginalized. If both the substrate is thicker than one-quarter wavelength and the conductor width is one-quarter wavelength or more, unwanted resonances and wave propagation issues will likely occur. Figure 4.4b illustrates a GCPW circuit. Even when the substrate thickness and conductor width are equal to one-quarter wavelength, resonance may be avoided due to the close coupling of the coplanar ground planes. The coplanar ground planes are adjacent to the center signal conductor and are grounded by means of plated through holes (PTHs). Certainly, tradeoffs are part of any choice of high-frequency circuit configuration, and GCPWs will suffer higher conductor loss than microstrips. However, depending upon the operating frequencies, this may not mean more overall insertion loss due to a GCPW circuit possibly exhibiting less radiation loss than a microstrip circuit (Coonrod 2013).

A high-speed digital signal is a square-wave signal made up of a combination of several sine waves. Comprising different frequency signal components, a high-speed digital signal actually includes a fundamental-frequency signal, a third-harmonic signal, a fifth-harmonic signal, a seventh-harmonic signal, and a host of other odd-harmonic signals of higher frequencies. Therefore, for a digital signal to maintain its integrity and the sharpness of its rise and fall times, it is necessary to transfer millimeter-wave signals with the lowest possible loss and distortion. For instance, any PCB material capable of handling analog signals of millimeter wavelengths, such as approximately 25–100 GHz, with low loss and distortion should be able to maintain the signal integrity of a high-speed digital signal at 10 Gbps. PCB material parameters that are critical to analog millimeter-wave circuit performance will also be important as guidelines for choosing PCB materials for high-speed digital circuits. The PCB parameters that can be used for guidelines when choosing circuit materials for high-speed digital applications include D_k, dissipation, loss, and dielectric thickness. The dielectric constant, D_k, of a PCB material has long been a guiding parameter for both analog and digital circuits since it is so closely related to the impedance of the circuits that will be fabricated on that material. Changes in a PCB material's D_k, whether as a function of frequency, as a function of temperature, or for other reasons, can adversely affect the performance of broadband high-frequency analog circuits as well as high-speed digital circuits because it will change the impedances of transmission lines in unexpected ways. In particular, these unwanted changes in D_k and impedance result in distortion to the higher-order harmonics, making up a high-speed digital signal, with loss of digital signal integrity (Rog Blog 2015).

In general, PCB materials with low and stable D_k values with frequency and temperature will support high-speed digital circuits with low distortion of the higher-order harmonic signal components. Dispersion is a PCB material characteristic closely related to D_k. All PCB materials exhibit some amount of dispersion, which refers to the change in D_k as a function of frequency. A circuit material with minimal change of D_k with frequency will exhibit minimal dispersion, a good characteristic for high-speed digital circuits. Dispersion can be caused by a number of different circuit material traits, including the polarity of the dielectric material, the loss of the material, and even how the surface roughness of the copper conductor affects the PCB material loss at higher frequencies. If a PCB material exhibits different D_k values for the different harmonic signal components comprising a high-speed digital signal, it will cause losses and even shifts in frequency for those harmonics, resulting in degradation of the high-speed digital signals. PCB signal losses at increasing frequencies, especially at the higher frequencies needed by a high-speed digital circuit's higher-order harmonic signal components, can suffer excessive losses to the amplitudes of those higher harmonic signals, resulting in distortion to those high-speed digital signals. Losses in a PCB can come from a number of different causes, including the dielectric material and the copper conductors. The length of a high-speed digital circuit on a PCB material can also have a great deal to do with maintaining the integrity of those high-speed digital signals. Circuit losses for any PCB material are a function of frequency and will increase with increasing frequencies. A PCB material with acceptable losses within a bandwidth closer to the fundamental-frequency tone of a high-speed digital circuit, such as 5 GHz, and perhaps even with low loss at the third-harmonic signal component, such as 15 GHz, may have excessive loss at the fifth- and seventh-harmonic signal components of that high-speed digital signal. In addition, signal losses are additive with length: a signal experiencing a loss of, for example, 0.5 dB per inch at 5 GHz for the first inch of a 10-inch-long high-speed digital circuit will suffer a loss of 5 dB at 5 GHz across the length of the circuit. Depending upon the circuit's dielectric losses and copper conductor losses, the total loss across the length of the circuit can be considerably higher for the high-speed digital signal's higher-order harmonic signal components than for the lower-order harmonic tones. For some circuit materials, the loss for a 10-in.-long circuit may be 10 dB or more at the fifth- and seventh-harmonic signal components of a high-speed digital signal, resulting in considerable distortion to the high-speed digital signal transferred across that PCB material (Rog Blog 2015).

Additionally, when working with PCBs for high-speed digital circuits, attention should be given to physical details. Such things as right-angle bends in transmission lines can affect performance. A right-angle bend represents a change in the effective width of the transmission line, resulting in an impedance discontinuity and an increase in the capacitance at that portion of the transmission line. The use of mitered 45-degree bends can minimize the impedance discontinuity and minimize the reflections of the signal passing through that junction. The choice of PCB material for high-speed digital circuits can be guided by the speed of those digital circuits, with material characteristics such as loss and dissipation factor (D_f) targeted for lower values at higher frequencies. Circuit materials with medium to low loss are suitable for digital circuits to 10 Gb/s, while lower-loss circuit materials are usable for digital circuits to approximately 25 Gb/s, and circuit materials considered to exhibit extremely low loss are well suited for the fastest digital circuits, such as operating at 50 Gb/s and faster. In terms of circuit material D_f, typical values might be 0.010–0.005 for applications to approximately 10 Gb/s, 0.005–0.003 for applications to approximately 25 Gb/s, and 0.0015 or less for circuit applications to 50 Gb/s and faster (Rog Blog 2015).

Therefore, for high-frequency transmission lines and transmission-line circuits, insertion loss is actually a total of various component losses: dielectric loss, conductor loss, radiation loss, and leakage loss. PCB substrate materials for high-frequency applications typically have high volume resistivity, with resulting minimal RF leakage losses. Dielectric losses are related to the tangent delta or dissipation factor of the circuit substrate material. These losses are affected by any additional substrate materials, such as a soldermask or prepreg/bonding layers. Soldermasks are typically not used at RF/microwave frequencies and especially at millimeter-wave frequencies, since these are very high-loss materials where a dissipation factor of 0.02 is not uncommon. Soldermasks are not typically characterized by a well-controlled dielectric constant (D_k), and using soldermasks can lead to impedance mismatches and their impact on increasing return loss and ultimately insertion loss (Coonrod 2013).

4.4.2 PCB Materials Used for High-Frequency Applications

PCBs are generally made of substrate, conductor (transmission line), solder mask, and silkscreen, which combine materials designed to conduct electrical signals along designated metal paths and insulate those metal paths from coming into contact with each other. Multilayer circuit boards use dielectric materials for a core and laminate, which serve as the foundation for the copper foil making up the metal paths. The dielectric materials used throughout the board have specific properties and characteristics that must be considered for the frequency and operating

Table 4.1 Typical ceramic-based PCB substrate materials (Essentra 2020)

Material	Typical application
Alumina (Al_2O_3) Most common ceramic PCB. Strong thermal dielectric with low expansion. Exceptional high-frequency performance. Operating temperature up to 350 °C	Cooling and heating; LED boards; Medical circuits; Sensor modules; High-frequency devices
Aluminum nitride (AlN) High thermal conductivity; high thermal strength with low expansion. Widely used as a substrate or package	High-power LEDs; Integrated circuits (ICs); Sensors; Cable television
Beryllium oxide (BeO) Highly toxic during machining. Used for its high thermal conductivity and low expansion	High-powered industrial microwave ovens; High-voltage amplifier ICs

environment of the PCB. Basic requirements for high frequency PCB materials include the following:

(a) Dimensional tolerance stability – The high frequency PCB should have materials that provide mechanical stability when experiencing different temperature ranges, vibrations, shocks, and electric surges.
(b) Superior thermal management – The PCB material should provide excellent heat transfer and dissipation without having the layers start to peel away, delaminate, or decompose at high rates.
(c) Enhanced signal performance – Signal performance should be constant across the board with little signal loss even when the frequencies increase. The materials should have a low D_f to provide this advantage.
(d) Tight impedance control – High-speed circuits will require tight control of the impedance routing to maintain a constant D_k when there are varying frequency ranges.
(e) Moisture and chemical resistance – Select materials that have low absorption rates of moisture and chemicals so there will be little changes to the electrical performance of the PCB.

4.4.2.1 PCB Substrate Materials

The PCB substrate materials selection should be based on the specific high-speed PCB application, their costs, and their manufacturability. As with many high-frequency components, PCB substrate materials are sorted and compared by a number of key parameters, including the relative dielectric constant (D_k or ε_r), dissipation factor (D_f), coefficient of thermal expansion (CTE), thermal coefficient of dielectric constant, and thermal conductivity. The base substrate for most PCB materials can be categorized as hard/rigid or soft/flexible (Essentra 2020).

1. Hard/rigid PCB substrate materials

Hard, or rigid materials, are used whenever the PCB has to retain the shape it was set up to be, for the device's lifespan. These PCBs are made out of a solid substrate material that prevents the board from bending. For example, a computer motherboard is perhaps the most common application for a rigid PCB. The motherboard is a multilayer PCB. It is designed to allocate electricity from the power supply while simultaneously enabling communication between all of the computer's parts, such as the CPU, GPU and RAM. Hard circuit materials are typically ceramic based. Generally, the advantages include exceptional high-frequency performance; high thermal conductivity, ~170 W/m-K; strong dielectric; high operating temperature >350 °C; low expansion coefficient <4 ppm/C; versatile packaging with an option to come in hermetic packages to prevent water absorption; high chemical erosion resistance; and limited or no outgassing. The most popular ceramic-based PCB substrate materials and their applications are shown in Table 4.1 (Essentra 2020)

2. Soft/flexible substrate materials

   To enable a PCB to flex and move, plastic is often used. Similar to rigid PCBs, flexible PCBs can be made in single, double or multilayer formats. They can be folded over edges and wrapped around corners. It's because of flexible materials that wearables are possible, allowing printed circuitry to insert into compact spaces. Flexible materials save on cost and weight, but they tend to cost more for fabrication. One advantage of flexible materials is that they can be used in areas with environmental hazards. Flexible materials can be waterproof, shockproof or corrosion-resistant, for example, which is not a feature that most rigid PCBs can offer. Soft circuit materials such as an epoxy or plastic form a coating around a filler, often a glass weave. This form of glass – or a ceramic filler – provides strength and rigidity to the plastic dielectric material. Table 4.2 shows three typical PCB base-substrate soft/flexible materials and their applications (Essentra 2020).

3. Flex-rigid PCBs

   There is a third option for the PCB substrate with a combination of flexible and rigid materials. Flex-rigid boards consist of multiple layers of flexible PCB material, such as polyimide, attached to a rigid PCB layer and are often used in aerospace, medical and military applications. Flexible circuit materials based on PTFE, such as RO3000 and RO4000, are popular because they are easy to machine and

Table 4.2 Typical PCB base-substrate soft/flexible materials (Essentra 2020)

Material	Typical application
Polytetrafluoroethylene (PTFE) The best-known PTFE-based brand name is Teflon. This material offers temperature stability and low dissipation factor	Antennas; Power amplifiers; Automotive cruise control; Aerospace guidance telemetry
Polyimide Good electrical properties, wide temperature range and high chemical resistance	Instrument panels; Under-hood controls; Cameras; Personal entertainment devices; Calculators
PEEK High melting point and resistant to chemicals, radiation and extreme temperatures	X-ray machines; Gamma ray devices (medical applications); Electronic systems on aircraft

Table 4.3 Flexible circuit materials based on PTFE such as RO3000 and RO4000 (Essentra 2020)

Material	Typical application
RO3000	Automotive radar; Cellular telecommunications systems; Remote meter readers; Direct broadcast satellites
RO4000	RF identification tags; Automotive radar and sensors; Cellular base-station antennas and power amplifiers

have low dielectric losses at microwave frequencies. Their typical applications are shown in Table 4.3.

Conventionally, flame retardant type-4 (FR-4) is widely used for PCBs. This material consists of epoxy laminate that is reinforced with woven glass. It is often used for low- to medium-speed circuit boards, as it is cost-effective for high-volume production runs. While it can be chosen for high-speed PCB applications, the standard FR-4 is susceptible to excessive heat, power, and voltage rates. It also does not have a stable D_k for high-speed frequencies as signal loss can become more frequent.

If FR-4 is not an option for a specific application, thermoset hydrocarbon laminates, PTFE laminates and polyimide may be acceptable alternatives. All these materials are readily available but at a higher cost:

(a) Thermoset hydrocarbon laminates – Thermoset hydrocarbon PPE and PPO offer good mechanical stability, age well and can be easily fabricated. It offers mechanical stability designed for high-speed PCBs. However, there may be an increase in dielectric loss at higher frequencies.
(b) PTFE laminates – Polytetrafluoroethylene (PTFE) materials are becoming standard alternatives to FR-4. This material has excellent dielectric properties for microwave frequencies and millimeter wave frequencies, especially for those now being used for 5G wireless communication systems. It also provides low signal loss, improved impedance control, and good signal performance.
(c) Polyimides: Polyimide materials are usually seen in flexible and rigid-flex PCB layers as they can be bent and folded around components without impacting their performance. This material also offers excellent thermal management, moisture resistance, and chemical resistance.

In addition, soldermasks are often guilty of thickness variations when applied to one circuit to the next or even within the same circuit, which can result in unwanted impedance variations. Soldermasks also typically have high moisture absorption characteristics, which can seriously degrade the performance of a PCB. Moisture, basically water, has a D_k of approximately 70 and a very high dissipation factor, both values much higher than the circuit material, so that as water or moisture is absorbed, a circuit material's D_k will rise, and its loss will increase. As a result, soldermasks should be used sparingly or not at all at millimeter-wave frequencies.

4.4.2.2 Conductors for High-Frequency PCBs

Most PCB boards are made of copper clad laminates- copper foil/adhesive/substrate…, referred to as cores. The core gives the boards structure and can influence other properties, such as thermal resistance and impedance.

The copper foil plays an important role in circuit boards and in building the final copper thickness. The quality of the copper conductors will also affect conductor losses. A copper conductor with a rough surface will exhibit higher conductor losses than a copper conductor with a smooth surface profile. As thinner substrates are used, as for millimeter-wave circuits, conductor loss becomes more of a concern; conductor loss also grows more significant with increasing frequency Copper with a roughened surface is often used for improved adhesion to the dielectric material in a PCB. However, this surface roughness can also result in higher loss. As a rule of thumb, when the skin depth for a frequency of interest is equal to or less than the copper surface roughness, the surface effects of the conductor will be significant. At millimeter-wave frequencies, the skin depth is commonly less than the conventional copper foil surface roughness. The rougher surface will create more parasitic inductance and cause a change in the surface impedance as well as an increase in insertion loss (Horn et al. 2010).

The plated finish applied to copper on final circuit production can also impact conductor loss, especially at higher frequencies. Unfortunately, many of the metals used as a finish for PCBs are less conductive than copper, and the addition of these finishes results in an increase in conductor loss. For example, electroless-nickel-immersion-gold (ENIG) finish is commonly used for PCBs, even though nickel is less conductive than copper and using an ENIG finish inevitably results in an increase in conductor loss. A typical conductor stack-up with an ENIG finish would start with the base copper for the circuit, then a barrier layer against copper oxide migration, which is the nickel, and gold on top of the nickel. In terms of thickness, immersion gold is a self-limiting process that typically produces a gold thickness of approximately 0.2 μm, while nickel has a thickness of 5.0 μm. For the skin depths at millimeter-wave frequencies, nickel and some of the gold will be used. At higher millimeter-wave frequencies, more of the gold finish will be used. However, gold is still less conductive than copper, so using this finish will result in a penalty in conductor loss (Coonrod 2013).

Pure silver is more conductive than pure copper, but the immersion silver process used to add silver as a PCB finish actually employs silver alloy and not pure silver. It is close to pure silver, so the conductivity of the silver alloy is very good and close to that of copper. The immersion silver process is self-limiting, so that silver is added to a thin coating, typically in the range of 0.2 μm. Unfortunately, silver will oxidize over time, in contrast to gold, which does not oxidize. Nevertheless, even though oxidation of silver will change its appearance, it apparently does not significantly impact the insertion-loss performance of a finished PCB (Coonrod 2013).

4.5 The Role of Materials in High Frequency PCB Fabrication

PCB fabrication processes to achieve circuits with repeatable, high-quality performance. A number of different mechanical processes are required as part of high-frequency PCB fabrication. In general, the most critical of these would be drilling, plated-through-hole (PTH) preparation, multilayer lamination, and assembly.

The drilling process is typically concerned with creating clean holes, which will later be metalized to form via holes for electrical connections from one conductive layer to another. Some concerns with the drilling process include smear, burring, and fracturing of the material. Smearing can be lethal to PCB fabrication using a PTFE-based material since there is no way to remove the smear. Fracturing can be fatal for some nonwoven glass hydrocarbon materials; however, most woven glass hydrocarbon materials do not have this concern (Coonrod 2012).

The PTH preparation process is relatively well defined and straightforward for most non PTFE materials, although special processing is required when forming PTHs for PTFE-based materials. Ceramic-filled PTFE-based materials offer PTH preparation options that are more forgiving. However, nonceramic-filled PTFE materials require a special process that can limit final circuit yields (Coonrod 2012).

Fabricating multilayer PCBs presents many challenges. One is the fact that dissimilar materials are often being bonded together, and these dissimilar materials can have properties that complicate drilling and PTH preparation processes. Additionally, a mismatch between certain material properties, such as the coefficient of thermal expansion (CTE), can lead to reliability problems when the circuit is thermally stressed during assembly. A goal of the material selection process is to find a good combination of circuit materials for a multilayer PCB that enable practical fabrication processing while also meeting end-use requirements. Designers and fabricators have many choices of materials used to bond together the copper-clad laminates that ultimately form a multilayer PCB. As Table 4.4 shows, the materials differ in terms of dielectric constant, dissipation factor, and processing temperatures. In general, lower lamination temperatures are preferred. If a PCB must undergo soldering or some other form of thermal exposure, it will be necessary to use a bonding material with a high reflow (remelt) temperature, one which is thermally robust and does not reflow at elevated processing temperatures. The greatest concern during PCB assembly is due to the effects of thermal stress from soldering. Other sources of thermal stress during PCB assembly are from solder rework or exposure to multiple thermal cycles. In terms of circuit materials, effects from thermal stress can typically be projected by comparing the CTE values for different materials, as shown in Table 4.5. In general, a circuit material with a lower CTE will be more robust and handle the thermal stress of PCB assembly better than a material with a higher overall CTE. This is one reason why multilayer PCBs typically use more than one type of circuit material. Materials that might provide good electrical performance may have characteristics (such as high CTE) that make them less than robust to handle the thermal stress of PCB assembly. By using a combination of materials, some with

Table 4.4 Substrate and bonding materials used for fabricating high-frequency multilayer PCBs (Coonrod 2012)

Substrate & bonding materials	Dielectric constant	Dissipation factor	Lamination temperature (°C)	Re-melt temperature (°C)
FEP	2.10	0.0010	296	271
Ceramic-filled PTFE, $D_k = 3$	3.00	0.0013	371	338
Ceramic-filled PTFE, $D_k = 6$	6.15	0.0020	371	338
LCP	2.90	0.0025	290	271
3001	2.30	0.0030	218	177
Ceramic-filled PTFE, $D_k = 10$	10.80	0.0023	371	338
Thermoset hydrocarbon	3.90	0.0040	177	–
FR-4	4.50	0.0180	182	–

Table 4.5 Typical CTE values for materials commonly used in high frequency PCBs (Coonrod 2012)

Material	CTE (ppm/°C)	Electrical performance
PTFE with micro-glass fiber	220	Excellent
PTFE with woven glass	200	Good
Ceramic-filled PTFE	25	Excellent
Ceramic-filled PTFE with woven glass	50	Good
Ceramic-filled hydrocarbon	20	Good
Ceramic-filled hydrocarbon with woven glass	35	Good
High performance FR-4	50	Poor

good electrical properties and others with a lower overall CTE, a robust multilayer PCB construction can be designed and assembled. Such a construction is known as a hybrid multilayer PCB, which can provide cost as well as performance benefits (Horn et al. 2010; Coonrod 2012).

In general, a circuit material with a CTE value of 70 ppm/°C or less is considered robust for PCB fabrication and assembly. As Table 4.5 shows, however, one of the materials with the best electrical performance may have the worst CTE, such as PTFE with micro-glass fiber. This is one reason why ceramic-filled PTFE laminates were formulated. They combine excellent electrical performance with very good CTE. Unfortunately, they exhibit poor dimensional stability since the material is soft and circuit dimensions can be easily distorted. To provide good electrical performance and CTE with improved dimensional stability, ceramic-filled PTFE laminates with woven glass reinforcement were developed. When making a choice in high-frequency circuit materials based on fabrication issues, the clear-cut favorite would be ceramic-filled hydrocarbon material with woven glass. These materials typically feature a low dissipation factor on the order of 0.003 and are robust in terms of most circuit fabrication processes. If better electrical performance is needed, the choice would be ceramic-filled PTFE with woven glass. These materials typically have a dissipation factor in the range of 0.002 and are generally fabrication-process friendly. The major concerns in fabricating PCBs with these materials relate to drilling and PTH preparation. For the best electrical performance, the choice is microfiber glass PTFE, although this material can be difficult in terms of fabricating more complex circuit constructions. The material, which is nearly pure PTFE, is often used for simple high-frequency circuitry such as microstrip filters and couplers. Additionally, this material is often used in a hybrid multilayer circuit, in which it supports critical functions, while other materials that are more friendly to circuit fabrication processes are used for the remainder of the multilayer PCB (Coonrod 2012).

Additionally, variations in copper plating thickness and the final plated-finish placed on copper surfaces can impact the performance of a mmWave circuit, and both processes must be tightly controlled for success with mmWave circuit fabrication. For laminated circuit materials, variations in the thickness of the raw copper foil on the laminates are typically held to a tolerance of ±10%. However, for circuits using plated through-hole (PTH) technology for circuit interconnections, the base copper on a laminate will be thicker in support of the PTH process. The copper plating process yields normal variations related to circuit design and the type of fabrication process. Copper is typically plated thinner in the middle of a processed circuit panel and thicker closer to the edges of the panel, but such variations in copper thickness can be a source of performance variations at mmWave frequencies. For mmWave circuits manufactured in large volumes, for example, variations in copper thickness can result in circuit-to-circuit variations in insertion loss and phase response (Coonrod 2012).

4.6 Material Issues Related to 5G Applications

Material considerations will be the top aspect that will have to be evaluated for designing and building the high frequency PCB stack-up. The 5G PCB will have to address all the specifications when carrying and receiving signal transmissions, providing electrical connections, and offering controls for specific functions. Additionally, challenges to PCB design will need to be addressed, such as maintaining signal integrity at higher speeds, dealing with thermal management, and preventing electromagnetic interference (EMI) between digital and analog boards (McMahon 2021).

4.6.1 Mixed Signal Acceptance Circuit Board Designs

Most wireless systems now are handling 4G and 3G PCBs. This means that components are transmitting and receiving frequencies that can range from 600 MHz up to 5.925 GHz and bandwidth channels of 20 MHz or 200 kHz for IoT systems. When designing PCBs for 5G network systems, the components will need mm-wave frequencies of 28 GHz, 30 GHz, and even 77 GHz based on the application. For bandwidth channels, 5G systems will be dealing with 100 MHz right below and 400 MHz right above 6 GHz frequencies. These higher speeds and higher frequencies will demand the appropriate materials within the PCB to capture and transmit both lower and higher signals at the same time without experiencing signal loss and EMI. In addition, an added problem is that the devices will become lighter, portable, and smaller. With strict weight, size, and space limitations, the PCB materials will have to be flexible and light while accommodating all the microelectronics along the board. Thinner traces and stricter impedance control will need to be adhered to for PCB copper tracings. The traditional subtractive etching processes used for 3G and 4G high-speed PCBs may be switched out for modified semiadditive processes. These modified semiadditive processes will provide more precise trace lines and straighter walls. Material substrates are also being redesigned, providing a dielectric constant as low as 3, as standard materials for lower speed PCBs are usually 3.5–5.5. Tighter fiberglass weaves, lower dissipation factor loss materials, and low-profile copper will also be options for high-speed PCBs used for digital signals to prevent signal losses and to promote greater signal integrity (McMahon 2021).

4.6.2 EMI Shielding Challenges

EMI, crosstalk, and parasitic capacitance are major problems with circuit boards. To deal with cross talk and EMI that will be present due to the analog and digital frequencies along the board, separating the traces is highly recommended. Using a multilayer board will offer greater versatility to decide how to place the high-speed traces so the routes for both analog and digital return signals will be kept away from each other while also keeping the AC and DC circuits apart. Adding shielding and filtering when arranging components should also lower the amount of natural EMI that will be present on the PCB (McMahon 2021).

4.6.3 Impedance Control and Signal Loss

Some high-frequency applications have very tight specifications for impedance control, and the D_k tolerance is a good indicator of how well this material may be suited for those applications. For an application that requires low-loss performance, a material with a lower D_f value would be a logical choice, although this choice should also be weighed against the ease or difficulty of PCB fabrication with that material (Coonrod 2012).

In addition to dielectric losses, conductor losses are important when comparing circuit materials. Especially for thin circuits, conductor losses can be more significant than dielectric losses. Conductor losses can be impacted by circuit design, circuit configuration, and the thickness of conductive metals, as well as the surface roughness of the copper conductor layers. For example, conductor losses are higher for materials with higher amounts of copper surface roughness than for materials with smoother copper conductor surfaces. When comparing measurements by this parameter, the surface roughness measurement of interest is the root-mean-square (RMS) roughness of the copper surface. A smooth copper conductor layer such as rolled annealed copper will typically have surface roughness RMS values of approximately 0.3 microns. A low-profile electrodeposited (ED) copper conductor layer will typically have a surface roughness of approximately 0.8 microns, with standard ED copper at approximately 1.8 microns and high-profile copper at approximately 3 microns (Coonrod 2012).

Another material property of interest is the thermal coefficient of the dielectric constant, or TcD_k. Often overlooked in material comparisons, this is a measure of how much the dielectric constant (D_k) will change with changes in temperature. Given the changes in the relative dielectric constant, ε_r, in parts per million (ppm) for changes in temperature (in °C), large values of TcD_k can be an indicator that circuits that perform well under ideal laboratory conditions may not fare as well under less controlled conditions, notably with large swings in temperature (Coonrod 2012).

4.6.4 Thermal Management Challenges

Higher signal speeds will lead to higher generated heat from the electricity passing through the PCB. PCB materials used for dielectric materials and core substrate layers will need to adequately handle the high speeds required for 5G technology. If the materials are inadequate, copper trace peeling, delamination, shrinking, and warping can result as problems that can cause deterioration of the PCB. To address these higher temperatures, fabricators will need to focus on material selections that address thermal conductivity and thermal coefficients. A material offering higher thermal conductivity, excellent heat transfer and consistent dielectric constants will be necessary to create a good PCB that will provide all the 5G functions that will be required for the application (McMahon 2021).

The thermal conductivity of a PCB material is important to consider for high-power applications in which a large amount of heat must be dissipated. A substrate with high thermal conductivity can assist thermal management issues with these applications. Many standard PCB materials have thermal conductivities in the range of 0.25 W/m/K. Additionally, some materials with typically good electrical performance, such as microfiber PTFE materials, also have thermal conductivities of approximately 0.20–0.25 W/m/K. By adding a ceramic filler, the thermal conductivity of a circuit material can be improved. This improvement can help solve many thermal management issues in high-power PCB designs (Coonrod 2012).

4.6.5 *Moisture Absorption*

Moisture absorption can also be important to consider for high-frequency applications. In an environment with high humidity, a circuit material that absorbs a high amount of moisture will exhibit increases in D_k and loss, both impacting PCB performance. Circuit materials with high moisture absorption may not suffer degraded performance in controlled environments, but performance can be quite variable in more hostile operating environments. Many standard PCB materials have moisture absorption in the range of 1%. However, most materials formulated for high-frequency applications are characterized by moisture absorption that is considerably less than 1%. For most high-frequency applications, laminates with moisture absorption values of less than 0.25% are considered acceptable (Coonrod 2012).

Therefore, there are many issues to consider when choosing a circuit material for high-frequency PCB applications. Some are related to fabrication issues for producing the most robust PCB possible, and some are related to achieving the best electrical performance possible for a given application. Because of various tradeoffs, the material for fabricating the most robust PCB may not be the same one for the highest electrical performance for an application. Multilayer hybrid PCBs represent one way to choose a blend of materials to combine robustness and good electrical performance. With the advent of 5G technology, printed circuit boards will be designed to offer optimal signal integrity, impedance control, and thermal management. Adopting and leveraging the new technology (such as additive manufacturing), while keeping PCB scope creep low to control costs, will be a delicate balancing act. However, as 5G technology becomes more mainstream, advanced techniques, materials, and fabrication equipment will be created to provide fabricators with greater capabilities to meet the high-speed PCB design and manufacturing demand. These technologies will also provide a more competitive field so that customers can find the right PCB design services that will meet their budgets (Coonrod 2012; McMahon 2021).

References

Bogatin E (2018) The physical basis of transmission lines. Pearson Education, Informit. https://www.informit.com/articles/article.aspx?p=2916283&seqNum=5

Coonrod J (2012) Selecting PCB materials for high-frequency applications. Microw Eng Eur. http://home.iitb.ac.in/~pradeepsarin/students/tether/generalectronicsfundaes/adv/Selecting-PCB-Materials-for-High-Frequency-Applications.pdf

Coonrod J (2013) Choosing circuit materials for millimeter wave applications. https://highfrequencyelectronics.com/Archives/Jul13/1307_HFE_CircuitMaterials.pdf

Coonrod J (2021) PCB design and fabrication concerns for millimeter-wave circuits. Printed Circuit Design & Fab. https://rogerscorp.com/-/media/project/rogerscorp/documents/articles/english/advanced-connectivity-solutions/pcb-design-and-fabrication-concerns-for-millimeter-wave-circuits.pdf

Essentra (2020) Your PCB substrate: a guide to materials. https://www.essentracomponents.com/en-gb/news/guides/your-pcb-substrate-a-guide-to-materials#:~:text=PCBs%20are%20generally%20made%20of,that%20offers%20so%20many%20choices

Horn AF, Reynolds JW, Rautio JC (2010) Conductor profile effects on the propagation constant of microstrip transmission lines. In: 2010 IEEE MTT-S international microwave symposium, pp 868–871.

Le JT (2019) Signal integrity optimization of RF/microwave transmission lines in multilayer PCBs. Master's theses. 5036. San Jose State University. https://doi.org/10.31979/etd.4ddq-zyef. https://scholarworks.sjsu.edu/etd_theses/5036

McMahon E (2021) High-speed PCBs and 5G technology considerations and challenges. https://www.epectec.com/articles/high-speed-pcbs-5g-technology-considerations-challenges.html

PCBWay (2018) High frequency PCB layout techniques. https://www.pcbway.com/blog/pcb_design_layout/high_frequency_pcb_layout_techniques.html

Pozar DM (2011) Microwave engineering, 4th edn. Wiley, pp 77–79

Practical EE (2019) Transmission lines. https://practicalee.com/transmission-lines/

Rog Blog (2015) Selecting PCB materials for high-speed digital circuits. Microw J.https://www.microwavejournal.com/blogs/1-rog-blog/post/23743-selecting-pcb-materials-for-high-speed-digital-circuits

Shamkhalichenar H, Bueche CJ, Choi J-W (2020) Printed Circuit Board (PCB) technology for electrochemical sensors and sensing platforms. Biosensors (Basel) 10(11):159

Chapter 5
Materials for High Frequency Filters

Abstract Different types of frequency filters (bandpass, bandstop, lowpass, and highpass) are used to obtain the desired frequencies and enable wireless communications. As 5G operates in higher frequency bands, new resonating structures are needed for frequency filters, which has been a significant challenge for industry. High percent bandwidth, high selectivity, low insertion loss, small form factor, increased rejection, and reliability under a broad range of temperatures are some of the important criteria for 5G filters. The most popular structures are acoustic wave filters, including surface acoustic wave (SAW) and bulk acoustic wave (BAW) filters, but SAW and BAW filters are only practical up to approximately 2.5 and 6 GHz, respectively. The piezoelectric layer used in BAW filters must have a precise crystal orientation to attain the finest electromechanical coupling. Aluminum nitride (AlN) is the most common material used for BAW filters, providing the greatest balance of performance, reliability, and manufacturability. For 5G, increasing the filter bandwidth requires enhanced piezoelectric coupling. Scandium-doped aluminum nitride (ScAlN) can solve this problem, but manufacturing becomes more challenging, motivating the transition to lithium niobate ($LiNbO_3$) materials. The precise deposition of the piezoelectric material, electrode configuration, acoustic reflector, and parasitic effects are key factors for enhanced coupling. Innovative structures such as XBAR and other options, including dielectric waveguides, on-chip filtering structures, cavity waveguides, and microstrips, can lead to filters operating at even higher frequencies. In the 5G era, the filter needs to be more miniaturized and better integrated with other devices. Great opportunities for stable ceramic materials with low dielectric constants/loss tangents and high quality factors (Q-values) exist. This chapter will provide a brief review of various material selections for different types of high-frequency filters used in 5G communication systems.

C. Tong, *Advanced Materials and Components for 5G and Beyond*, Springer Series in Materials Science 327, https://doi.org/10.1007/978-3-031-17207-6_5

5.1 The 5G Effect on Filter Technologies

High-frequency filters are increasingly essential components within wireless products, especially as those wireless products continue to compete for limited frequency spectra. Various types of RF/microwave filters help wireless radio transmitters and receivers operate with their proper signals while shielding them from interference caused by out-of-band signals. As millimeter-wave (mmWave) frequency bands are increasingly used for high-speed links in next-generation wireless systems, filters will be essential to minimize interference. Fifth-generation (5G) wireless technology targets more signal traffic, increased capacity, reduced latency, and lower energy consumption than its predecessors. To achieve these goals, networks will need to increase their bandwidths through carrier aggregation and a push into the mmWave spectrum while improving their spatial efficiency with base station densification, massive multiple-in multiple-out (MIMO) antenna technology, and beam-forming antenna arrays. These enabling technologies will place new demands on the underlying RF front-end components, particularly the vast number of filter designs required across a heterogeneous network of base stations (of varied cell sizes) and mobile devices. Systems offering large bandwidths through carrier aggregation and ubiquitous coverage through massive overlapping of microcells must manage both in-band and out-of-band interference. Likewise, implementation of massive MIMO will require compact filtering technology that mitigates the adverse impact of out-of-band interference on the uplink sum rate of maximum-ratio-combining (MRC) receivers. Therefore, it is necessary to explore the filter challenges brought on by moving into relatively uncharted frequency spectra and adopting these new technologies, as well as the factors driving the physical, electrical, and cost restraints for 5G filters and the supporting simulation technology that will help designers physically realize these components (Cadence 2018).

5.1.1 Current Status of Mobile Device Filter Technologies

Today's Fourth-Generation (4G) Long Term Evolution (LTE) smartphones support in excess of frequency 30 bands, requiring over 60 filters, many in the form of multiplexers. This number of filters consumes significant space and commands the largest share of the RF expense budget in the mobile ecosystem, putting considerable cost pressures on component manufacturers to meet performance goals for low costs. The majority of these filtering components are based on surface acoustic wave (SAW) or bulk acoustic wave (BAW) technology. BAW filters fall into two general architectures, solidly mounted resonators (SMRs) and film bulk acoustic resonators (FBARs). Additionally, both SAW and BAW devices have temperature compensated (TC) versions and will often see TC used as a descriptor prefix, e.g., TC-SAW (Mahons 2017; Cadence 2018).

Electrical filters fundamentally work by storing enough of a signal over time to determine the rate of change of the signal, allowing one frequency to be differentiated from another. The easiest filter to conceptualize is a digital signal processing (DSP) filter since the storage is digital memory. DSP filters do this by converting the signal with an A/D converter to a digital sequence stored in local semiconductor memory. The math processor can then operate on the data, removing the unwanted frequencies and allowing the desired frequencies to pass. Analog filters effectively do the same thing, storing the signal, not in digital words, but in stored energy. A conventional RLC filter stores the energy in the charge on capacitors and current in inductors, and BAW/SAW filters store the signal in acoustic resonators. Waveguide and cavity filters store the signal in the electromagnetic (EM) resonance in the transmission lines or the cavity (Mahon 2017).

Printed-circuit board (PCB)-based filters can be designed with various responses, including bandpass, bandstop, lowpass, or highpass filters, and from a number of different transmission-line technologies, including microstrip, stripline, or coplanar-waveguide (CPW) transmission lines. In general, filter designers should start with a printed-circuit-board (PCB) material selection, targeting optimum characteristics for RF/microwave filters. The choice of circuit material can impact not only a filter's performance but also the size of a printed circuit filter. The function of a filter is to shape part of the frequency spectrum, ideally stopping unwanted signals while passing desired signals with minimal loss or attenuation. Each filter system performs these functions by means of different spectral regions: stopbands, passbands, and transitions between a stopband and a passband. For example, a lowpass filter has one passband in the lower-frequency portion of its frequency range and one stopband in the upper-frequency part of its frequency range, with one transition region between them. A highpass filter is the opposite, with one passband in the upper-frequency part of its range, one stopband in the lower-frequency part of its range, and one transition region between them. A bandpass filter has a passband, lower and upper stopbands, and two transition regions. A band-reject filter can be thought of as the opposite, with a stopband with transition regions linking upper and lower passbands (ROG Blog 2016).

Different transfer functions describe a filter's transition regions. A Chebyshev filter, for example, is characterized as having an abrupt transition from the passband to the stopband, i.e., very little spectrum is required to make the change from the lowest signal loss to the highest signal attenuation. On the other hand, a filter with a Butterworth or binomial function makes a more gradual transition from the passband to the stopband. It requires a greater frequency spectrum to make the transition from filter regions, but it can also achieve a passband with low loss and very little ripple compared to a Chebyshev filter with its shorter transitions (ROG Blog 2016).

A filter's frequency response is truly a composite of the responses of its different spectral regions, with the transfer function having a major influence on the loss characteristics of the passband and stopband regions. A Chebyshev filter is capable of a quick, clean transition from a passband to a stopband, but at the cost of some amplitude variations or ripples in the passband insertion-loss response. A Butterworth filter can achieve a much flatter passband insertion-loss response but less attenuation of signals at frequencies closer to the passband than a Chebyshev filter (ROG Blog 2016).

A PCB filter design aims to achieve a set of responses for a desired frequency range but also minimize transmission and reflection losses at the filter's input and output ports by means of impedance matched junctions. The input and output ports are often coaxial connectors and most typically at a characteristic impedance of 50 Ω. When sorting through PCB material options prior to a design, a filter designer usually starts with dielectric constant (D_k) evaluation. PCB filters are typically formed of resonant circuit structures, such as the quarter-wave or half-wavelength resonators used in edge-coupled microstrip bandpass filters. The D_k of the dielectric material will determine the dimensions of the transmission lines required for specific resonator characteristics and frequencies. Circuit materials with higher D_k values will yield smaller filter resonator structures for a given wavelength and frequency, when miniaturization of a filter design is an important goal. In any case, for predictable, repeatable filter and resonator performance, the D_k of a circuit material choice should be as consistent as possible, held to the tightest tolerance possible (ROG Blog 2016).

5.1.2 The 5G Filter Performance Challenges

Surface acoustic wave (SAW) and bulk acoustic wave (BAW) filters cover a range of frequencies up to 6 GHz, come in small sizes and offer good performance-to-cost tradeoffs, making them the dominant off-chip approaches in mobile devices at the present time. Unfortunately, analogous filtering options for the mmWave spectrum have issues in viability, performance, size, availability, and so forth. In past decades, each generation of mobile systems – from 1G analog systems to 2G digital standards to 3G mobile broadband capabilities to 4G LTE and LTE-Advanced networks – has overcome a unique set of challenges. Leaps in technology are necessary to enable these advancements in performance. To tackle the challenge of increased data rates, the Shannon-Hartley theorem explains the maximum amount of error-free digital data that can be transmitted over a channel of a given bandwidth in the presence of noise (Matthews 2019):

$$C = M \times B \times \log_2\left(1 + \frac{S}{N}\right) \tag{5.1}$$

where C is the channel capacity, M is the number of channels (e.g., the MIMO order), B is the bandwidth, and S/N is the signal-to-noise ratio. To increase channel capacity (or data rates), one of the primary factors of focus for 5G is bandwidth, as well as how much bandwidth is available in a given part of the spectrum.

5.1.2.1 The 5G Frequency Spectrum

High frequencies suffer from range limitations and path loss through air, objects and buildings. However, mmWave signals require much smaller antennas, which can be tightly packed together to create single, narrowly focused beams for point-to-point communication with greater reach. In the U.S., RF/microwave/mmWave bands have been proposed for 5G between 3.5 and 6 GHz, 27 and 40 GHz, and 64 and 71 GHz. Frequency bands at approximately 27, 40 and 71 GHz are the main candidates for 5G mmWave, having demonstrated directional antenna, beamforming and beam tracking performance in multipath environments. Each band will undoubtedly present its own set of issues and solutions for components in the radio design. As a result, this significantly expanded spectrum is expected to result in a greater diversity of filter solutions than those serving the current mobile communication bands (Cadence 2018; Matthews 2019).

When allocating the band spacing for a new wireless communications standard, the third Generation Partnership Project (3GPP), a telecommunication association responsible for the maintenance and development of system specifications, must strike a balance between the efficient use of the available spectrum and the current capabilities of radio technology, including the state of filter design with regard to performance, cost, and size. With 5G, the need for bandwidth has motivated the 3GPP to push for advances in radio access technology into the mmWave spectrum as well as select unused bands between frequencies that have been authorized for public safety and defense applications (Cadence 2018).

To solve the practical issue of how to build an actual mmWave-capable base station and implement high-performance RF filtering, many experiences from other industries can be learned and evaluated. mmWave technology has been employed for decades in various fields and functions. For example, it has a long history in military, aerospace and satellite communication applications such as K-Band inter-satellite communication and ranging and Ka-Band high-resolution radar. In other industries, the automotive field is using both 24 GHz in short-range and 77 GHz in long-range radar-based advanced driver assistance systems (ADAS) to scan a vehicle's environment for driver support or automated driving. In the U.S., 36–40 GHz is currently licensed for high-speed microwave data links between a cellular base station and a base station controller, and the unlicensed 60 GHz band is used in short-range data connections and IEEE 802.11ad WiGig for high bandwidth streaming (Matthews 2019).

5.1.2.2 The 5G Filter Requirements

High-performance filtering is critical, as spectral crowding increases the need for interference mitigation, and the consumption of bandwidth drives the need to reduce or even eliminate guard bands. Supporting technology will require very low-loss, narrowband filters with exceptionally steep filter skirts (high selectivity), high rejection, and very little temperature drift. In addition to these stringent requirements, increased parasitics and substrate losses associated with the filtering device and its packaging (laminate) at mmWave frequencies will most certainly degrade performance unless properly addressed. Specific filter performance metrics must address the following (Cadence 2018; Matthews 2019):

1. Percent bandwidth

   Filter technology must not limit the radio access system bandwidth. For the 3.5–6.0 GHz 5G bands, the frequencies are close enough to the current mobile highband frequency channel that 5G systems can employ a similar set of radio solutions. While the higher (6-GHz) frequencies will challenge the performance levels of current off-the-shelf components, the basic radio architectures employed in current wireless communications systems are expected to work effectively for the higher 5G frequencies. From a filter perspective, the incremental higher frequency will be an additional challenge to the use of SAW filters, which already struggle at the 2.5-GHz band. This leaves the field open for BAW and TC-BAW components. However, the performance degradation that current BAW filter technology experiences at higher frequencies may also disqualify these types of filters.
2. Selectivity

   High selectivity enables designers to make good use of available bandwidth.
3. Insertion loss

   Power is a system cost driver, and on the receiver side, insertion loss impacts the overall noise figure of the receiver.
4. Size and packaging

   For 5G applications, filter technology will also be driven by size and integration concerns, which will be influenced by the system architecture. In phased arrays, the antenna elements must be sufficiently close together to avoid generating grating lobes; a half-wavelength spacing for mmWave frequencies amounts to only a few millimeters. As shown in Fig. 5.1a, b, phased arrays often used in mmWave applications use a plank architecture, in which the gold areas indicate the antennas mounted on a printed circuit board (PCB) (green area) and the blue areas indicate the circuit "planks" extending 90 degrees from the array (where space for filters is already minimal). For base station manufacturing, however, flat-panel architectures are also considered, where the circuitry is implemented on the backside of the antenna PCB, requiring even denser spacing for filtering and other functional blocks. Two alternative radio architecture approaches gaining attention include software-defined radio (SDR) architectures (cognitive and reconfigurable) and tunable filters. With SDR, all filtering is performed after the analog-to-digital conversion on the receive side and before the digital-to-analog conversion on the transmit side. The current state of the art in silicon semiconductor integration can address the filtering requirements, although this active filter approach consumes significant power (tens of watts) in contrast to passive filters, which consume zero power. In addition, a front-end amplifier in an SDR would be vulnerable to any potentially strong out-of-band signals, and the ADC would be converting the entire received spectrum. A tunable band-select filter

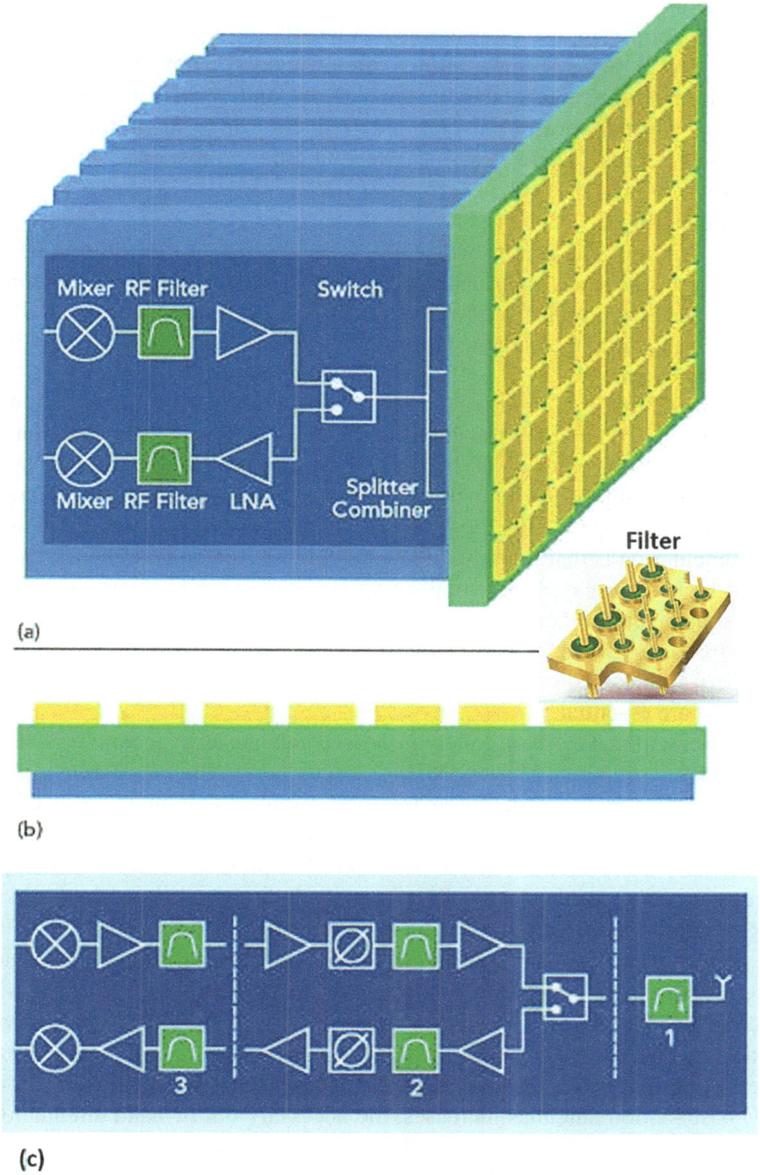

Fig. 5.1 Phased array architectures (Adapted from Matthews (2019), Credit: Microwave Journal): (**a**) plank; (**b**) flat panel; and (**c**) potential radio filter locations

before the LNA would address out-of-band signals, while a tunable anti-aliasing filter before the ADC would greatly improve the power efficiency.

5. Temperature Stability

   To make efficient use of available bandwidth, the filter must meet its specifications over a range of temperatures. Small-scale systems may be deployed in exposed environments that experience extremes in temperature and temperature variation. Furthermore, the overall size reduction in systems leads to densely populated boards, where heat from surrounding components can affect filter stability.
6. Power handling

   The ability of the filter to withstand large amounts of transmitting RF power is mostly a concern in traditional macrocell use cases below 3 GHz. In mmWave 5G use cases, the transmit power is spread over the individual array elements, and the transmit range is reduced as well.
7. Filter placement

   As shown in Fig. 5.1c, there are several practical locations: at the antenna element (Position 1), behind the amplifiers closest to the antenna (Position 2) and on the high-frequency side of the mixers (Position 3). Note that the dashed lines represent potential 1: N branching in a beamforming architecture, showing a simplified single path from the mixer to an antenna. The advantage of Position 1 implementation is that linearity and bandwidth design constraints for the subsequent blocks are eased, potentially reducing the overall system cost. Filtering placed at Position 1 suppresses noise far from the channel of interest across a wide frequency range, making wideband suppression and shallow insertion loss key performance features. Given that each subarray has a filter, a small size and low cost are essential.

5G networks will make use of greatly improved antenna array technology, requiring additional filter solutions. Massive MIMO, which may contain 100 or more antenna elements, may offer an order of magnitude improvement in spectral efficiency (60–110 b/s/Hz/cell, under ideal conditions, compared to ~3 b/s/Hz/cell) over a 4 × 2 MIMO. One concern for massive MIMO implementation is the complexity and quantity of components per RF chain, including broadband high-resolution analog-to-digital and digital-to-analog data converters (ADCs and DACs), highly linear power amplifiers (PAs) with linearizing control circuits, and a large number of filters with strong out-of-band suppression to address any interfering signals. Reducing interference as early as possible in the receiver chain is a favorable approach to achieving the objectives of MIMO antennas, simultaneously increasing interference robustness while decreasing power consumption. As the number (M) of MIMO base-station antennas increases, the necessary out-of-band attenuation provided by the Berkeley Packet Filters (BPFs) increases proportionately to the square root of M. This implies a practical limit on the number of base-station antennas due to the increase in BPF design complexity and power consumption. Insufficient out-of-band attenuation would result in aliasing of the filtered out-of-band interferers into the useful band at the output of the ADC, thereby corrupting the received baseband signal (Cadence 2018).

5.1.2.3 Physical Design and Emerging Solutions for the 5G Filters

SAW and BAW filters have long dominated off-chip filtering applications in mobile devices because of their excellent performance specs, tiny footprints, and affordable prices compared to other options. Unfortunately, the nature of their design – using interdigital transducers (IDTs) to process signals as acoustic waves – exhibits degradation in selectivity at frequencies greater than 6 GHz, making SAW and BAW technology unfeasible for mmWave applications (Matthews 2019).

The film bulk acoustic resonator (FBAR) filter is a form of BAW filter that can operate from 5 to 20 GHz but still below desired mmWave ranges. Due to their high Q factor, these filters offer low insertion loss for decent system performance and can be integrated with monolithic-microwave-integrated-circuit (MMIC) and other technologies to reduce cost, size, and power consumption (Cadence 2018).

As frequencies increase toward the mmWave range, signal wavelengths are small enough that filters based on EM techniques are feasible. Waveguide and cavity filters are the two most common high-performance filter types between 20 and 80 GHz. These filter types have dimensions in centimeters rather than millimeters. There have been many efforts to miniaturize these filters at mmWave frequencies. However, the wavelength size for the EM wave being filtered is still large with respect to the filter's physical size requirements, so it is likely that these mmWave filters will be larger than lower-band acoustic filters, which may be permissible if a different radio architecture can reduce the quantity of filters needed. Otherwise, an alternative construction approach must be developed for EM-based filtering for mmWave radio architectures (Cadence 2018).

Likely candidates for filtering at mmWave frequencies are filters based on substrate integrated waveguides (SIWs). They offer a planar construction that can be easily incorporated into MMIC, RFIC, and printed circuit board (PCB) substrates with existing interconnect structures. They have also been demonstrated using standard complementary-metal-oxide-semiconductor (CMOS) technology. Gallium arsenide (GaAs) and indium phosphide (InP) technologies offer better high-frequency performance than CMOS because of higher breakdown voltages, higher electron mobility, high cutoff frequencies, and superior noise performance. GaAs MMICs also offer high-quality passive circuit elements. While attractive for SIW's wideband properties, good isolation, and lower losses, challenges for mmWave include radiation leakage between plated through-holes (PTH), difficulties designing SIW transitions and dimensional variations in the PTH sidewalls (which are detectable at mmWave frequencies). Microelectromechanical systems (MEMS) offer the potential for tiny, tunable RF filters created using conventional semiconductor fabrication processes. The methodology is interesting due to its dimensional accuracy, high component density and low cost at high volumes; however, current designs are still in the research phase or are limited to submmWave ranges (Cadence 2018; Matthews 2019).

Figure 5.2a summarizes the frequency ranges of commonly available filters at present. At higher mmWave frequencies, acoustic filtering is not as practical, so many developers have turned to electromagnetic (EM) solutions, including

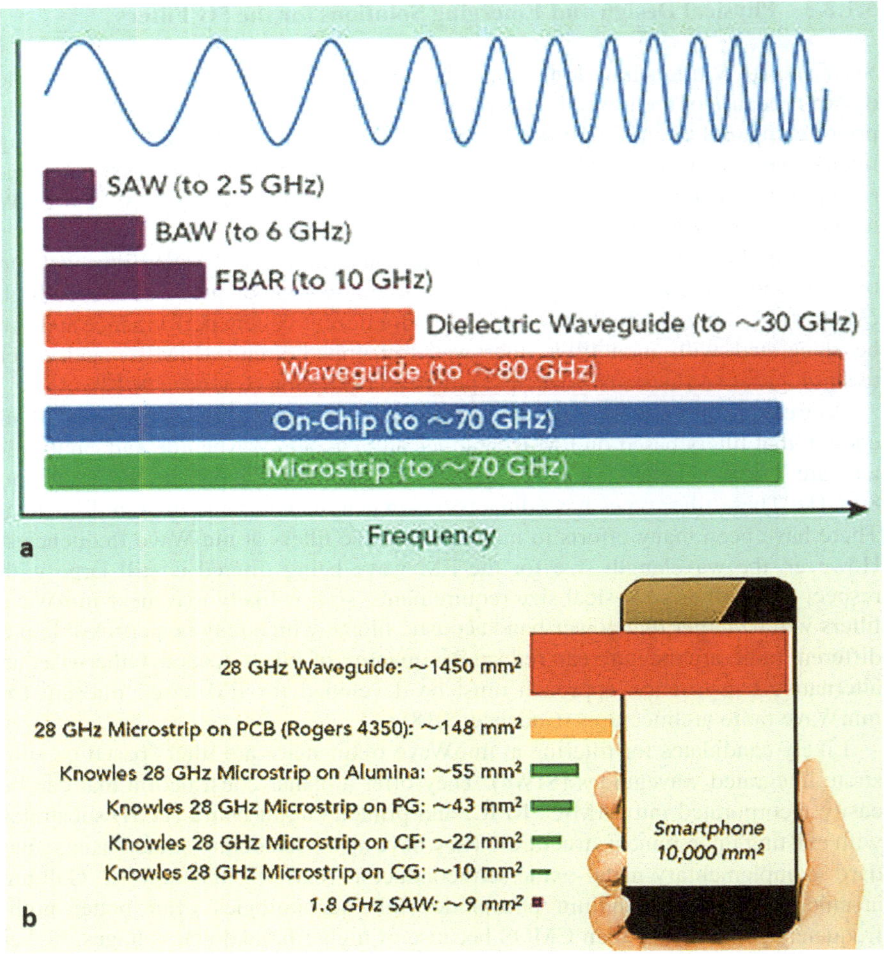

Fig. 5.2 Common standard filter technologies (Adapted from Adapted from Matthews (2019), Credit: Microwave Journal): (**a**) Frequency ranges; (**b**) Sizes of mmWave filters compared to typical SAW filters

dielectric and cavity waveguide, on-chip, and microstrip (or planar thin film) filters. Waveguide filters are hollow or dielectric-filled cavities within metal or dielectric tubes that act as BPFs, blocking certain wavelengths while allowing others to pass. Characterized by high-power handling and low loss, waveguide filters are widely used for mmWave frequencies from 20 to 80 GHz in military, radar, satellite and broadcasting areas. Unfortunately, waveguides typically have dimensions in the centimeter range (see Fig. 5.2b). The $\lambda/2$ array element spacing at 28 GHz in free space is 5.35 mm. Until manufacturers can sufficiently reduce waveguide sizes while still meeting electrical performance requirements, this solution may not be practical for an array antenna system (Cadence 2018).

On-chip filters using semiconductor technologies are attractive because of the short circuits, tight tolerances, and the capability for integration with other devices to form systems on chip (SoC) solutions. However, significant performance issues regarding the Q-factor, losses and noise fig. (NF) occur with the production of on-chip devices having reduced dimensions for mmWave frequencies. Challenges arise from various factors, including the physical characteristics of the semiconductor material and the cost of implementation. For example, GaN circuits are made as thin as possible to increase heat dissipation; however, filter Q is proportional to dielectric substrate thickness, so the high-power advantage of using GaN devices works in opposition to integrating filters with high Q. Additionally, a filter structure in GaN occupies valuable real estate on the wafer that would be better allocated to active devices. Building on-chip high-Q filter structures to serve in a front-end application is currently impractical (Cadence 2018).

Microstrip filters have been considered for mmWave applications but are commonly dismissed for various performance issues. Currently, there are at least three different form factors (Cadence 2018):

1. Microstrip on PCB
2. The microstrip is a multilayer, low-temperature cofired ceramic (LTCC) package
3. Microstrip is a small form factor, single-layer package

Microstrip filters printed on PCB are appealing because of their simple construction. Nevertheless, high-performance PCB solutions generally reach centimeter-range sizes, which are much larger than the subwavelength dimensions required for mmWave antennas. Variations caused by the PCB manufacturing processes also limit performance, resulting in higher insertion loss and lower suppression values.

Another option is surface mount technology (SMT). SMT assembly has been long used in commercial systems and is now being adopted in mmWave military systems for potential cost savings. Unlike chip and wire solutions, SMT filters have standardized form factors to reduce the overall assembly time and require no post-tuning. Low-temperature co-fired ceramic (LTCC) filters come in SMT form factor and are similar to multilayer capacitors, in which multiple layers of very thin ceramic tape are printed with different passive elements and then stacked together to prevent substrate warping. Prototypes of LTCC technology for mmWave applications are being developed, potentially offering a method of including both filters and antennas on the same component with a very small footprint. Unfortunately, since the metal coatings are screen printed, dimensional precision is not as high as other thin film solutions, and the unpolished substrate can lead to high losses. Additionally, suppression is generally limited to 30 dB and below (Cadence 2018).

Another type of SMT filter assembly is a single-layer microstrip, which is printed with distributed transmission lines to create high-performance resonant structures. With a careful choice of filter topology and dielectric materials, high rejection, low loss filters that are temperature stable from −55 to 125 °C can be produced. These filters provide performance similar to their lower frequency SAW and BAW counterparts (see Table 5.1). High performance and small form factor devices are possible within the constraints of 5G New Radio (NR) systems. For instance, a low-loss

Table 5.1 Comparing SAW/BAW filter performance with SMT microstrip (Cadence 2018)

Performance goals	Typical SAW/BAW performance (<6 GHz)	SMT microstrip performance (<6 GHz)
Low insertion loss	<3 dB	<3 dB
Excellent rejection	>30 dB out of band	>50 dB out of band
Broad bandwidth	<100 MHz	3 MHz
Small size	~9 mm^2	<9 mm^2
Good temperature stability	~3 ppm/°C	~3 ppm/°C

26 GHz filter with greater than 50 dB of suppression can fit in a 4 mm × 1.6 mm footprint. This size is significantly smaller than half a wavelength, enabling integration into both plank and planar architectures (Cadence 2018).

5.2 Materials and Design for Acoustic Filters

5.2.1 Current Application and Band Allocation of Acoustic Filter Technology

Acoustic wave technology has attracted great attention as an effective solution for achieving high-performance RF filter operations while offering low cost and small device size. At present, the most mainstream implementation of RF filters is dominated by surface acoustic wave (SAW) filters and bulk acoustic wave (BAW) filters.

The SAW propagates in the lateral direction with the shape and center frequency of the passband determined by the pitch, line width and thickness of the interdigital transducers (IDT) (Fig. 5.3). Because they are fabricated on wafers, SAW filters can be created in large volumes at low cost; filters/duplexers for different bands can be integrated on a single chip with little or no additional fabrication steps. Their key advantages are low cost, wide relative bandwidth, and flexible port configurations. At the lower frequency range, SAW filters meet the requirements for low insertion loss and excellent rejection, covering broad bandwidths at a fraction of the size of traditional cavities and even ceramic filters. However, due to the degradation in selectivity at higher frequencies, SAW filters have limited use above ~2 GHz; at those higher frequencies, they are mostly used for applications with modest performance requirements, such as global systems for mobile communications (GSM), code division multiple access (CDMA), and third generation (3G) wireless receiver front ends, duplexers, and filters. SAW devices are also highly sensitive to temperature: the stiffness of the substrate material decreases with higher temperatures, resulting in a diminished acoustic velocity and degraded RF performance. Typically, SAW filters operate in the mobile environment at signal frequencies from 600 MHz to 2 GHz, whereas BAW filters operate between 1.5 and 6 GHz, putting them in the range of the lower 5G bands (Matthews 2019).

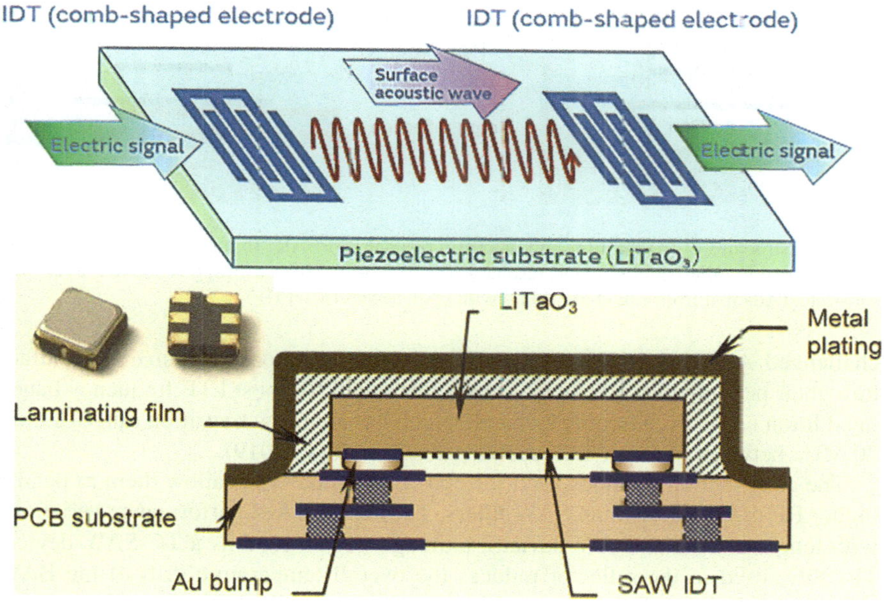

Fig. 5.3 Illustration of surface acoustic wave (SAW) filter and its electronic package (IDT-Interdigital transducer). (Modified from Murata (2022), Credit: Murata Manufacturing Co., Ltd)

Temperature-compensated SAW (TC-SAW) filters are fabricated using a more complex and costly layer structure to increase the substrate stiffness at higher temperatures and extend their operating range. Since the temperature-compensating process doubles the number of required mask layers, TC-SAW filters are more expensive to manufacture than standard SAW filters, but they are still less expensive than bulk acoustic wave (BAW) filters. In comparison, BAWs require approximately 10 times more processing steps than SAW filters. While BAW filter technology yields approximately 4 times more parts per wafer, it still has a higher cost-per-filter compared to SAW filters (Mahon 2017).

BAW filters fall into two general architectures: solidly mounted resonators (SMRs) and film bulk acoustic resonators (FBARs), as shown in Fig. 5.4. With BAW filters, an electric field excites an acoustic wave that travels in a vertical direction through the body of a piezoelectric substrate. The resonant frequency is determined by the thickness of the piezoelectric layer, which must be precisely controlled to extremely tight tolerances. The result is a device with lower loss, higher quality factor (Q), better power handling, and sharper corners (greater selectivity) compared to SAW filters operating at the same higher frequencies (Matthews 2019).

In the case of FBAR filters, the resonator is surrounded by an air interface created through etching or micromachining. In contrast, acoustic reflectors below the bottom electrode of a BAW-SMR filter allow them to be optimized for wideband performance in frequency regions where FBAR filters are more technically

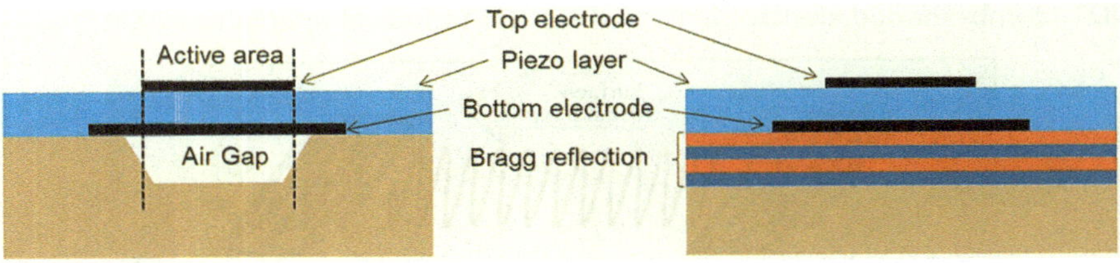

Fig. 5.4 Structure of (**a**) the film bulk acoustic resonator (FBAR) formed over an air gap on a substrate and (**b**) the solid mounted resonator (SMR) formed over a Bragg reflection layer on a substrate. (Adapted from Liu et al. (2020) with open access (MDPI))

challenged. Although BAW-SMR and FBAR filters are more expensive to manufacture, their performance advantages are better suited for most LTE frequency bands in addition to the PCS frequency band, which has a narrow transition range of only 20 MHz between transmit and receive paths (Matthews 2019).

The construction methods used for both BAW filter types allow them to handle higher RF power levels than SAW filters. They exhibit less performance variations with temperature than SAW devices, although not as good as a TC-SAW device. The SiO_2 used in the reflector reduces the overall temperature drift of the BAW significantly below what either traditional SAW or FBAR filters can achieve. Since the BAW-SMR resonator sits on a solid substrate, it can dissipate heat more effectively than an FBAR device, which dissipates heat laterally through a much smaller edge surface. This allows BAW devices to achieve higher power densities, allowing compact devices to handle upward of 10 W and ample power-handling capability for small-cell base station applications (Matthews 2019).

The current market application and band allocation of acoustic RF filter technologies are presented in Fig. 5.5. Compared with surface acoustic wave (SAW) resonators, bulk acoustic wave (BAW) resonators have more potential in fabricating high-quality RF filters because of their lower insertion loss and better selectivity in the middle and high frequency bands above 2.5 GHz. The high-quality factor (Q-factor) of the BAW makes filter skirts steeper, while high acoustic velocity and thermal conductivity enable high power handling. Additionally, the low power consumption, high isolation, and complementary metal oxide semiconductor (CMOS) compatibility of BAW filters make it a dominant device in the field of RF communications. However, the emergence of 5G is a challenge for data rates, coverage, and minimization of acoustic RF filters. Considering the entire RF system, the decreasing module size and increasing input power are inevitable trends in the future. The BAW resonator, as a basic unit of RF circuits, significantly affects the performance of BAW filters. AlN is still the preferred material for bulk production of BAW devices, but the most limiting property is probably the coupling coefficient. A new design of laterally excited bulk-wave resonators (XBAR) with ~25% strong piezoelectric coupling and RF integration technology are proposed for further improving the properties of BAW devices, thus enhancing the availability of BAW resonators and filters in RF communication (Ruppel 2017; Liu et al. 2020).

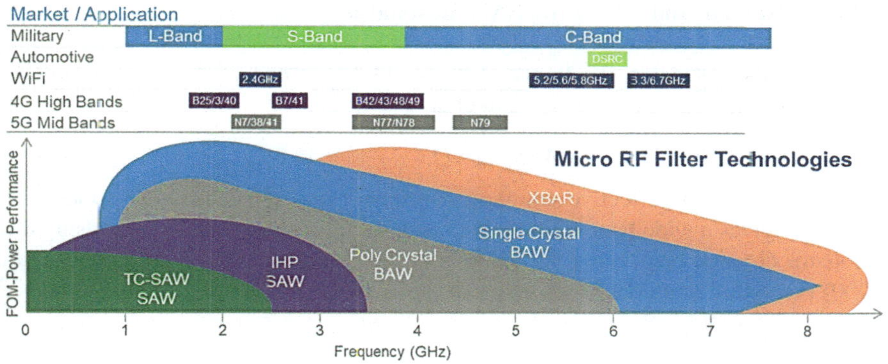

Fig. 5.5 Current market application and band allocation of acoustic radio frequency (RF) filter technologies. (Adapted from Liu et al. (2020) with open access (MDPI)). *SAW:* Surface acoustic wave, *TC-SAW:* Temperature-compensated surface acoustic wave, *IHP SAW:* Incredible high-performance surface acoustic wave, *BAW:* Bulk acoustic wave, *XBAR:* Literally excited bulk-wave resonator

5.2.2 *Basic Working Principle of the BAW Filter*

5.2.2.1 Structure of the BAW Resonator

The BAW resonator is a piezoelectric stack structure consisting of a piezoelectric film sandwiched between two metallic electrodes. The area where the upper and lower electrodes and the piezoelectric layer overlap in the thickness direction is defined as the active area of the resonator. The FBAR (film bulk acoustic resonator) and SMR (solid mounted resonator) are two types of BAW resonators (Fig. 5.4). The fundamental resonance occurs when the wavelength of the mechanical wave generated by the excitation is twice the thickness of the piezoelectric layers. Regardless of the electrodes, the resonant frequency depends on the acoustic velocity and the thickness of the piezoelectric film (Liu et al. 2020):

$$f_0 = \frac{\upsilon}{2d} \tag{5.2}$$

where f_0 and v are the resonant frequency and acoustic velocity, respectively, and d is the thickness of the piezoelectric film. Additionally, the thickness of the electrode can affect the resonant frequency. An electrode of different thicknesses as a mass loading can generate a frequency shift, which is often used to form the filter passband.

5.2.2.2 Key Parameters of the BAW Resonator

For the BAW resonator, the figure of merit (FOM) is one of the most important parameters and can be defined as (Liu et al. 2020):

$$FOM = k_{\mathrm{eff}}^2 \times Q \tag{5.3}$$

where k_{eff}^2 is the effective coupling coefficient and Q is the quality factor. A larger k_{eff}^2 gives a wide bandwidth, which is desired for the 5G band. Increasing the resonator Q-value (especially such that both the Q_s and Q_p values are greater than 2000) is a significant goal of resonator design.

1. Effective coupling coefficient

The piezoelectric material-coupling coefficient k_t^2 and effective coupling coefficient k_{eff}^2 are two different expressions. It is not difficult to design a resonator with the required k_{eff}^2 from a material having an adequate k_t^2 value. In addition, electrode configuration, acoustic reflector, and parasitics are other major factors that can significantly influence k_{eff}^2. The calculation of k_{eff}^2 via the series resonant frequency f_s and parallel resonant frequency f_p is given as follows (Bi and Barber 2008):

$$k_{\mathrm{eff}}^2 = \frac{\pi^2}{4} \times \frac{f_s}{f_p} \times \left(\frac{f_p - f_s}{f_p} \right) \tag{5.4}$$

When AlN is deposited on the substrate, a transition region with poor crystal quality will be formed at the interface between the AlN and the substrate. The thinner this layer is, the better the overall layer coupling will be. For a given material, there is always an optimum thickness ratio that maximizes k_{eff}^2 (Fig. 5.6a). For electrodes, high acoustic impedance materials can boost the achievable k_{eff}^2 due to the matching of the shape of the stress field and the electric field (Fig. 5.6b). Similarly, reflector layers can cause the loss of the stress field, thereby reducing the coupling (Kaitila 2006; Fattinger 2008).

2. Quality Factor

The Q of a resonator is defined by the device's loss mechanism. Various loss mechanisms are possible, including acoustic losses, ohmic losses of the electrodes and viscoelastic losses. In particular, shear waves and laterally leaking waves can constitute a major loss mechanism for BAW resonators. The electrode of the BAW resonator is usually not equipotential; thus, eddy currents are produced, leading to I^2R losses and decreased Q_s. Because of the existence of the acoustic reflection layer, SMR-BAW may have more losses, such as material loss, incomplete suppression of energy loss, which can flow through the reflection layer, and resistive losses from the interconnections. There are several approaches to estimate the Q of a resonator. Among them, methods of 3-dB bandwidth and phase derivation are sensitive to the quality of the measured data and can just characterize the Q-value at f_s and f_p. The Bode method, or using the stored energy divided by the dissipated energy per cycle, can derive a Q that is not only at the two resonance frequencies but also over

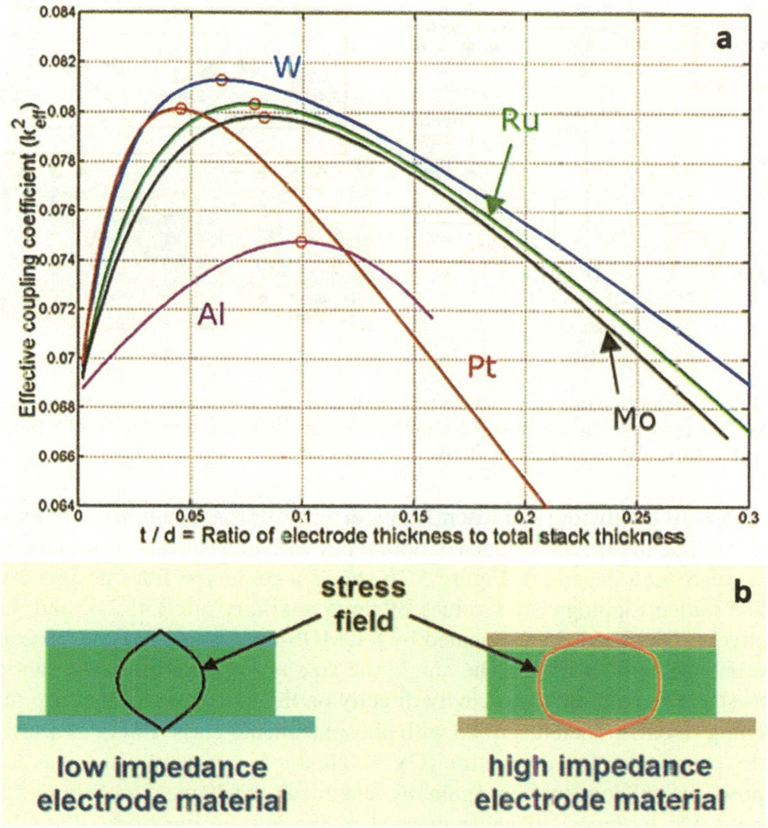

Fig. 5.6 Relationship between keff² and the ratio of electrode thickness to total stack thickness for an AlN resonator with different electrode materials (**a**) (Adapted from Kaitila (2006) with permission from IEEE); and Stress field distribution throughout a bulk acoustic wave (BAW) resonator with low impedance and high impedance electrode materials (**b**). (Adapted from Fattinger (2008) with permission from IEEE)

a whole frequency range in the vicinity of those frequencies (Ruby 2006; Bi and Barber 2008).

5.2.2.3 Topology of the BAW Filter

The BAW filters can be divided into ladder and lattice topologies, as shown in Fig. 5.7. Generally, ladder-type filters present a steep rejection near the passband but a poor out-of-band rejection. The conventional ladder-type filter has been modified to improve the out-of-band rejection. By adding a series/shunt inductor, for example, a new pair of transmission zeros will appear in the transmission response. Moreover, a modified lattice network with two auxiliary inductors is shown in Fig. 5.7a, and the

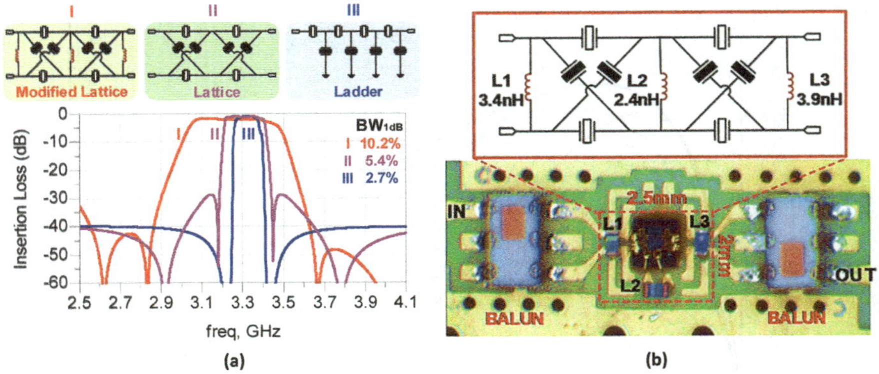

Fig. 5.7 Topologies of ladder type, lattice type, modified lattice with two auxiliary inductors, and their electrical responses (**a**); and board assembly of the filter with balun circuits (**b**). (Adapted from Yang et al. (2016) with open access (MDPI))

measured 3-dB bandwidth can reach 12.3% at 3.25 GHz. A combined topology (ladder-lattice) that ally the advantages of both types with high selectivity and high isolation has also been proposed. Figure 5.7b shows a prototype filter designed in this modified lattice topology. The values of three auxiliary are 3.4, 2.4, and 3.9 nH, respectively. The filters are fabricated by a CMOS-compatible MEMS process in a high-resistivity silicon wafer. The single die size is 1 mmx 1mm. The fabrication process starts by etching an air cavity directly on the silicon wafer by deep reactive ion etching (DRIE). Then it is filled with phosphosilicate glass (PSG), as a sacrificial layer, by chemical vapor deposition (CVD). Next the deposited wafer was reduced to expose the silicon surface again by chemical mechanical polishing (CMP). Afterward Mo is deposited and patterned as the bottom electrode. Then follows depositing AlN and Mo by physical vapor deposition (PVD), successively, to form the piezoelectric layer and top electrodes, respectively. The top electrode is then patterned by plasma etching, while AlN is etched by KOH wet etching. Finally, gold (Au) is evaporated by PVD and patterned by lift-off, serving as the electrical connection and pads, which is followed by release of the sacrificial layer in diluted hydrofluoric acid solution (HF/H2O, 1:10). After the wafer is fabricated, the AlN-based BAW filter dies are identified through the on-wafer test and picked up for further assembly (Yang et al. 2016; Liu et al. 2020).

5.2.3 *Materials for the BAW Resonator*

5.2.3.1 Piezoelectric Materials

At present, there are three main kinds of piezoelectric materials used in BAW resonators: AlN, PZT, and ZnO. For PZT, although it has a high k_t^2 and large dielectric constant, the inherent loss of PZT is too large, the speed of sound is too low, and the preparation process is difficult to keep the composition uniform. The k_t^2 of ZnO is also larger than that of AlN, but its longitudinal sound velocity is much smaller than that of AlN. Furthermore, zinc reduces the carrier lifetime during processing and is not compatible with CMOS technology (Liu et al. 2020).

AlN is a hexagonal wurtzite structure with lattice constants $a = 0.322$ nm and $c = 0.498$ nm. The maximum piezoelectric response (piezoelectric strain coefficient d_{33} is approximately 5.5 pC/N) can be obtained in the longitudinal direction. Simultaneously, it has high stiffness C_{33} (3.67×10^{11} N/m^2), a relatively high k_t^2 (6.5–7%), relatively high longitudinal sound velocity v (11,354 m/s), and a low temperature coefficient of frequency (−25 ppm/°C or less). Other factors, such as the relative dielectric constant ε_r (approximately 10.2), the intrinsic Q of the material, and compatibility with CMOS processes, are also consistent with the design requirements of BAW resonators and filters (Akiyama et al. 2008; Liu et al. 2020).

AlN thin films can be deposited by magnetron sputtering, molecular beam epitaxy, pulsed laser deposition, and metal organic chemical vapor deposition. Among them, magnetron sputtering is the most commonly used deposition method for preparing piezoelectric AlN films. It is generally performed at a lower temperature (typically 200–300 °C), and the maximum temperature is up to 400–500 °C during the reaction, which makes deposition processes compatible with CMOS technology and reduces deposition cycles. In addition, the method has the advantages of high efficiency, low cost, large deposition area, etc. Rapid thermal annealing (RTA) after sputtering is another effective way to improve the quality of the AlN film. In addition, the quality of AlN can also be affected by the thickness of the films (Bhugra and Piazza 2017; Liu et al. 2020).

Compared with polycrystalline AlN, single crystal AlN has quantifiable higher crystal quality, fewer defects, and more stable chemical properties. Improved crystal quality in the single crystal AlN results in improvements in acoustic velocity and potentially improved piezoelectric coefficients. Moreover, single crystal AlN-based BAWs are superior to the polycrystalline AlN devices. Defects existing in polycrystalline AlN will cause absorption or scattering of bulk acoustic waves and increase acoustic transmission loss. However, there are still many problems that directly hinder the commercialization of single crystal AlN-based BAWs. For example, the area of single crystal AlN that can be obtained is usually small and normally based on a 2-inch substrate; the thickness is small, usually only 100–200 nm; and the film deposition is expensive, slow, and conditional (Wang et al. 2014).

Doping other elements into AlN is another effective method to improve the piezoelectric properties of AlN. The elements include Sc, Ti, Ta, V, Y, Mg, and Er. The codoped elements include Mg + Zr, Mg + Hf, Mg + Nb, and Mg + Ti. Sc, Ta, Y, and Er increase the piezoelectric properties of the film within a certain doping concentration range. Ti, V, Mg, Zr, Hf, and Nb may deteriorate the piezoelectric properties of the film. Moreover, controlling the ratio of multielement doping, such as Mg + Zr/Hf/Nb/Ti, can also greatly improve the piezoelectric properties of the film. On the other hand, although doping elements such as Sc can improve the piezoelectric properties of piezoelectric materials and increase the k_{ef}^2 of the BAW resonator, the Q-value of the resonator is generally reduced. The reason is that the doping element will increase the dielectric loss, thus deteriorating the quality of the piezoelectric material. Therefore, it is important to control the doping element concentration in the AlN film to obtain good performance (Akiyama et al. 2008; Zywitzki et al. 2017).

5.2.3.2 Electrode Materials

In addition to the influence of piezoelectric materials, the choice of electrode material is also important. The ideal electrode material should have high acoustic impedance (high acoustic impedance difference with piezoelectric material is beneficial to improve the effective electromechanical coupling coefficient of the resonator), low resistivity (the lower the resistivity, the smaller the R_s, which is beneficial to increase the Q-value of the resonator), and low density (higher density means higher mass loading, which is not conducive to the high frequency of the resonator). Moreover, since the piezoelectric material is directly deposited on the electrode material, the type and quality of the electrode directly affect the quality of the piezoelectric material. Good orientation, low surface roughness (reducing the scattering loss of sound waves), and low lattice distortions with piezoelectric material are beneficial to improve the quality of the piezoelectric film and improve the performance of the resonator (Liu et al. 2020).

Many electrode materials have been investigated, including Pt, Mo, W, Ru, Ir, Ta, Al/Au/Cu, CNTs (carbon nanotubes), graphene, and ITO (indium tin oxide). Among them, Mo, W, Ru, and Ir have high Young's moduli, but the density of W is relatively high. Al, Au, and Cu have lower resistivity, but their acoustic impedance is too low. Mo has a relatively moderate acoustic impedance, density, and resistivity, so it is the material most used as an electrode. CNTs and graphene are novel electrode materials that have the advantages of low density and low electrical resistivity, and the preparation of electrodes is currently difficult (Kamohara et al. 2008; Liu et al. 2020).

5.2.4 Temperature Compensation

Since the sound velocity of materials (such as AlN) decreases as the temperature rises, the operating frequency of a BAW resonator decreases as the temperature increases. Generally, the method to improve the temperature coefficient of a BAW resonator is to insert a material that has a positive temperature coefficient frequency (TCF), such as silicon dioxide (SiO_2), although it is known that deposition of SiO_2 will decrease the k_{eff}^2 and Q-value. This approach can achieve temperature compensation without any feedback control and external components (Liu et al. 2020).

Many factors influence the effectiveness of temperature compensation when using SiO_2 is a compensated layer, such as the deposition position and the ratio between SiO_2 and negative TCF layers. In the case of the position of the SiO_2 layer, it is usually deposited on the topmost layer in a piezoelectric stack, which can protect resonator from moisture and contamination. However, this method will increase the mass loading sensitivity, making it difficult to trim the layer to obtain a desired frequency. On the other hand, to achieve a zero-drift frequency of the BAW resonator, a large amount of SiO_2 (large ratio between the SiO_2 and piezoelectric film) is usually used, which can deadly deteriorate the electrical character of BAW. Putting SiO_2 in the high-stress region of BAW is an excellent method; thus, a minimum amount of SiO_2 is used without large k_{eff}^2 reduction and Q-value loss. Furthermore, a higher positive temperature compensation can be operated by using boron-doped SiO_2 or SiOF as the temperature compensation layer. Compared with FBAR, SMR-BAW shows a huge potential in temperature compensation when using a positive material as the low impedance layer (Liu et al. 2020).

5.2.5 *Frequency Tenability*

In the fabrication process of the BAW device, the resonant frequency is usually different from the target value due to the deviation in the process. Furthermore, the frequency offset caused by the temperature change, electromagnetic disturbances and aging also needs to be considered. On the other hand, 5G frequency allocations for mobile phones vary widely among the world's regions, so it is difficult for a filter with a single center frequency to cover all the bands. Therefore, tunable BAW resonators and filters that can operate at different frequencies have been developed. At present, BAW resonator tuning technology mainly includes the following (Liu et al. 2020):

1. Changing the physical characteristics of the piezoelectric film by bias voltage;
2. Tunable components such as passive elements;
3. A ferroelectric material with an electrostrictive effect;
4. A microheater element (by adjusting the temperature of the FBAR through a heater, but a relatively large electric power and time are needed to change the resonant frequency of the FBAR).

Table 5.2 shows some tunable BAW resonators using ferroelectric materials such as $SrTiO_3$, $BaTiO_3$, and $Ba_xSr_{1-x}TiO_3$ (BST) to replace the piezoelectric layer of the devices. These materials exhibit strong piezoelectric and/or electrostrictive properties. By applying a DC voltage, the electromechanical coupling coefficient and sound speed of the device are changed, thereby achieving a frequency shift. In particular, because the BST thin film is the paraelectric phase, no resonance is observed at zero bias. However, with increased DC bias, resonance develops and can be clearly tuned. Obviously, although those devices show some tuning abilities, the low Q-value inherent to ferroelectric materials limits their practical use (Noeth et al. 2010).

In fact, tuning frequency is complicated in acoustic resonators. Most of the current methods for tunable devices involve applying a variable electric field. The battery voltage of mobile phones is typically 2–3 V. In case a higher voltage is required for tuning, an intricate and space-consuming system is needed. Even if a suitable

Table 5.2 Performance of some BAW resonators based on tunable materials (Liu et al. 2020)

Material	Frequency (GHz)	Tuning ability[a] (%)	Max bias (V)	Q^b	Max k_{eff}^2 (%)
$SrTiO_3$	5.87	f_r:1	7.5	78 (Q_s)	2
$BaTiO_3$	~4	f_r:1.3/f_a:4.0	10	30	6.2
$Ba_{0.25}Sr_{0.75}TiO_3$	~4	f_r:1.2/f_a:0.8	10	120	0.5
$Ba_{0.25}Sr_{0.75}TiO_3$	~5	f_r:0.9	25	250 (Q_s) 350 ($Q_{s\text{-}de}$)	–
$Ba_{0.25}Sr_{0.75}TiO_3$	5.46	f_r:2	50	50 (Q_s) 140 (Q_p)	4.4
$Ba_{0.25}Sr_{0.75}TiO_3$	5.76	f_r:3.8	25	130 (Q_s)	7.1
$Ba_{0.25}Sr_{0.75}TiO_3$	5.46	f_r:1.9	50	260 ($Q_{s\text{-}de}$)	3.5
$Ba_{0.3}Sr_{0.7}TiO_3$	2.85	f_r:2.4/f_a:0.6	615/ kV/cm	200 (Q_s) 120 (Q_p)	4.4
$Ba_{0.5}Sr_{0.5}TiO_3$	5.45	f_r:2.4	50	250 ($Q_{s\text{-}de}$)	7.5
$Ba_{0.7}Sr_{0.3}TiO_3$	5.44	f_r:1.3	6	352.6 (Q_s) 746 (Q_p)	17.15

[a]Tuning ability at resonance frequency f_r and antiresonance frequency f_a
[b]Q-factor at series resonance Q_s, parallel resonance Q_p and series resonance after de-embedding $Q_{s\text{-}de}$

approach to tuning frequency can be found, there are many other conditions that need to be fulfilled for a tunable filter with any practical value (Liu et al. 2020).

5.2.6 *Lithium Niobate and Laterally Excited Bulk-Wave Resonators (XBAR)*

Lithium niobate (LN) appears to be a potential material with the form of crystalline films due to the large value of its coupling coefficient for the fast shear mode. In fact, the traditional BAW technology cannot cover the wider bandwidths required due to the relatively small piezoelectric coupling. Therefore, a novel device named laterally excited bulk-wave resonators (XBAR) is developed (Fig. 5.8). The resonator with interdigital transducers (IDTs) on a suspended LN plate can excite standing bulk acoustic shear wave resonance by applying a lateral electrical field parallel to the crystalline y-axis. The curve in the cross-section diagram represents the shear deformation in the LN layer, and small arrows indicate the direction and magnitude of atomic motion. The direction of acoustic energy flow of the primary shear acoustic mode is substantially orthogonal to the surface of the piezoelectric plate (Plesski 2020).

As shown in Fig. 5.8c, the main shear resonance is at 4.55 GHz, and the third harmonic is near 13 GHz. The relative bandwidth for the shear bulk wave fundamental (A1) mode is ~11%, and the coupling coefficient $k_{eff}^2 \approx 25\%$, which

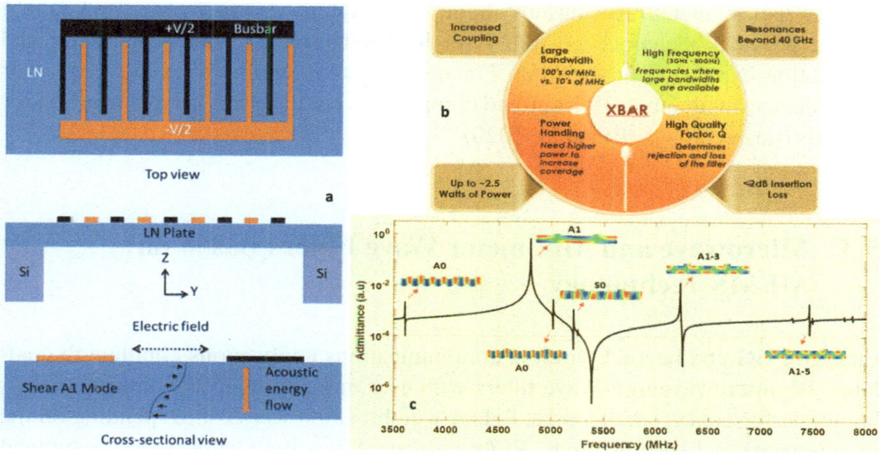

Fig. 5.8 XBAR and its characteristics (Adapted from Yandrapalli et al. (2019) with permission from IEEE): (**a**) The structure of XBAR with crystalline z-axis perpendicular to the platelet surface; shear wave resonance (A1 Mode) by applying lateral electrical field parallel to the crystalline y-axis. (**b**) Key characteristics of the XBAR resonator addressing 5G requirements. (**c**) Admittance curve from 2D periodic simulation indicating main XBAR mode and spurious mode

absolutely satisfies the requirement for mobile phones at 4–6 GHz frequencies and even has huge potential at 10–25 GHz without harsh lithography conditions. In addition to the exciting vertical shear vibration A1 mode, higher-order harmonics of Lamb waves appear in the horizontal direction (A0, S0, A1–3, A1–5). Because the horizontal A1–3 harmonic is close to the right of anti-resonance, it may adversely affect the filter design. Changing the value of pitch and mass loading of metal electrodes can not only tune the frequency but also adjust the relative position of the A1–3 mode and anti-resonance (Yandrapalli et al. 2019).

A typical XBAR may have hundreds, possibly thousands, of parallel fingers in the IDT. The IDT of an XBAR differs substantially from the IDTs used in SAW resonators, in which the pitch p of the IDT is typically 2–20 times the width w of the fingers. The width of the IDT fingers in an XBAR is not constrained to one-fourth of the acoustic wavelength at resonance. However, there are still many problems to be solved for XBAR, such as poor mechanical reliability (large-size film hanging on the cavity), poor temperature coefficient of frequency, poor heat dissipation, and low power capacity (Liu et al. 2020).

In brief, acoustic resonators/filters based on BAW technology can be used in base stations, automotive, electronics, radar, and wireless communications. AlN-based BAW devices are still one of the best choices for Sub66 GHz due to their irreplaceable superiority in performance and manufacture. High frequency, wide bandwidth, small size, and high-power capacity are new challenges in 5G communication networks. It is difficult to complete a high-performance filter based on conventional BAW resonator technology. Doped AlN, single AlN, and LN are preferred materials to improve the inherent performance of the resonator. Research on resonator characteristics, including spurious mode, temperature compensation, and tuning ability, provides technical feasibility for high-performance, stable operation and application expansion of the filter. For mobile phone applications, integration of BAW devices with other RF front-end components could be an optional solution for miniaturization targets (Liu et al. 2020).

5.3 Microwave and Millimeter Wave Filters Based on MEMS Technology

Emerging 5G and beyond wireless communications applications continue to challenge RF/microwave/mm-Wave filters with ever more stringent requirements, e.g., higher performance, smaller size, lighter weight, and lower cost. Depending on the requirements and specifications, RF/microwave filters may be designed as lumped element or distributed element circuits, may be realized in various transmission line structures, e.g., microstrip, waveguide cavities, coplanar, coaxial, stripline, and can be implemented using a variety of fabrication techniques. As the operating frequency of communication systems is extending to the mm-wave spectrum, conductor loss, dielectric loss, and radiation loss increase rapidly, which makes it challenging to develop high-Q mm-wave filters. Three-dimensional (3D) waveguide filter structures exhibit excellent RF performance at mm-wave frequencies and have been widely employed in high-performance RF systems. Nevertheless, as the operating frequency increases to the mm-wave frequency, the physical sizes of the waveguide filters become miniature in size, impeding the use of postfabricated tuning elements to compensate for the manufacturing tolerances of traditional machining technologies. MEMS technology has the potential to develop very accurate miniature 3D mm-wave filters. MEMS filters can be classified into micromachined filters and tunable filters. Micromachined filters can be classified into the following groups: microstrip, coplanar, micromachined cavities, filters supported by membranes and coaxial. For tunable filters, switches or varactors have been integrated on a filter topology to achieve reconfigurability (Jaimes-Vera et al. 2007; Jia 2021).

5.3.1 Micromachined Filters

Micromachining techniques can be generally compartmentalized into two main categories (Jia 2021): surface micromachining and bulk micromachining. In surface micromachining (Fig 5.9a), all fabrication processing is performed on top of the substrate by alternatively depositing and patterning the dielectric and conductor layers on the surface of the substrate. The additive layers on top of the substrate can have a total thickness from several nanometers to several micrometers. Normally, surface micromachining techniques are used to develop planar transmission line

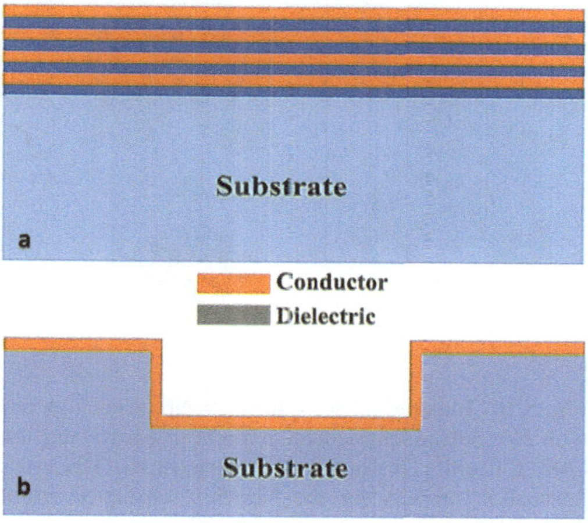

Fig. 5.9 Illustration of micromachining techniques (Adapted from Jia (2021), Credit: University of Waterloo): (**a**) Surface micromachining; and (**b**) bulk micromachining

structures. If using superconductors in surface micromachining, compact high-Q planar transmission line resonators can be realized. In bulk micromachining (Fig. 5.9b), the substrate is etched by using wet chemical etching or plasma etching techniques. Several etched substrates can be stacked together to form a waveguide that can have a height from several micrometers to several millimeters. Currently, bulk micromachining is widely used to build wafer-level cavity resonators for high frequency applications.

Different transmission line structures and topologies have been investigated with the general objective of producing low-cost micro/millimeter-wave high-performance devices. The filters are designed with different structures, such as cavities, coplanar, microstrip or coaxial transmission lines (Jaimes-Vera et al. 2007).

5.3.1.1 Surface Micromachining Superconductor Filters

Surface micromachining techniques are normally employed to develop lumped-element and planar transmission-line resonators. These two resonator types have compact physical sizes in comparison to 3D cavity and dielectric resonators. However, lumped-element and planar transmission-line resonators have low-Q values due to high ohmic losses when using ordinary conductors. By combining surface micromachining and superconducting techniques, highly miniature high-Q filters can be realized (Jia 2021).

A multilayer Nb-based superconductor bandpass filter operating over the frequency range of 30–60 GHz has been developed. The filter is realized in a coplanar waveguide (CPW) structure, where crossovers in the multilayer structure are employed to reduce the filter size. The filter is fabricated using the MIT Lincoln Laboratory (MIT-LL) 8-niobium (Nb)-layer fabrication process shown in Fig. 5.10. In this process, there are 8 niobium layers (M0–M7) and silicon dioxide (SiO_2) layers in between. Each Nb layer has a thickness of 200 nm, and all the Nb layers can

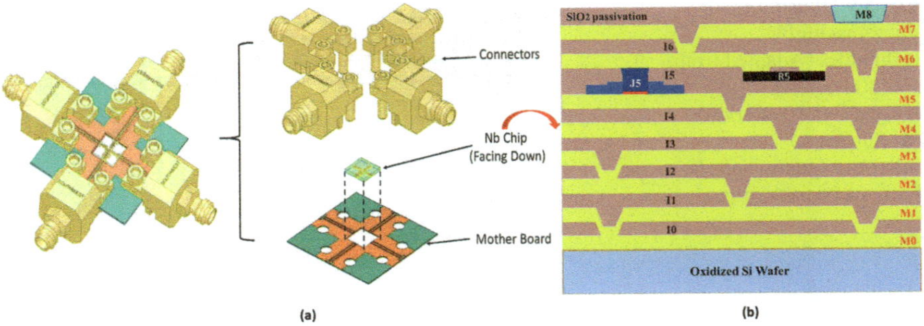

Fig. 5.10 Illustration of multi-layer Nb-based superconductor bandpass filter: (**a**) The triplexer with Nb chip assembled on a mother board using the flip-chip technology (Adapted from Jia (2021), Credit: University of Waterloo); (**b**) The cross section of the MIT-LL 8-niobium-layer fabrication process for the Nb chip. (Adapted from Tolpygo et al. (2015) with permission from IEEE)

be connected using vias (I0–I6). The height of each via is 200 nm, except via I5, which has a height of 280 nm. Moreover, Layer J5 is used to realize Josephson junctions, and the lossy conductor layer, R5, with a thickness of 40 nm, can be used to realize resistive terminals. The gold layer, M8, with a thickness of 250 nm, can be used for landing pads (Tolpygo et al. 2015).

5.3.1.2 Planar Microstrip Filters

LIGA (a German acronym with an English translation of lithography, electroforming, and molding) process has been used to fabricate planar millimeter-wave and microwave transmission lines and filters. In the LIGA process, structures with near-vertical sidewalls are fabricated by means of deep X-ray diffraction, metal electrodeposition, and mechanical grinding. This process enables the creation of thick structures or molds with near-vertical sidewalls from two-dimensional patterns. These structures are between 10 μm and 1 mm in height with aspect ratios as high as 100:1. A variant of the standard LIGA fabrication process is employed, utilizing the intermediate metallic structures and carrier substrate as the final microwave circuitry and dielectric, respectively, as shown in Fig. 5.11a. LIGA-micromachined transmission lines will enable several important advancements in microwave and millimeter-wave integrated circuits. The tall metal transmission lines, when bounded by a good thermal conductor, are perfect for high-power applications due to the increased conduction interface, which allows the development of high-power monolithic circuits for commercial transmitter applications. High-aspect conductor sidewalls possessing a nearly vertical slope are ideal for devices that require very high coupling levels that are impossible to attain with conventional integrated transmission lines (Willke and Gearhart 1997).

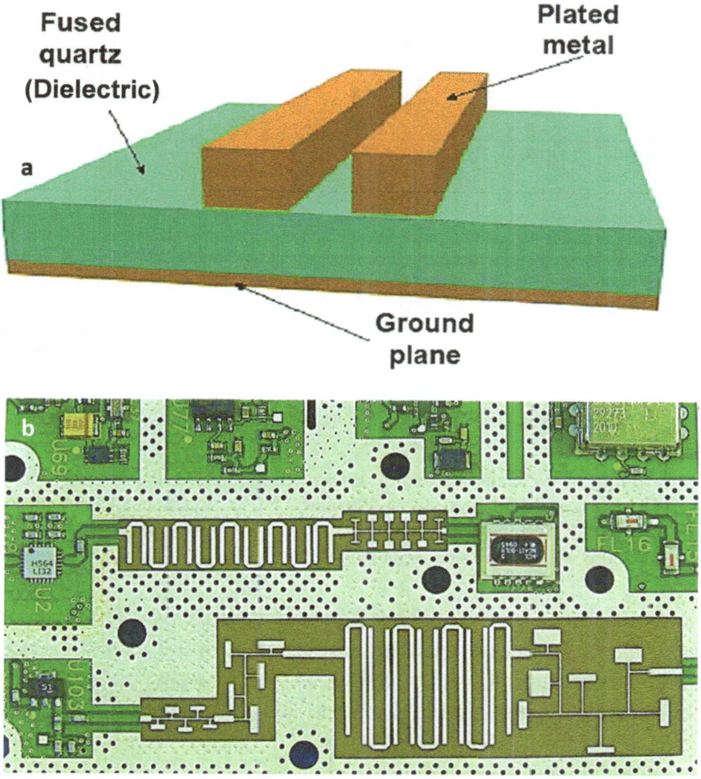

Fig. 5.11 Planar microstrip filters: (**a**) LIGA fabricated two 200 μm thick coupled LIGA microstrip lines (Adapted from Willke and Gearhart (1997) with permission from IEEE); (**b**) The PCB inside a 20GHz Agilent N9344C spectrum analyzer showing various microstrip distributed element filter technology elements. (Credit: https://commons.wikimedia.org/wiki/File:Microstrip_Distributed_Element_Filter_Technology.jpg)

Another approach is to design a high-frequency filter using microstrip distributed elements consisting of copper tape etched onto a PCB. Figure 5.11b demonstrates the PCB inside a 20 GHz Agilent N9344C spectrum analyzer showing various microstrip distributed-element filter technology elements.

5.3.1.3 Coplanar Waveguide Filters

A conductor-backed coplanar waveguide (CPW) filter was fabricated using the LIGA-fabricated process. In thick-metal CPWs, the electric field is confined primarily to the gap between the signal and ground lines (Fig. 5.12a). This geometry distributes the surface current over the full height of the metal, reducing ohmic losses relative to standard thin-metal CPWs while simultaneously drawing the

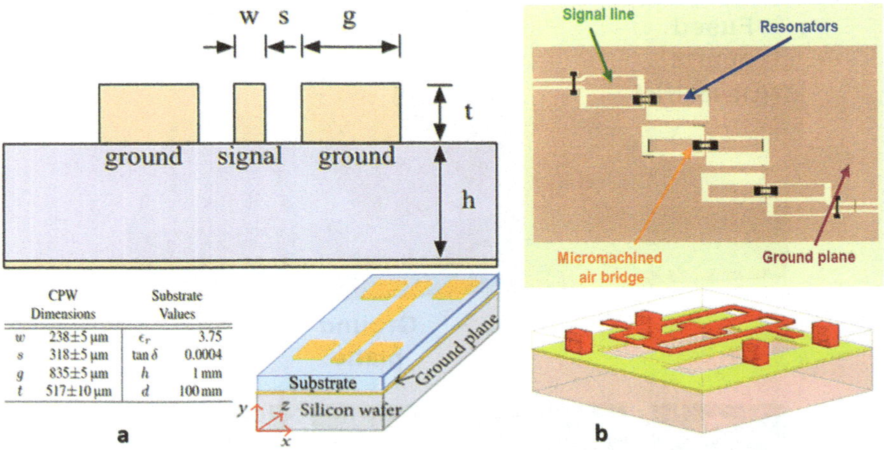

CPW Dimensions		Substrate Values	
w	238±5 μm	ϵ_r	3.75
s	318±5 μm	$\tan\delta$	0.0004
g	835±5 μm	h	1 mm
t	517±10 μm	d	100 mm

Fig. 5.12 Illustration of coplanar waveguide filters: (**a**) Conductor-backed coplanar waveguide made with LIGA process (Adapted from Forman (2006), Credit: IEEE 2006 Asia-Pacific Microwave Conference); (**b**) Layout of a coplanar 94 GHz frequency coupled-line filter. (Adapted from Rius et al. (2003) with permission from IEEE)

electric field out of the substrate and reducing dielectric loss. The reduced interaction between the propagating fields and the substrate minimizes dispersion, radiation loss, and the excitation of substrate modes. Additionally, the ability to fabricate microwave structures with variable thickness adds an additional design parameter, increasing design flexibility (Forman 2006).

Figure 5.12b shows a third-order, 5% fractional bandwidth coupled-line waveguide filter with a center frequency of 94 GHz. The insertion loss and return loss are 4 and 20 dB, respectively, for a center frequency of 91.5 GHz. Compared to traditional microstrip technology, the flexibility of coplanar technology is an important advantage for the design of filters in the high frequency range. Choosing suitable cross-section dimensions avoids dispersion and allows the use of analytical quasi-TEM models. The association of such models with specific design rules for describing electrical discontinuity constitutes a very convenient method for passive device design (Rius et al. 2003).

5.3.1.4 Micromachined Dielectric Waveguide Resonate Filters

To further reduce the physical size of mm-Wavewave filers, dielectric loaded double-ridge waveguide resonators have been developed as shown in Fig. 5.13, which consist of three parts. The top and bottom parts can be fabricated by applying Si-DRIE on standard Si wafers (a standard 4-inch Si wafer is 525 μm thick). A 254-μm thick alumina wafer is then sandwiched between the two processed Si wafers. Figure 5.14 shows the 3D layout of a 4-pole dielectric loaded double-T-septum resonator bandpass filtersfilter operating at the center frequency of 25.5 GHz with a

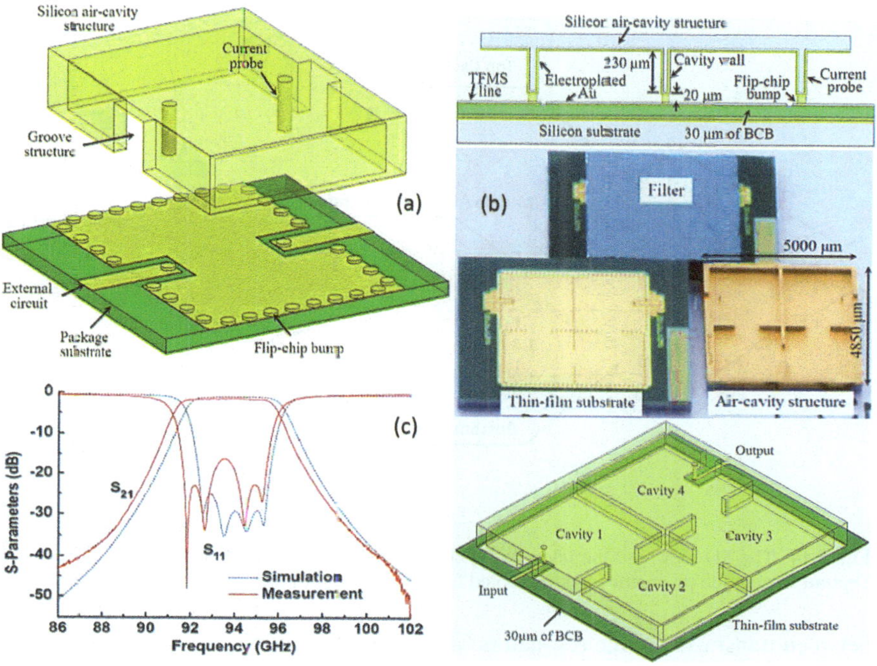

Fig. 5.13 Micromachined cavity resonator (**a**) and filter (**b**) and its simulated and measured S-P parameters vis frequency (**c**). (Adapted from Song et al. (2010) with permission from IEEE)

fractional bandwidth of 11.7%. This filter has a compact physical size of 4.7 mm × 4.7 mm × 1.3 mm. The top and bottom parts of the filter are fabricated on a standard 4-inch Si wafer by using DRIE, and a 3 μm thick gold layer is coated on the processed Si wafer. The middle part of the filter is fabricated on a 254 μm thick alumina substrate. The 50 Ohm CPW lines are used as the input and output ports of this filter (Jia 2021).

5.3.2 *Micromachined Tunable Filters*

Micromachined tunable filters present circuit reconfigurability, e.g., the frequency of the passband or rejection band can be varied by adjusting its components. MEMS switches or varactors have been integrated on filter topologies to achieve reconfigurability. There are two different types of frequency-tuning methods for MEMS filters: analog and digital. Analog tuning is performed with MEMS varactors and provides continuous frequency variation between 5% and 15%. Digital tuning is made with MEMS switches, where discrete center frequencies and wide tuning ranges of 20–60% are possible. Additionally, MEMS switches can be used to switch

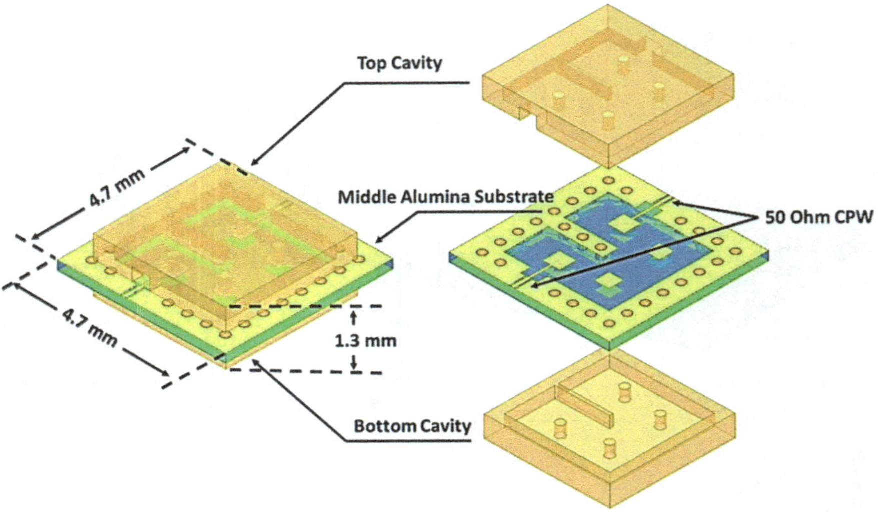

Fig. 5.14 The 3D view and the layer-by-layer view of an mm-Wave dielectric loaded double-T-Septum waveguide resonator filter. (Adapted from Jia (2021), Credit: University of Waterloo)

between transmission line resonators of different lengths to provide more than one filter center frequency (Jaimes-Vera et al. 2007).

5.4 Metamaterial and Metasurface Filters for 5G Communications

A metamaterial is engineered to have a property that is not found in naturally occurring materials, which is usually arranged in repeating patterns, at scales that are smaller than the wavelengths of the phenomena they influence. Metamaterials derive their properties not from the properties of the base materials but from their designed structures. Their precise shape, geometry, size, orientation and arrangement give them their smart properties capable of manipulating electromagnetic waves by blocking, absorbing, enhancing, or bending waves to achieve benefits that go beyond what is possible with conventional materials (Tong 2018).

Metamaterials in filter applications are based on resonant-type metamaterial transmission lines. This helps to obtain a propagating medium with controllable characteristics. Additionally, the host transmission line combines subwavelength resonators with elements such as series capacitances or shunt inductances. Some examples of different kinds of subwavelength resonators used by resonant-type metamaterial transmission lines are split-ring resonators (SRRs), complementary split-ring resonators (CSRRs), double-slit complementary split ring resonators (DS-CSRRs) and double-slit split ring resonators (DS-SRRs). Characteristic

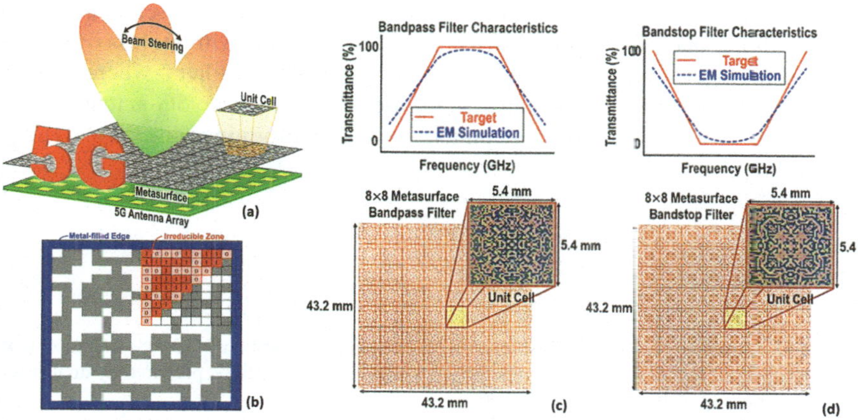

Fig. 5.15 Designed metasurface filter integrated with a 5G antenna array (Adapted from Hong et al. (2021) with open access (IEEE)): (**a**) Single-layer metasurface combined with a beam-steerable 5G antenna array; (**b**) illustration of a unit cell metasurface filter configuration including irreducible zone and metal-filled edge; (**c**) bandpass metasurca filter and (**d**) bandstop metasurface filter

impedance and phase shift are among the chief properties of such transmission lines. As these resonators are of very small dimensions, they result in compact transmission lines. Once the transmission line has been designed to show certain desired characteristics, its size can be reduced (Jindal and Sharma 2012).

Microstrip filters based on resonant-type left-handed and composite right/left-handed metamaterial transmission lines have been developed for ultrawide bandpass and band stop implementations. A metamaterial structure behaves as an effective (continuous) medium with left-handed wave propagation in the pass band. Due to the small electrical length of the resonators employed, metamaterials have become very attractive for the design of compact planar microstrip filters. These microstrip filters using metamaterials are advantageous over conventional microwave filters, as they result in improved performance as well as compact sizes (Jindal and Sharma 2012).

In the case of the mm-wave band, metasurface-based spatial filters can be used in place of existing on-chip and off-chip filters. Metasurfaces are defined as arrays of spatially varying geometries used to generate desirable electromagnetic wavefronts. The target frequency response can be achieved using metasurfaces with subwavelength thickness, enabling the development of conformal arrays. Figure 5.15 illustrates an approach to designing a metasurface filter integrated with a 5G antenna array, where an ultrathin (50 μm) metasurface bandpass filter (or bandstop filter) is placed on a printed circuit board (PCB) carrying an antenna array. Beam-steering circuits are printed on the other side. A 5G antenna array is implemented in conjunction with the metasurface bandpass filter to provide low loss within the passband and high suppression within the out-of-band while maintaining angular stability and polarization insensitivity. Meanwhile, the 5G antenna array with the metasurface bandstop filter manifests a complementary performance, maintaining angular stability and polarization insensitivity. Furthermore, without loss of generality, this approach can be extended to the design of spatial filters, where mutual coupling with the embedded antenna is considered to maximize 5G antenna array performance (e.g., gain, radiation pattern, and beamforming diversity) (Hong et al. 2021).

References

Akiyama M, Kamohara T, Kano K, Teshigahara A, Takeuchi Y, Kawahara N (2008) Enhancement of piezoelectric response in scandium aluminum nitride alloy thin films prepared by dual reactive cosputtering. Adv Mater 21:593–596

Bhugra H, Piazza G (2017) Piezoelectric MEMS resonators. Springer, Berlin/Heidelberg

Bi F, Barber B (2008) Bulk acoustic wave RF technology. IEEE Microw Mag 9:65–80

Cadence (2018) RF filter technologies for 5G wireless communication systems. https://resources.system-analysis.cadence.com/i/1326172-filter-technologies-for-5g-wireless-communication-systems/0?

Fattinger GG (2008) BAW resonator design considerations – an overview. In: Proceedings of the 2008 IEEE International Frequency Control Symposium, Honolulu, HI, USA, 19–21 May 2008, pp 762–767

Forman MA (2006) Low-loss LIGA-fabricated coplanar waveguide and filter. In: 2006 Asia-Pacific microwave conference, pp 1905–1907. https://doi.org/10.1109/APMC.2006.4429780

Hong YP, Hwang IJ, Yun DJ, Lee DJ, Lee IH (2021) Design of single-layer metasurface filter by conformational space annealing algorithm for 5G mm-Wave communications. IEEE Access 9:29764–29774. https://doi.org/10.1109/ACCESS.2021.3059019

Jaimes-Vera A, Llamas-Garro I, Corona-Chavez A, Zaldivar-Huerta I (2007) Review on microwave and millimeter filters using MEMS technology. In: 17th international conference on electronics, communications and computers (CONIELECOMP'07). https://www.inaoep.mx/~altafrec/papers/2007_conielecomp.pdf

Jia H (2021) High-Q millimeter wave RF filters and multiplexers. PhD thesis, University of Waterloo, Ontario, Canada

Jindal S, Sharma J (2012) Review of metamaterials in microstrip technology for filter applications. Int J Comput Appl 54(3):48–54

Kaitila J (2006) 3C-1 review of wave propagation in BAW thin film devices – progress and prospects. In: Proceedings of the 2006 IEEE Ultrasonics Symposium, Vancouver, BC, Canada, 3–6 October 2006, pp 120–129

Kamohara T, Akiyama M, Kuwano N (2008) Influence of molybdenum bottom electrodes on crystal growth of aluminum nitride thin films. J Cryst Growth 310:345–350

Liu Y, Cai Y, Zhang Y, Tovstopyat A, Liu S, Sun C (2020) Materials, design, and characteristics of bulk acoustic wave resonator: a review. Micromachines 11:630. https://doi.org/10.3390/mi11070630

Mahon S (2017) The 5G effect on RF filter technologies. IEEE Trans Semicond Manuf 30(4):494–499

Matthews P (2019) Approaching the 5G mmWave filter challenge. https://www.microwavejournal.com/articles/32228-approaching-the-5g-mmwave-filter-challenge

Murata (2022) Innovation in technological development realizes high-performance surface acoustic wave devices. https://corporate.murata.com/en-eu/more_murata/techmag/metamorphosis20/productsmarket/ihp-saw

Noeth A, Yamada T, Muralt P, Tagantsev AK, Setter N (2010) Tunable thin film bulk acoustic wave resonator based on $Ba_xSr_{1-x}TiO_3$ thin film. IEEE Trans Ultrason Ferroelectr Freq Control 57:379–385

Plesski V (2020) Transversely excited film bulk acoustic resonators for high power applications. US Patents 16/578,811, January 16

Rius E, Prigent G, Happy H, Dambrine G, Boret S, Cappy A (2003) Wide- and narrow-band bandpass coplanar filters in the W-frequency band. IEEE Trans Microw Theory Techn 51(3):784–791

ROG Blog (2016) Filtering through material choices. https://www.microwavejournal.com/blogs/1-rog-blog/post/27004-filtering-through-material-choices

Ruby R (2006) 11E-2 Review and comparison of bulk acoustic wave FBAR, SMR technology. In: Proceedings of the 2006 IEEE Ultrasonics Symposium, Vancouver, BC, Canada, 3–6 October 2006, pp 1029–1040

Ruppel C (2017) Acoustic wave filter technology – a review. IEEE Trans Ultrason Ferroelectr Freq Control 64:1390–1400

Song S, Yoo C, Seo K (2010) W-band bandpass filter using micromachined air-cavity resonator with current probes. IEEE Microw Wirel Compon Lett 20(4):205–207. https://doi.org/10.1109/LMWC.2010.2042552

Tolpygo SK, Bolkhovsky V, Weir TJ, Galbraith CJ, Johnson LM, Gouker MA, Semenov VK (2015) Inductance of circuit structures for MIT LL superconductor electronics fabrication process with 8 Niobium layers. IEEE Trans Appl Supercond 25(3):1–5. Art no. 1100905. https://doi.org/10.1109/TASC.2014.2369213

Tong CT (2018) Functional metamaterials and metadevices. Springer, New York

Wang W, Yang W, Liu Z, Lin Y, Zhou S, Qian H, Gao F, Yang H, Li G (2014) Epitaxial growth of homogeneous single-crystalline AlN films on single-crystalline Cu (111) substrates. Appl Surf Sci 294:1–8

Willke TL, Gearhart SS (1997) LIGA micromachined planar transmission lines and filters. IEEE Trans Microw Theory Techn 45:1681–1688

Yandrapalli S, Plessky V, Koskela J, Yantchev V, Turner P, Villanueva LG (2019) Analysis of XBAR resonance and higher order spurious modes. In: Proceedings of the 2019 IEEE International Ultrasonics Symposium (IUS), Glasgow, Scotland, UK, 6–9 October 2019, pp 185–188

Yang Q, Pang W, Zhang D, Zhang H (2016) A modified lattice configuration design for compact wideband bulk acoustic wave filter applications. Micromachines 7:133

Zywitzki O, Modes T, Barth S, Bartzsch H, Frach P (2017) Effect of scandium content on structure and piezoelectric properties of AlScN films deposited by reactive pulse magnetron sputtering. Surf Coat Technol 309:417–422

Chapter 6
EMI Shielding Materials and Absorbers for 5G Communications

Abstract Control of electromagnetic interference (EMI) is critical in 5G systems, which adopt massive multiple-input multiple-output (MIMO) techniques and consist of arrays of antenna elements. Traditional shielding approaches become less effective at high frequencies as the amount of EMI increases in 5G technology. Absorbers are needed to reduce the cross talk of the antenna array and eliminate reflections that may interfere with the desired signal. Additionally, when a 5G signal radiates, the desired transmitted signal may be interfered with since objects near the antenna can reflect the signal. These reflections can be reduced or eliminated by the application of absorbers. Module and board-level shields are also vital parts of EMI mitigation. When a conductive enclosure is larger than half the wavelength, energy can resonate inside the cavity and affect circuit operation, resulting in increased shielding leakage. An absorber material inside the shield can be useful to absorb energy, and high permittivity/permeability materials are in high demand. This chapter will provide a brief review of materials and component design for 5G EMI shielding technologies and absorbers, including EMI shielding in 5G package modules, board-level shields, advanced metallic composite-based EMI shielding materials, and emerging polymer-based EMI shielding and absorber materials.

6.1 EMI Shielding Design Principle in 5G Systems

EMI can disrupt electronic devices, equipment, and systems in terms of consumer electronics (such as smartphones) and automotive, medical, or aerospace electronics. When designing and manufacturing modern electronic devices, there are strict electromagnetic compatibility (EMC) standards that must be followed to ensure that the device will not be impacted by external electromagnetic waves, nor will it be affected by its own radiated electromagnetic waves. Beyond this, it must not interfere with external devices or cause harm to human health, so it is important to block the path of electromagnetic waves. A common way to do this is through EMI shielding, and with 5G it is no different.

C. Tong, *Advanced Materials and Components for 5G and Beyond*, Springer Series in Materials Science 327, https://doi.org/10.1007/978-3-031-17207-6_6

5G system hardware designers are challenged to implement better power solutions that help all necessary electronics to fit into form factors close to those of existing 4G base-station enclosures while maintaining a low EMI. Increased board component density in 5G systems demands space-saving solutions with higher efficiency and lower EMI than traditional discrete DC-DC ICs with external inductor-based solutions. This will challenge designers to create a careful layout of components for filters to minimize the radiated and conducted EMI generated via power converter and inductor switching currents. DC-DC converters are notorious for conducted EMI coming from magnetic fields emanating from the current loop path. That path comprises the output switching node to ground as well as the input capacitor to ground. Radiated EMI is also generated from the MOSFET switching node to the connection at the inductor. This creates a high dV/dT because it continuously switches from the high input voltage level to the ground as well as from the inductor-generated electromagnetic fields. This system design may lead to multiple, costly EMI testing to meet the standard. An integrated module with EMI shielding may be a better option (Taranovich 2021).

In general, a nested approach is one where EMI shielding is applied at each of the lowest possible levels of a product design. For example, EMI shielding is first applied to individual ICs, small areas of PCBs such as package-level shielding, followed by entire PCBs, subassemblies, and finally complete products. A nested EMI shielding approach is one that results in the lowest overall cost to manufacture a quality product, on time, and within performance specifications. The main reasons to implement EMI shielding at the lowest possible levels (individual ICs, package level/small areas of the PCB, and the PCB level) include the following (MacArthur 2019):

(a) Enclosure shielding does not help attenuate interference between individual ICs located on a PCB, whereas PCB level shielding does help attenuate interference between individual ICs.
(b) From a practical/cost-efficiency level, typical enclosure shielding technology is incapable of providing significant attenuation performance at higher (GHz) frequencies, whereas PCB level shielding does provide this performance.
(c) The cost and weight of shielding at higher levels is minimized through the effective use of shielding at the PCB level.
(d) From a susceptibility standpoint, modern ICs with their ever-shrinking silicon features, faster rise-times, and lower noise margins can be made to function dependably in the noisy atmosphere that they are often required to operate in simply by employing shielding at the PCB level.
(e) Integration of intentionally noisy wireless communication modules within products can cause harmful inference to other sensitive analog and digital components located in close proximity. This noise can also be mitigated through the use of PCB level shielding.
(f) Enclosure shielding is often compromised to a point of total ineffectiveness due to the need to have holes and slots added for the penetration of input/output

cables, displays, ventilation, access to removal media, etc. This situation becomes less of a problem when PCB level shielding is utilized.

(g) Effective enclosure shielding usually requires substantial filtering of all cables that pass in and out of the product, right at the point where they penetrate the enclosure shield. It is possible to lessen the need for this extra filtering when PCB level shielding is utilized.

EMI shielding applications in a 5G system can range from smartphones, healthcare, and vehicle-to-everything (V2X) to smart appliances, public safety, and much more. However, one improper shielding device failure could cause a larger communications snafu that possibly affects a broader system disruption. Designers must avoid open apertures as well as nonconductive regions that diminish the overall electronic enclosure. They allow EMI and RFI to exit the design areas to the outside or enter the system design area to possibly cause the system to malfunction or even fail. One of the main open apertures and nonconductive regions is the cable entry, which is particularly vulnerable near connectors. Power and signal access panels could contribute to EMI problems. Viewing windows, as well as heating, ventilation, and air conditioning (HVAC) vents, can access areas that EMI may enter or exit. Some other potentially vulnerable use cases include the following (Taranovich 2021):

(a) 5G in the military

Military enclosures will need to pass MIL-STD-461, possibly requiring 60-dB, 80-dB, or even higher attenuation levels to achieve compliance. To achieve such levels of electromagnetic shielding attenuation, open apertures and nonconductive areas may not be larger than a few thousandths of an inch.

(b) 5G in the medical arena

EMI in some sensitive medical applications could harm a patient or even cause death. Medical facilities must properly shield operating theater equipment, as well as other areas that need protection, such as where imaging devices are in use.

6.2 Component Package-Level EMI Shielding for 5G Modules

The convergence of advanced RF/mm-wave communication capability, miniaturization and package-level integration underscore the requirement for novel approaches to EMI shielding. This has traditionally been managed at the board level, which alone will no longer address the challenges of integrated packages, such as SiPs. Currently, compartment and conformal shielding materials offer two solutions, as shown in Fig. 6.1, which can be used independently of one another or together as a comprehensive EMI system. Compartmental shielding materials provide adaptable designs for internal component-to-component shielding with in-package partitioning, while conformal shielding materials provide external

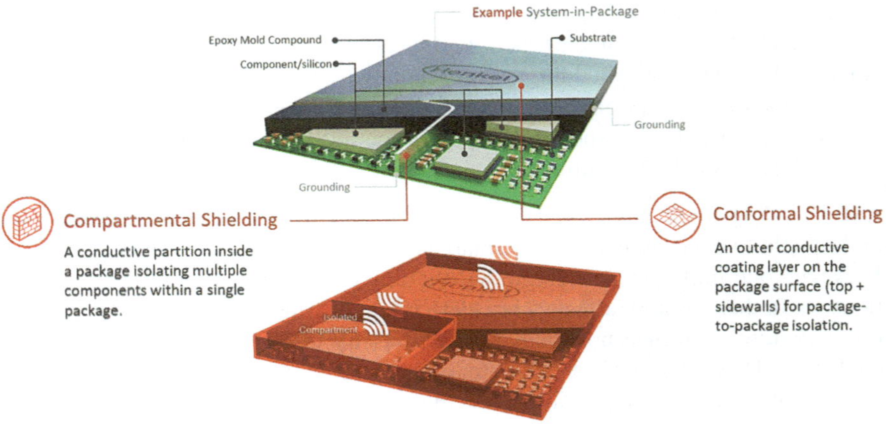

Fig. 6.1 Package – level EMI shielding solutions. (Adapted from Henkel 2017; credit: https://dm.henkel-dam.com/is/content/henkel/webinar-henkel-package-level-emi-shielding-externalpdf)

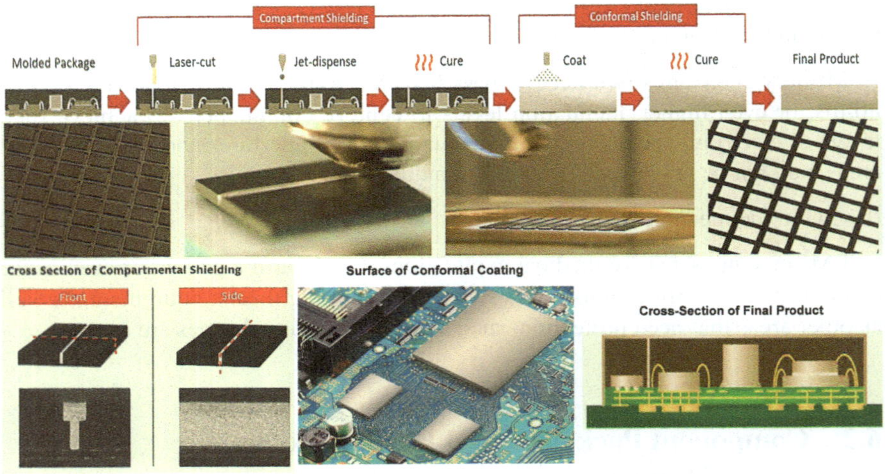

Fig. 6.2 Package – level EMI shielding implementation process. (Adapted from Henkel 2017; credit:https://dm.henkel-dam.com/is/content/henkel/webinar-henkel-package-level-emi-shielding-externalpdf)

package-to-package shielding with an ultrathin, conductive outer coating layer (Henkel 2017).

With multiple ICs, antennae, and MEMS sensors integrated into a single module, in-package effective RF protection between the various baseband, RF, analog, and power management components is required to protect performance. As shown in Fig. 6.2, compartment shielding is a package-level solution that allows a shielding partition to be created using laser ablation on the package mold to isolate chips within the package. This is followed by jet dispensing or 3D printing of highly conductive material into the trench. If desired, trench filling can be followed by conformal sputter or spray coating of the entire package. Conformal coating is usually applied via PVD (such as sputtering), plating or spray coating equipment, enabling a high-UPH process that results in an ultrathin package coating to protect against electromagnetic interference. By applying state-of-the-art masking techniques, a conformal shield can be applied to select areas of the package, if needed (Henkel 2017).

More advanced compartmental shielding techniques have been developed that utilize sophisticated wire bond technologies to create a wire fence, a wire cage or a vertical wire structure within the molded package, as shown in Fig. 6.3. Strip grinding or laser ablation processes are used to reveal the encased wire. Conformal shielding is then applied to create a Faraday cage effect, whereby the wire structure serves to block electric fields and electromagnetic waves.

In 5G connectivity applications, different frequency transmissions require integrating an antenna in the SiP module so that the product designer can develop the products as compact as possible for specific applications. The 3D System in Package (3D SiP) includes double-side molding technology, and the antenna in package (AiP) is a combination solution. The AiP structure can use different antenna patterns to simplify the manufacture of the antenna and shielding structure. Therefore, AiP technology has been widely applied in IoT, RF module and wireless transmission applications. For 5G mmWave generation, whose frequency is higher than 28 GHz, the antenna wavelength is suitable for small package sizes, which creates various challenges in AiP design, such as EMI shielding technology, partition design, low loss substrate material selection and multilayer substrate antenna design capability. Partition EMI shielding technology is used for the isolation of different function blocks in the SiP module. The advanced SiP with partition EMI process flow and sequences is shown in Fig. 6.4. When the substrate is in-coming, the ICs

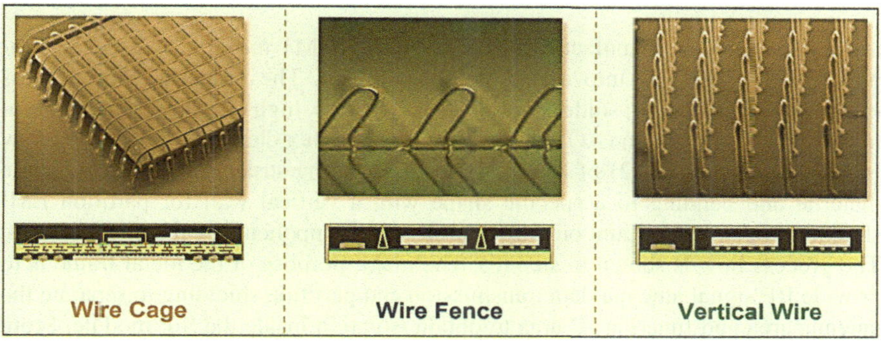

Fig. 6.3 EMI shielding techniques for double-sided, molded ball grid array (DSMBGA) packages to achieve improved system performance. (Adapted from Zwenger 2021; credit: https://chipscalereview.com/wp-content/uploads/2021/09/ChipScale_Sep-Oct_2021-digital.pdf)

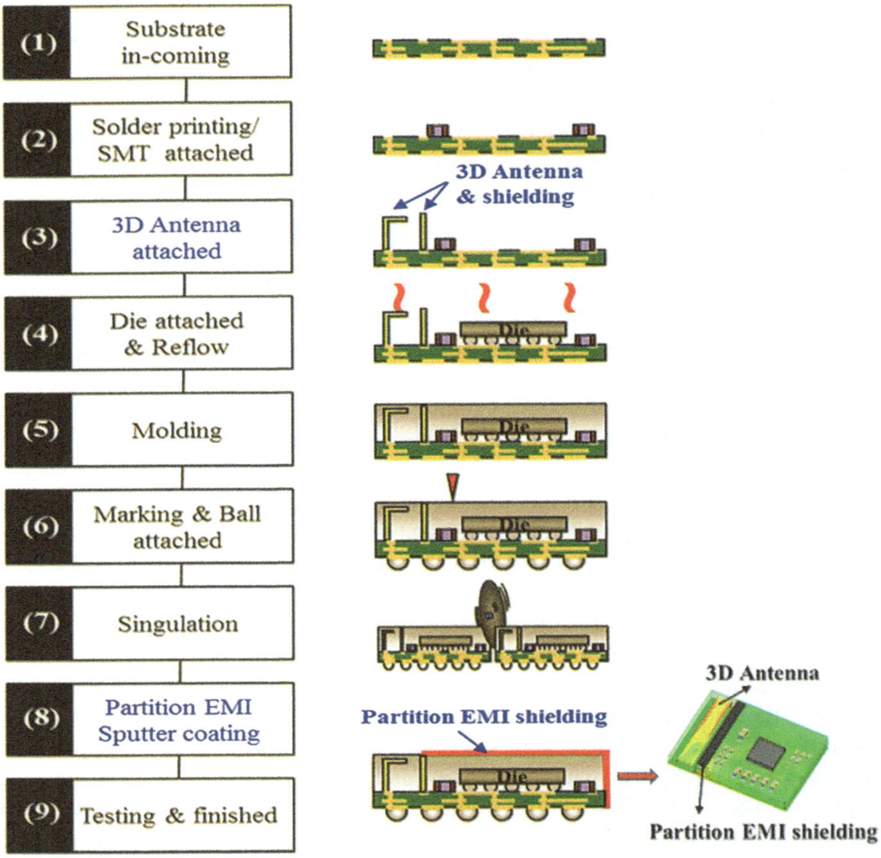

Fig. 6.4 Advanced SiP with partition EMI process flow. (Adapted from Tsai et al. 2019 with permission from IEEE)

and components are manufactured by a high-speed SMT machine and with a compact SMT design rule into a small product layout. The standard SMT process includes solder printing, solder paste inspection (SPI), high-speed placement, automated optical inspection (AOI), thermal reflow and flux cleaning; the process flow is shown in steps (1)~(2) of Fig. 6.4. The 3D antenna structure uses metal frame material and bending to a specific shape with a vertical wall for partition EMI shielding together to mount on the substrate with components by the SMT process. The process flow is shown in step (3). The major purpose of the metal frame is to provide RF signal antenna data transmission and partition shielding to separate the antenna area and function IC area to obtain isolation inside the SiP module. From the die attachment process to control the good bonding accuracy and through reflow for solder melting with the 3D antenna, refer to step (4) of Fig. 6.4. After finishing the entire top-side SMT process and following up by the molding process, laser marking, ball placement and the singulation process, refer to steps (5)~(7), respectively. For the molding process for advanced SiP with 3D antenna structure, the major key factor is focused on mold flow control to achieve mold void free performance, also need to optimize the EMI shielding material to get good adhesion on 3D metal antenna material without any delamination problem (Tsai et al. 2019).

Consequently, the conformal EMI shielding methodology, such as metal frame, sputter coating or spray coating technology, is used to prevent the disturbance of the circuit and even stop the function. The typical metal frame shielding solution has a higher coating thickness of approximately 100~150 μm and poor reliability performance; hence, the Cu/SUS sputter coating is a widely accepted coating solution in the assembly industry, showing a thinner coating thickness of approximately 3~5 μm, ideal shielding effectiveness and better particle adhesion capability on the coating surface. Nevertheless, package miniaturization, high performance and heterogeneous integration are the key design factors for SiP technology development. To achieve a more flexible shielding design between the different functional blocks, compartment shielding technology is used to isolate the signal noise between the digital and RF/analog circuits inside SiP modules. Multiband RF SiP modules, such as CDMA, LTE and dual band Wi-Fi integration systems, or EMI sensitive SiP modules, such as AP, PMIC and RF integration modules, in terms of compartment technology, wire bonds, metal lids and epoxy filling are developed in the SiP industry, and each partition methodology is required for a particular design purpose and applications. Metal frame partition design has the advantages of simple assembly, lower cost and good shielding effectiveness for low-frequency spectra, but the shielding effectiveness (SE) may degrade in the mmWave range due to the jagged solder pad design. Hence, epoxy filling has become an increasingly popular shielding design that supports flexible partition shapes and capable SE performance in mmWave signals (Li et al. 2020).

6.3 Board Level EMI Shielding for 5G Systems

The main difference in package-level and board-level EMI shielding is shown in Fig. 6.5. The goal of an EMI shield is to create a Faraday cage around the enclosed RF noisy components. The board level shield is usually made using the six sides of a metallic box. The top five sides are created using a shielding cover or metal can, while the bottom side is achieved by using the ground plane within the PCB (Fig. 6.6). In an ideal enclosure, no emissions would enter or exit the box. Unwanted emissions from these shields do occur, such as from holes perforated into soldered cans that allow thermal heat transfer during solder reflow. These leaks can also occur from imperfections along an EMI gasket or solder attachments. Noise can also escape from the spaces between grounding via-holes used to electrically connect the shielding cover to the ground plane. The typical attributes of PCB level shield cans are as follows (MacArthur 2019): small footprints; low-profile configurations; two-piece design (fence and cover); through-hole or surface mount;

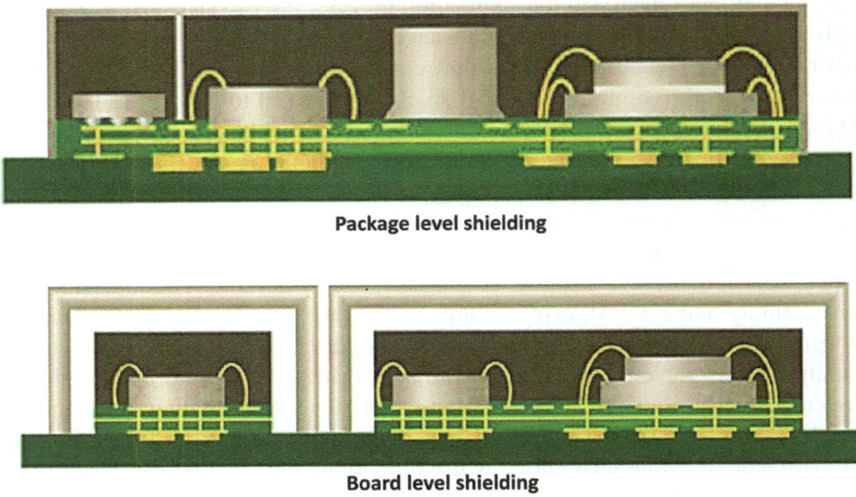

Fig. 6.5 Comparison of board level shielding and package level shielding. (Adapted from Henkel 2017; credit: https://dm.henkel-dam.com/is/content/henkel/webinar-henkel-package-level-emi-shielding-externalpdf)

multicavity patterns (isolate multiple components using the same shield); virtually limitless design flexibility; ventilation holes; removable covers for quick access to components; I/O holes; connector cutouts; enhanced shielding with RF absorbers; ESD protection with insulator padding; reliable protection from shock and vibration using secure locking features between the frame and cover.

A wide range of materials are generally available for shielding, including brass, nickel silver and stainless steel. The most common types are: tin-plated cold rolled steel (cheapest option), tin-plated copper, nickel silver, stainless steel, and tin-plated phosphorous bronze. In general, tin-plated steel is the best choice for shielding below 100 MHz, while tin-plated copper is best above 200 MHz. Tin plating allows for the best soldering efficiency possible. Because aluminum on its own is not easily soldered to a ground plane with its heat-sinking properties, it is not generally used for PCB level shielding. Depending on the regulatory burden of the end-product, all materials used for shielding may need to be RoHS compliant. In addition, if a product is intended for hot and humid environments, galvanic corrosion and oxidation may be of concern. While every application is unique and shielding effectiveness is a function of shield and board design, board level shields generally provide 60 dB isolation for frequencies up to 10 GHz. At higher frequencies, board level shields are susceptible to a loss of isolation between cavities and leakage between the cover and fence (MacArthur 2019).

Board level shields are generally stamped out of flat metal sheets. Some designs enable 'fully drawn' metal shields where the metal stretches and results in a shield with no gaps. Other designs bend the metal and the corners of the shield fold closely together. This can create a gap that can be straight or somewhat jagged to improve

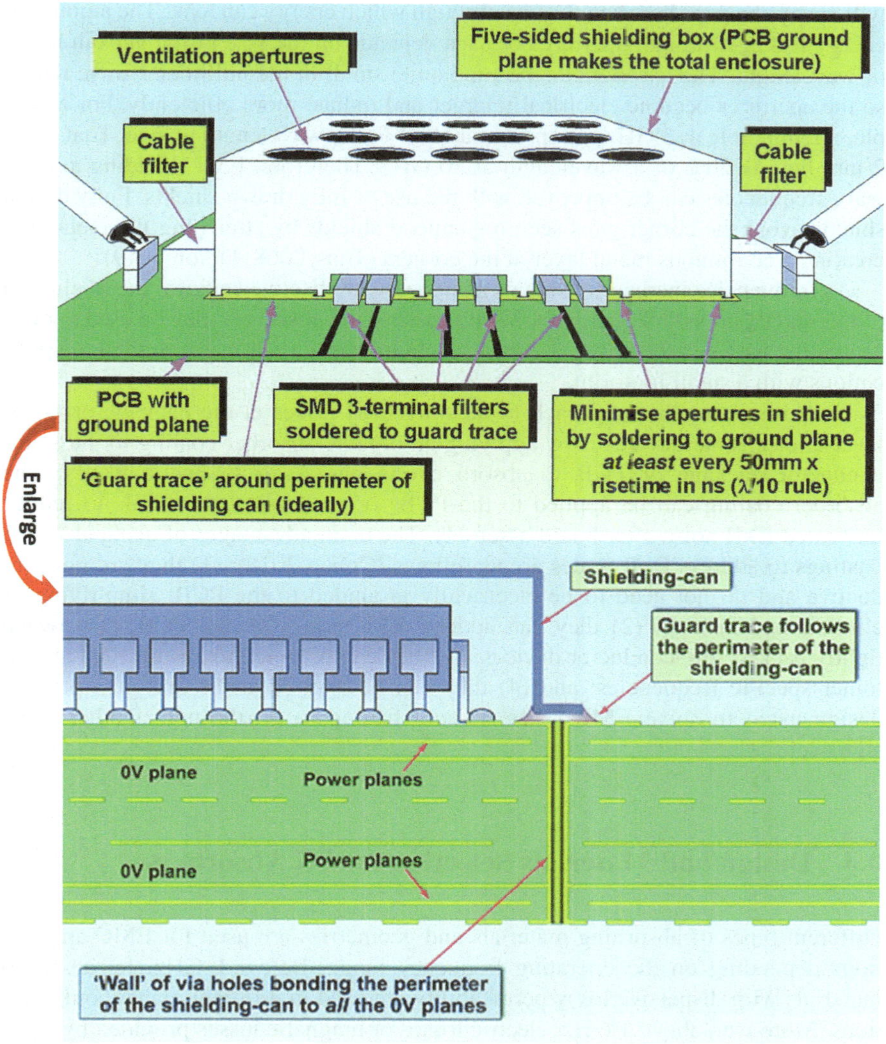

Fig. 6.6 Sketch of board level shielding and grounding in the PCB. (Modified from Armstrong 1999; credit: https://www.emcstandards.co.uk/files/design_techniques_for_emc_1999_part_4_shielding.pdf)

EMI performance. Other designs include holes in the shield to allow thermal energy to escape. Another type of design uses two metal parts with a fence and a lid. This type is used if access to the components inside is desired following the solder reflow process for tuning or repair. The lid then snaps onto the fence to create the shield. Other gaps in a shield arise from the need for vias to connect the shield to the PCB ground plane (Fig. 6.6). The vias through the PCB provide the connection, but there will always be gaps between the vias through which energy can leak. The amount of energy leakage through apertures or gaps depends on the size of the gaps in terms of wavelength. The wavelengths become quite small in the millimeter wave range, so the apertures become electrically larger and radiate more efficiently. For example, a 2 mm hole at 3 GHz is approximately 2% of a wavelength in size. That same 2 mm hole is 20% of a wavelength at 30 GHz. Therefore, EMI shielding at mm-wave frequencies can be improved with the use of fully drawn shields. Fully drawn shields avoid the corner gaps seen in stamped shields by stretching the metal and creating a continuous metal layer at the corners (Tong 2008; Dixon 2019).

As frequencies increase in shields designed with discrete cavities, the likelihood of intercavity cross-talk increases. While gaskets and absorbers may be used to offer grounding between the cover and internal cavity walls, their use is limited to applications with a small footprint.

In addition, another approach to tackling EMI issues at the package or board level in electronic devices is by applying an organic absorber coating to the semiconductor package or PCB to absorb extraneous electromagnetic waves. The absorber coating can be applied to the PCBs or sections of the PCBs to reduce unwanted EMI noise after board assembly. The main advantages of using absorber coatings to address EMI issues are as follows (Cohen 2018): (1) they are nonconductive and do not need to be electrically grounded to the PCB, simplifying or eliminating masking; (2) they can address EMI issues on the PCB, in-between tightly packed semiconductor devices; (3) they can be tuned to absorb EMI at customer specific frequencies; and (4) they can be applied at the end of a product design cycle to solve EMI issues without having to go through another redesign cycle.

6.4 Design and Materials Selection for 5G Absorbers

Different types of absorbing materials and geometries are used for EMC applications depending on the operating frequency range. Below 1 GHz, ferrite-based materials with dispersive lossy permeability are used in different planar configurations. Moreover, above 1 GHz, electric losses or magnetic losses provided by lossy dielectric or lossy magnetic materials, respectively, can be used. To cover both of the above ranges, hybrid absorbers constructed by combining the ferrite tiles and pyramidal absorbers are a solution, as shown in Fig. 6.7a, b). At higher microwave/THz frequencies, metamaterial absorbers can be used (Raad 2021).

With the development of fifth-generation (5G) communication technology utilizing massive multiple-input multiple-output (MIMO) phased array antennas for high-speed communication, the use of millimeter-wave absorbers (30–300 GHz) with high-power handling capability is needed. Convoluted (egg-carton) absorbers with the typical absorption of 50 dB at normal incidence at 30 GHz are primarily useful in millimeter-wave bands (Fig. 6.7c). Moreover, pyramidal structures with vent holes and heat sinks are developed. Additionally, the millimeter-wave

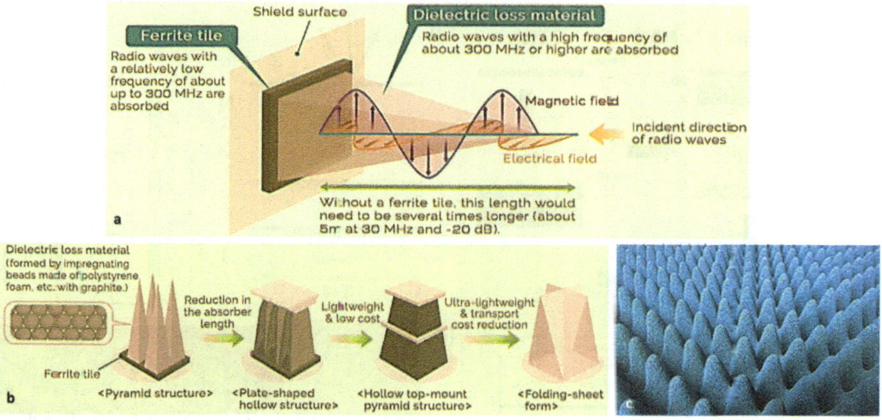

Fig. 6.7 Different types of RF/microwave/mm-wave absorbers. (Modified from TDK 2020; Credit: https://www.tdk.com/en/tech-mag/ferrite02/011): (**a**) Hybrid structure of a composite absorber in which a ferrite tile and a dielectric loss material are combined; (**b**) Evolution of different absorber structures; (**c**) Convoluted (egg-carton) absorber structure

metamaterial absorber bandwidth is widened by embedding one of the resonators inside another in the unit cell, which is fabricated by a standard optical photolithography process. Furthermore, the optical transparency requirement for some special EM protections, demanding large light transmittance and high absorptivity, can also be satisfied. In addition, naturally thin and transparent graphene material has great potential in the millimeter-wave absorber design, as shown in Fig. 6.8. Optically transparent broadband absorbers operating in the millimeter wave regime achieved by stacking graphene bearing quartz substrates on a ground plate. Broadband absorption is a result of mutually coupled Fabry–Perot resonators represented by each graphene-quartz substrate. Using a repeated transfer-and-etch process, multilayer graphene was processed to control its surface resistivity. Millimeter wave reflectometer measurements of the stacked graphene-quartz absorbers demonstrated excellent broadband absorption of 90% with a 28% fractional bandwidth from 125 to 165 GHz. The absorber operation can also be extended to microwave and low-terahertz bands with negligible loss in performance (Wu et al. 2014; Raad 2021).

In addition, metamaterials have enabled the design of thin electromagnetic wave absorbers. Compared to conventional millimeterwave absorbers, which are physically thick and where frequency performance is dictated by the inherent complex permittivity and permeability of the bulk material, metamaterial-based absorbers can be tailored for frequency response through the geometry of metallic inclusions called meta-atoms or metamolecules, which can be either periodically or randomly organized and layered with or embedded into dielectric materials. Metamaterial absorbers are frequency selective and have broadband responses over a wide range of frequencies spanning the microwave, terahertz (THz), infrared (IR), and optical spectra. The design of absorbers at millimeterwave or higher frequencies using metamaterial-based approaches is frequency scalable compared to conventional

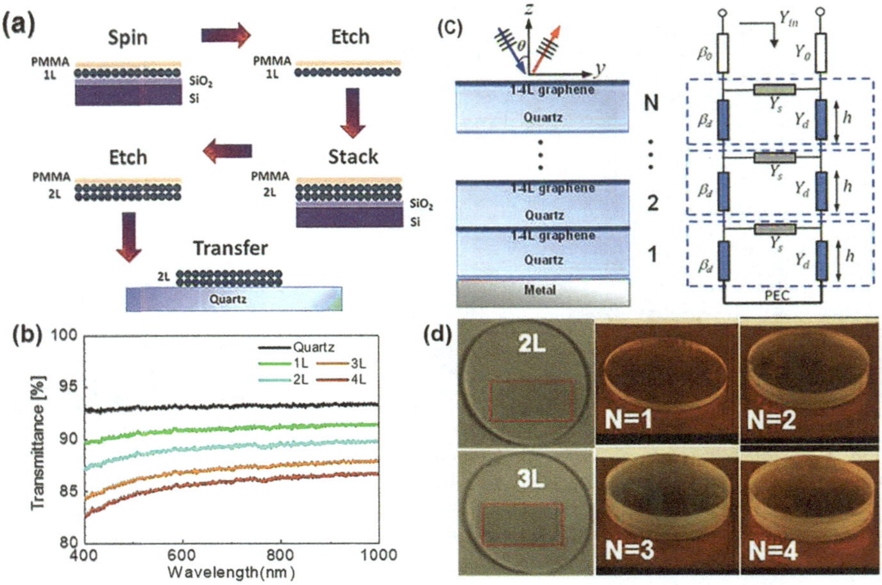

Fig. 6.8 Schematic and optical images of multilayer graphene on quartz and stacked graphene-quartz absorbers. (Adapted from Wu et al. 2014 with open access (Springer Nature)): (**a**) Schematic of the multiple transfer-etch processing for a 2 L device; (**b**) Typical UV-Vis spectra for the 1.3 mm thick bare quartz and 1–4 L graphene samples; (**c**) Schematic of the N-unit stacked absorber and the equivalent transmission-line circuit model; (**d**) Optical images of 2 L and 3 L absorbers and N 5 1–4 stacked graphene-quartz structures backed with a ground plate (N is the number of stacked graphene-quartz units)

bulk material-based absorbers. However, the bandwidth of a metamaterial-based absorber relies on the resonance bandwidth of the resonator structure, which is typically narrowband. To meet the broad bandwidth application requirements, several approaches have been investigated to design absorbers beyond single resonance. One example is that of a dual band absorber demonstrated at millimeterwave frequencies and at THz frequencies. In another application, a broadband absorber at 2.4 GHz has been investigated using stacked multifrequency resonators that achieved an absorption bandwidth of 700 MHz. An ultrabroadband absorber over a frequency range of 6–18 GHz with an absorption level of 90% has been demonstrated using a frequency selective resistive surface that relies on the tuning of material resistivity. An absorber at 30 GHz has also been investigated for wideband absorption using multiple resonators. Broadband absorbers at IR wavelengths designed using multiplexed metal patches have been shown to achieve 98% absorption over 3.40–3.55 μm. A broadband IR absorber is also realized using plasmonic nanoantennas with multiple resonance and dual resonance absorption responses. (Singh et al. 2013).

Figure 6.9a illustrates a metamaterial-based absorber with broad bandwidth at millimeter wave frequencies designed using two resonators, one embedded inside

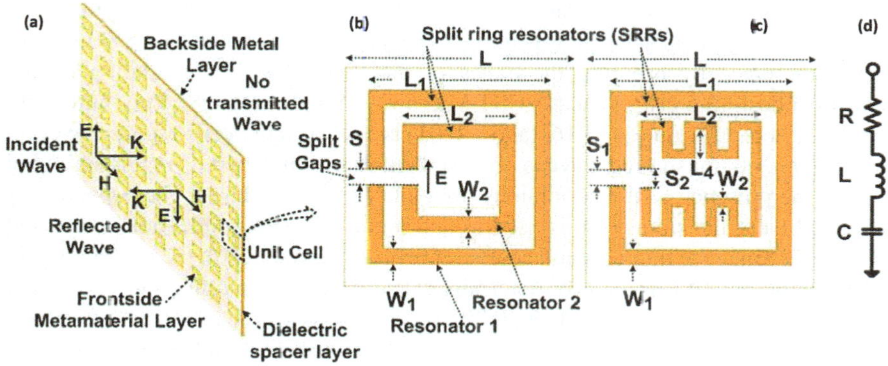

Fig. 6.9 Schematic representation of metamaterial absorber. (Adapted from Singh et al. 2013, reproduced courtesy of The Electromagnetics Academy): (**a**) The incident wave is reflected back and there is no transmitted wave due to backside continuous metal layer. The wave is absorbed at resonance frequency when electric field is parallel to the gap of the resonator; (**b**) layout of the unit cells for dual band absorber using two resonators; (**c**) inner resonator is folded to increase length of the resonator for decreasing second resonance frequency and bringing that close to the first resonance frequency; and (**d**) circuit based (RLC) model of the single SRR resonator

another. The top layer of the absorber consists of a 2D periodic array of split ring resonators (SRRs) made of metals such as gold patterned on a polyimide dielectric spacer layer. The back side of the absorber is coated with a continuous thin metal layer. The simultaneous interactions of electric (in the gap of SRR) and magnetic (between SRR and backside metal) fields of incident wave with the metamaterial result in the wave impedance ($\eta = E/H = \sqrt{(\mu/\varepsilon)}$) matching of absorber to the free space at the resonance frequency eliminating reflection. This incident wave is absorbed as metallic and dielectric losses at the resonance frequency. Waves at frequencies other than resonance are reflected back primarily due to the presence of the backside continuous metal layer; this also implies zero transmission through the absorber. A split ring resonator ('C' shaped) is used as the metamaterial unit cell, as shown in Fig. 6.9b, c, with a resonance frequency of 77 GHz. Another split ring resonator is inserted inside the first resonator to achieve the second resonant absorption frequency. The absorption frequency of the resonator can be modeled using the equivalent inductance (L) and capacitance (C) of the resonator, as shown in Fig. 6.9d. The resonant absorption frequency of the resonator, $f = 1/(2\pi\sqrt{(LC)})$, can be varied by changing its dimensions. The increase or decrease in the inductance value of the resonator is related to the increase or decrease in the length of the resonator, respectively. Similarly, the value of capacitance can be increased or decreased by increasing or decreasing the split gap capacitance of the SRR. Since the second resonator is placed inside the first resonator, without changing its design, it will have a smaller length and will therefore exhibit resonance at a much higher frequency than the first resonator. To enhance the absorption bandwidth at a single resonant frequency, the goal would be to bring the second resonance close to the first resonance so that the combination achieves broad absorption. To achieve this goal, different parameters have been varied for the second resonator. Increasing the effective length of the resonator through meandering was found to be more effective than trying to increase the capacitance by bringing the edges of the SRR closer or extending inside. Absorber designs are simulated in CST Microwave Studio software. Polyimide with a thickness of 125 μm ($\varepsilon = 3.16$, $\tan\delta = 0.01$) is used as a dielectric spacer layer. In this design, no transmitted wave is observed either in simulation or in measurement due to the presence of the continuous backside conductive layer. The S-parameter obtained from the simulation is used to determine the power reflection coefficient ($R = S^2_{11}$). Since the transmission coefficient is zero ($T = S^2_{21} = 0$), the power absorption coefficient (A) is directly determined from the power reflection coefficient ($A = 1 - R - T = 1 - R$) (Singh et al. 2013).

Further optimization of the design through numerical simulation can be performed to achieve perfect absorption over the designed frequency band. To demonstrate an absorber with a broadened bandwidth, two design approaches have been used. In design 1, the resonance frequency corresponding to the outer resonator is selected to be 77 GHz and that corresponding to the inner resonator is 80 GHz, as shown in Fig. 6.10a. However, in design 2, the inner resonator frequency is selected to be much closer to the first resonator (<80 GHz) so that the combined response appears similar to the single resonator but with enhanced absorption bandwidth, as shown in Fig. 6.10b. In design 2, the number of folds in the inner resonator is three compared to the two in design 1 to scale down the second resonance frequency. These designs are optimized to achieve perfect absorption peaks at 77 GHz and 80 GHz in the case of design 1 and peak absorption at 77 GHz with maximum

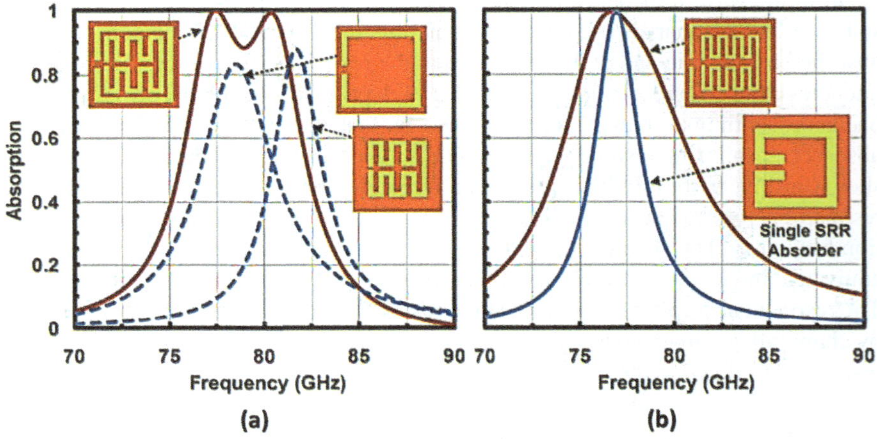

Fig. 6.10 Two designs of metamaterial absorber using dual SRRs for the enhanced bandwidth. (Adapted from Singh et al. 2013, reproduced courtesy of The Electromagnetics Academy): (**a**) Design 1, two very close resonance absorption frequencies (77 and 80 GHz) achieved using two additional folds in the inner resonator; (**b**) Design 2, both resonator frequencies are close enough to appear as a single resonance absorber with enhanced bandwidth, inner resonator has three additional folds in this design

bandwidth in the case of design 2. Figure 6.10a shows that the combination of two resonators results in perfect absorption at 77.45 GHz (absorption, 99.9%) and at 80.35 GHz (absorption, 99.0%), although the absorption level of the resonators individually is not perfect. The simulated bandwidth for an absorption level of 90% is 4.2 GHz for design 1 and 2.7 GHz for design 2 in comparison to 1 GHz for the absorber designed using a single resonator, as shown in Fig. 6.10b, indicating that a large enhancement of bandwidth is possible using a dual resonator structure. The absorber samples shown in Fig. 6.10 (designs 1 and 2) are fabricated using a standard optical photolithography process. Silicon wafers with a 4 inch diameter are used as the supportive substrate. Polyimide film as a dielectric spacer layer of the metamaterial absorber is prepared on a silicon wafer. Metamaterial patterns are transferred onto the polyimide using the lithography process. Chromium and gold metals serving as the metamaterial layer are deposited using a DC sputtering system with thicknesses of 20 nm and 200 nm, respectively. The polyimide film is peeled off from the silicon wafer after fabrication of the metamaterial layer. Chromium and gold layers are then coated on the back side of the sample to form a continuous conducting back plane (Singh et al. 2013).

The above designs show an approach to orient the resonators in the same direction such that mutual coupling is minimized and no significant hybridization effects are seen. Implemented absorbers are ultrathin (1/30th of the wavelength) and flexible. These metamaterial-based absorbers can be scaled to other frequencies from RF to THz, covering various applications in WiFi, automotive radars, radomes, and point-to-point 5G wireless communications (Singh et al. 2013).

6.5 Advanced Metallic Composite Materials for High-Frequency EMI Shielding

The most commonly used materials for EMI shielding are metals such as copper, aluminum, and stainless steel. While these metals provide adequate shielding against EMI, their susceptibility to corrosion, bulkiness, and difficulties in forming to the required shape create problems when these materials are to be used in modern electronic devices. To overcome this, some emerging metallic composite materials have been developed to modify existing metals used in shielding to eliminate their weaknesses or in different ways to eliminate one or more of their shortcomings while trying to maintain the same level of shielding effectiveness.

6.5.1 Hollow and Porous Metal-Based EMI Shielding Materials

One of the main advantages of synthesizing a material as foam is the substantial reduction of its weight. Since one of the primary drawbacks of metals used in shielding is their high specific weight, creating foams is one of the best ways to avoid this problem. Another advantage of foam material designed for EMI shielding is that the creation of pores inside the material could help to increase multiple reflections of electromagnetic waves, leading to higher attenuations and increased shielding effectiveness (Wanasinghe and Farhad 2019).

Figure 6.11 compares the shielding effectiveness of the foam materials designed for EMI shielding with a frequency range from 0 to 18 GHz, including Al foam, Cu–Ni-CNT composite, and PVA-Fe_2O_3 composite. When comparing these porous metal-based materials designed for EMI shielding, foam Al seems to show a higher SE than other materials under a frequency of less than 1.8 GHz (Xu and Hao 2014).

The Cu–Ni-CNT foam composite is fabricated using foamed polyurethane (PU) as the precursor material. The PU foam was subjected to electroless copper coating followed by electroplating of nickel. The subsequent heat treatment removed the PU, leaving behind a hollow metal structure. The final composite was fabricated by the electrophoretic deposition of carbon nanotubes (CNTs) on a hollow metal structure. Composites with different porosities specified as pores per inch (PPI) have been fabricated by utilizing PU foams with different porosities. A specimen with

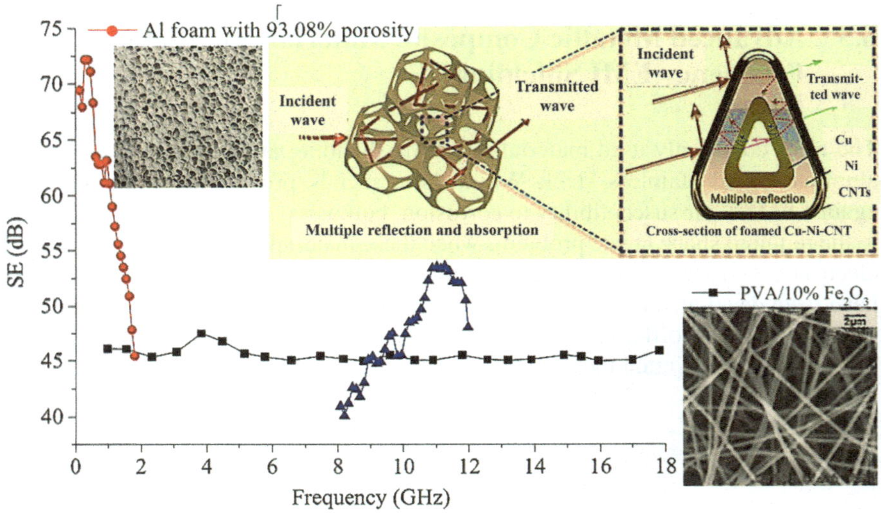

Fig. 6.11 Comparison of EMI shielding effectiveness (SE) properties of porous metal-based materials. (Modified from Wanasinghe and Farhad 2019 with permission from Elsevier)

110 PPI and a thickness of 1.5 mm showed an overall shielding effectiveness of approximately 40–55 dB within the 8–12 GHz frequency range. The addition of a CNT layer to the metal structure increased the shielding effectiveness by approximately 14 dB. The total shielding provided by the composite results from the reflection, scattering, and absorption mechanisms. Because of the open pore structure of these composites, they can be used in spaces such as air vents where EMI shielding is extremely difficult to achieve (Ji et al. 2014).

A PVA-Fe_2O_3 fibrous composite was fabricated using electrospun Ag-decorated poly(vinyl alcohol) (PVA) fibers. The fibers were first fabricated by mixing Fe_2O_3 nanoparticles into PVA solution followed by electrospinning. The obtained fibers were then coated with Ag by using a physical vapor deposition technique. A maximum shielding effectiveness of approximately 45.2 dB was obtained by the composite with 10% Fe_2O_3 nanoparticles prepared in mixed solvent and coated with a 200 nm Ag coating, which showed very stable shielding effectiveness within the tested frequency range of 0.5–18 GHz. These composites are used in aircraft, spacecraft, and automobile applications (Kim et al. 2012).

6.5.2 Metal-Based EMI Shielding Composites with Frequency-Selective Transmission

In some specific applications, such as the glass used in microwave ovens or other transparent windows, a certain amount of transparency to visible light is required while providing shielding from microwaves. Achieving this goal is problematic since visible light is also a form of electromagnetic wave. Since most EMI is caused by electromagnetic waves with lower frequencies than visible light, the shielding material should be transparent to visible light while providing shielding against lower frequency electromagnetic waves (Wanasinghe and Farhad 2019).

Figure 6.12 compares the EMI shielding effectiveness of some transparent metal-based materials, including IZO/Ag multilayer thin film, SSF/silicon composite, EHD printed Ag mesh, and hierarchical graphene/metal film. An EMI shielding IZO/Ag multilayer thin film was fabricated using layers of Ag alloy sandwiched between indium–zinc oxide (IZO) layers for optically transparent applications. Each layer was created using a radio frequency (RF) magnetron sputtering technique with a thickness varying between 7 and 100 nm. The composite specimens could produce a shielding effectiveness above 45 dB in the 30 MHz to 1 GHz frequency range. The optical transmission has been above 70% with smooth metal layers, indicating that the composite specimens have adequate optical properties to be used in many EMI shielding applications requiring optical transparency (Kim et al. 2005).

The SSF/silicon composite was fabricated by distributing an SSF with a diameter of 4 μm and a length of 1 mm in a silicon matrix. The maximum shielding effectiveness of approximately 27–36 dB is produced by the composite with 1.25 wt% SSF

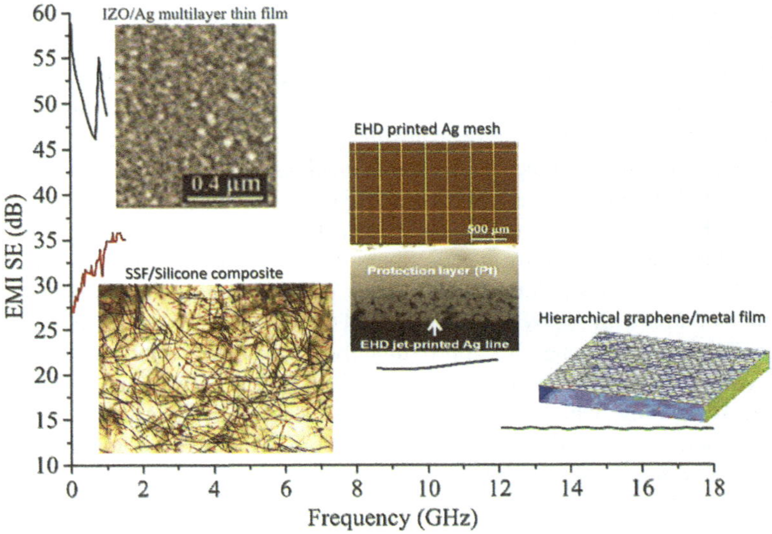

Fig. 6.12 Comparison of EMI shielding effectiveness properties of transparent metal-based materials. (Modified from Wanasinghe and Farhad 2019 with permission from Elsevier)

in the frequency range of 300 MHz to 1.5 GHz. The visible light transmittance of the composite with 1.25 wt% SSF is approximately 67%. This composite could be used in applications where the EMI shielding effectiveness requirement is below 20 dB, such as in mobile phones, since it has proven to be more cost-effective than composites with carbon fibers or carbon nanotubes. Using SSFs with smaller diameters and longer lengths may achieve better shielding effectiveness while maintaining the required visible light transmittance values (Jin et al. 2014).

EHD-printed Ag mesh was fabricated using electrohydrodynamic (EHD) jet printing to fabricate a metal mesh that is invisible to the naked eye and provides sufficient EMI shielding effectiveness. The specimen with a 300 µm pitch showed relatively high transmittance and was able to produce an average shielding effectiveness of approximately 20–21 dB in the 8–12 GHz frequency range. This method could be utilized in the manufacturing of displays where EMI shielding is required (Vishwanath et al. 2014).

As an EMI shielding visible light transparent composite, a hierarchical graphene/metal film was fabricated using a graphene layer on a metal mesh embedded quartz substrate with the crack-template-based lift-off method. The composite is able to produce a shielding effectiveness of approximately 14 dB in the frequency range of 12–18 GHz, transmitting close to 90% visible light. The addition of the graphene layer has led to the expansion of the conductive network of the composite that has increased the absorption as well as the reflection of electromagnetic waves, increasing the overall shielding effectiveness of the composite (Han et al. 2017).

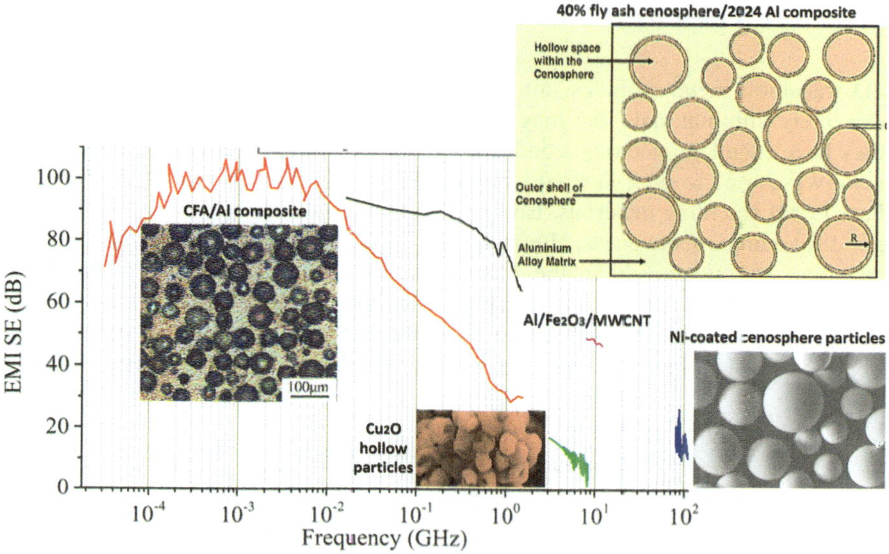

Fig. 6.13 Comparison of EMI shielding effectiveness properties of particle-based metallic materials. (Modified from Wanasinghe and Farhad 2019 with permission from Elsevier)

6.5.3 Particle-Based EMI Shielding Metallic Composites

The use of particles for the fabrication of EMI shielding composites has been widely investigated, as particles act as effective absorbers of electromagnetic waves and help to enhance their EMI shielding properties while improving their mechanical properties and reducing their specific weight. Figure 6.13 compares some metal-based particle-incorporated materials, indicating a wide variation in the shielding effectiveness produced by these materials. At the lower frequency end of the graph, the CFA/Al composite showed high shielding effectiveness, which drastically decreased with increasing frequency. The 40% fly ash/Al composite showed very similar behavior to the CFA/Al composite. This indicates that the addition of fly ash alone to the composite is insufficient to create a stable shielding effectiveness in metallic materials. The Al/ferrite/MWCNT composite showed a moderate shielding effectiveness compared to materials at a frequency of approximately 10 GHz. The shielding effectiveness produced by hollow copper oxide particles was the lowest, with a substantial fluctuation in shielding effectiveness. In the higher frequency range, Ni-coated cenospheres produced low but somewhat stable SE, which might be improved with additional development. Therefore, particle-based EMI shielding metallic composites need further investigation to assess their suitability for EMI shielding applications over a wide range of frequencies (Wanasinghe and Farhad 2019).

6.5.4 MXene-Based EMI Shielding Composites

2D transition metal carbides, nitrides, and carbonitrides known as MXenes are extremely thin materials that may be used in EMI shielding applications because they have excellent electrical conductivity and the ability to attenuate electromagnetic waves due to their layered structure. Figure 6.14 shows two of the MXene-based EMI shielding materials. Both of these materials have shown comparatively good shielding effectiveness, although the $Ti_3C_2T_x$ layered material exhibits higher shielding effectiveness than the MXene foams. Moreover, MXene foams have shown hydrophobic properties, which have led to a longer lifetime of the material. Furthermore, the shielding effectiveness produced by the MXene foam shows much higher stability in the tested frequency range than the layered composite. Since the

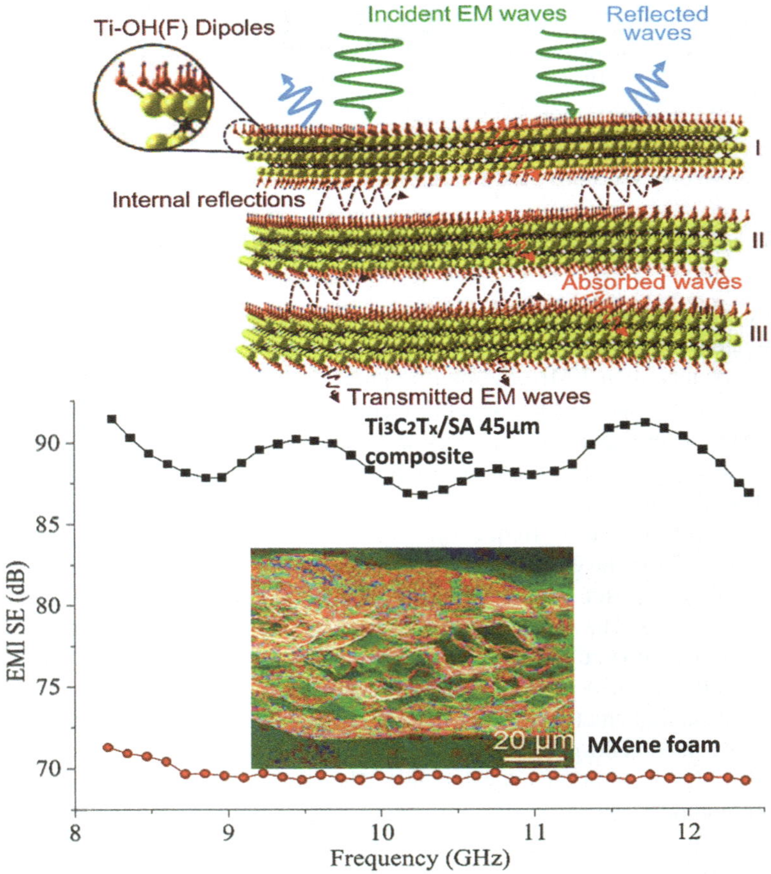

Fig. 6.14 Comparison of EMI shielding effectiveness properties of MXene based materials. (Modified from Wanasinghe and Farhad 2019 with permission from Elsevier)

EMI shielding effectiveness of the MXene foam is approximately 20 dB lower than that of the layered composite, the addition of a secondary wave absorber could increase the shielding effectiveness of the foam. When comparing the thickness of the two materials, the layered composite showed an extremely low thickness value to produce this shielding effectiveness, while it was considerably high for the foam. Overall, MXene is a promising material that can be developed to have high EMI shielding effectiveness and even replace existing metallic shields (Wanasinghe and Farhad 2019).

6.5.5 *Metal-Based Flexible EMI Shielding Materials*

Metal-based flexible EMI shielding materials have been developed for flexible, stretchable or wearable electronics, including electronic textiles. For example, one of the easiest ways to incorporate EMI shielding metals into clothing is to weave fabric by mixing metal filaments and traditional clothing fibers. Another way of fabricating flexible shielding materials would be to create flexible layers or foams of materials that are already good at EMI shielding. Having flexibility for EMI shielding materials enables them to be used in miniature electronic devices that require complex shapes.

Figure 6.15 compares the EMI shielding effectiveness properties of some flexible metal-based materials. The best shielding effectiveness is produced by the CPAN NF/Ag NP composite. The primary reason for this material to show such high levels of SE is the high conductivity shown by the material owing to the high conductivity of Ag and PAN fibers. Moreover, this material has been able to

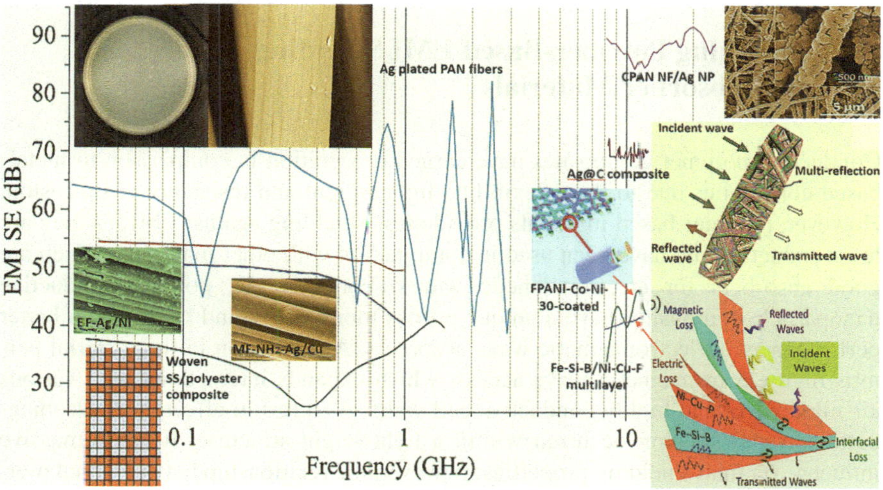

Fig. 6.15 Comparison of EMI shielding effectiveness properties of flexible metal-based materials. (Modified from Wanasinghe and Farhad 2019 with permission from Elsevier)

maintain high shielding effectiveness within a wide range of frequency bands, making it suitable for use in high-frequency telecommunication-related applications. Ag@C particles have shown moderate but stable shielding effectiveness in the frequency range it was tested. Again, the primary reason for these particles having high shielding effectiveness is the high conductivity produced by their components. On average, the PANI-Co-Ni composite and the Fe–Si–B/Ni–Cu–P multilayer composite showed the same amount of shielding effectiveness. However, their shielding effectiveness has been very low compared to most of the materials shown in Fig. 6.15. Materials that have been tested in the lower frequency range have shown wide variation in their shielding effectiveness properties. The Ag-plated PAN fibers have shown somewhat good shielding effectiveness, but their shielding effectiveness has varied largely in this frequency range (Wanasinghe and Farhad 2019).

In summary, although numerous alternative materials for EMI shielding have been investigated to address various shortcomings of metals, metal sheets continue to be used due to their high shielding effectiveness, as shown in Table 6.1. Out of all these materials, arc thermal metal spray (ATMS) Zn–Al has been shown to be extremely effective against EMI and has been applied over existing walls as a thin coating. The other materials that have shown some amount of potential against EMI shielding are the CFA/Al and fly ash/Al composites, which have shown very high shielding effectiveness, but unfortunately, the shielding effectiveness produced by the CFA/Al composite decreases with increasing frequency of electromagnetic waves. MXene-based materials have also shown many potentials to have high shielding effectiveness with extremely low thicknesses. However, most of these novel metal-based materials have considerable variations in their shielding effectiveness, which requires further research to make them suitable for practical applications, especially for 5G communication systems.

6.6 Emerging Polymer-Based EMI Shielding and Absorber Materials

Conducting polymer composites have earned recognition in comparison to metal-based composites due to their flexibility, light weight and resistance to corrosion. However, polymer-based materials provide less shielding against EMI, as they are less conductive and have been used in combination with other materials, which are good absorbers for electromagnetic waves. For example, polymer-conductive nanoparticle composites have a unique porous morphology and have shown better performance as electromagnetic wave absorbers. Air insertion in the material permits high electromagnetic wave access, which expands interactions with various air-filled pores and highly conductive cell walls. As a result, effective electromagnetic wave dissipation occurred within a lightweight structure. It has a massive influence on final shielding properties; however, the relationship between final morphology and foaming conditions differs and is not predictable. Table 6.2

Table 6.1 Typical metallic shielding materials (Wanasinghe and Aslani 2019)

Material-filler	Material – Matrix	Frequency	Specimens thickness	Effect of shielding
Aluminium/ Nickel on copper plating	Plastic	1 kHz to 20 GHz	–	Apertures found in practical enclosures is more important than the wall material
Ag coating	Fe_2O_3 mixed PVA	0.5–18 GHz	200 nm Ag coating	45 dB
Aluminum foams	–	130–1800 MHz	10 mm	25–75 dB
CNT	Cu–Ni	8–12 GHz	1.5 mm	54.6 dB
Indium–zinc oxide/Ag multilayers	–	30 MHz–1 GHz	3 mm	>45 dB
Stainless steel fibers	Silicon	300 MHz to 1.5 GHz	2.5 mm	27–36 dB (1.25 wt%)
Electrohydrodynamic jet printed Ag metal mesh	–	8–12 GHz	150–300 μm	20–25 dB
Graphene layer on Gr–Ag mesh	Quartz substrate	12–18 GHz	~4 μm	~14.5 dB
Fly ash cenospheres	2024 Aluminium	0.03 MHz to 1.5 GHz	2 mm	Absorber, −65 to −95 dB
Fly ash microparticles	2024 Al	30 kHz to 1.5 GHz	2 mm	30–110 dB
Ni coating	Fly ash cenosphere particles	80–110 GHz	60 mm	Absorber, −11 to −27 dB
Fe_2O_3+MWCNT	A1-LM6	8–12 GHz	–	46.8–49.3 dB
Fe-Co nanoparticles	Porous carbon decorated with graphene	2–18 GHz	2.5 mm	Absorber, –52.9 dB (max at 10.1 GHz)
$ZnFe_2O_4$ nanoparticles	Carbonyl iron (Fe) flakes	2–18 GHz	1.5 mm	Absorber, ~ −24 dB (max at ~ 8.5 GHz)
Ti_3C_2/Fe_3O_4	PANI	2–18 GHz	2.5 mm	Absorber, –40.3 dB (max at 15.3 GHz)
Cu_2O	–	3–8.5 GHz	~1 mm	1.5–17.8 dB
Fe_3O_4 quantum dots	Carbon sphere	2–18 GHz	1.5 mm	Absorber, −26.2 dB (max at ~ 9 dB)
MWCNT-Fe_3O_4@Ag	Epoxy	8.2–12.4 GHz	2 mm	35 dB
Siliconboron carbonitride (SiBCN)	graphene@Fe_3O_4	8.2–12.4 GHz	2.58 mm	Absorber, −66.21 dB (max at 11 GHz, sample oxidated at 600 °C)
Fe@Carbon dual-shells	–	1–18 GHz	3 mm	Absorber, −54.4 (max at 8.8 GHz)

(continued)

Table 6.1 (continued)

Material-filler	Material – Matrix	Frequency	Specimens thickness	Effect of shielding
Zeolitic imidazolate frameworks	Paraffin	2–18 GHz	1.9 mm	Absorber, −50.9 dB (max at 17 GHz)
$Ti_3C_2T_x$ [2D transition metal carbides (MXenes)]	Sodium alginate (SA)	8.2–12.4 GHz	0.0045 mm	57 dB
MXene foam	–	8.2–12.4 GHz	60 μm	Absorber, –70 dB
$Ti_3C_2T_x$ MXene	Epoxy	8.2–12.4 GHz	2 mm	41 dB
MXene	polystyrene	8.2–12.4 GHz	2 mm	53–62 dB
MXene	poly(ethylene terephthalate) textiles	8.2–12.4 GHz	1.3 mm	90 dB
Zn–Al Arc thermal metal spray	Sandblasted tempered glass	1 kHz to 10 GHz	100–700 μm	50–120 dB
Metallic multilayers	Plastic substrate	30 MHz to 3 GHz	–	Al- 45 dB, Ni- 40 dB
MnO_2/Fe_3O_4/CNT	PDMS	2–18 GHz	1 mm	42.2 dB (7.1 GHz)
Electroless Ag coating	PET	0.045–1 GHz	–	64.5 to 54.2 dB
Fe_3O_4/thermally annealed graphene aerogel	Epoxy	8.2–12.4 GHz	3 mm	35 dB
Fe_3O_4-loaded multiwalled carbon nanotubes	Polyazomethine (PAM)	8.2–12.4 GHz	1.8 mm	Absorber, –40.6 dB (max at 11.5 GHz)
Metal filaments	Polyester fibers	30 MHz to 1.5 GHz	–	25–40 dB
Electroless Ni coating	Bamboo fabric	30–1000 MHz	–	>40 dB
Electroless Cu coating	Modal fabric	100–1000 MHz	–	>35 dB
Electroless Ag coating	Polyacrylonitrile (PAN) fibers	30 kHz to 3 GHz	–	40–80 dB
PANI/Co–Ni	Lyocell fabric	8.2–12.4 GHz	–	34–46 dB
Electroless Ni–Cu–P coating	Fe–Si–B	8–12 GHz	0.1 mm	~40–43 dB
Ag/Gu/Ni nanoparticles	PAN nanofibers	8–12.4 GHz /12–18 GHz /18–26.5 GHz	–	85–93 dB
Carbon	Ag nanowires	8.2–18 GHz	3 mm	70 dB
Cu coating	Bacterial cellulose	10 MHz to 1.5 GHz	–	61–66 dB

summarizes emerging polymer-based composites for EMI shielding and absorber applications. The shielding effectiveness depends on the thickness and amount of filler content, where their increasing proportions can enhance the effectiveness. Polymer-based composites show efficient performance with graphene and iron

Table 6.2 Typical emerging polymer-based composites for EMI shielding and absorber applications (Ayub et al. 2021)

Material	Thickness	Loading	Methods	Frequency	Shielding effectiveness
PVC/PANI/GNP	–	5 wt.%	Solution processing method	18–20 GHz	51 dB
GNP@PANI	1.5 mm	–	In-situ growth	12 GHz	Absorber, −14.5 dB
TiO_2/PANI/GO	3.12 mm	–	In-situ growth	2–18 GHz	Absorber, −51.7 dB
Graphene@PANI@TiO_2	1.5 mm	–	1. In-situ growth; 2. Hydrothermal method	2–18 GHz	Absorber, −45.4 dB
Graphene@Fe_3O_4@PANI@WO_3	4 mm	–	1. Hydrothermal method; 2. Chemical oxidation polymerization	9.4 GHz	Absorber, −46.7 dB
Graphene@Fe_3O_4@SiO_2 @ polyaniline	2.5 mm		Dilute polymerization	12.5 GHz	Absorber, −40.7 dB
PANI/GO/Fe_3O_4	3.91 mm	–	Hummers method	2–18 GHz	Absorber, −53.5 dB
Ag@Graphene/PANI	–	5 wt.%	In-situ growth	0.4–1.6 GHz	29.33 dB
Fe_3O_4/PANI rod/RGO	3.5 mm	–	Facile method	2–18 GHz	Absorber, −33.3 dB
Graphene@NiO@PANI@Ag	3.5 mm	–	1. Hydrothermal method; 2. In-situ growth	2–18 GHz	Absorber, −37.5 dB
G-PANI	1.5 mm	–	In-situ growth	2–18 GHz	32.5 dB
γ-Fe_2O_3/RGO/PANI	2.5 mm	–	1. Chemical oxidation polymerization; 2. In-situ growth	8–12 GHz	51 dB
$Ti_3C_2T_x$/Fe_3O_4@PANI	12.1 μm	–	Co-precipitation method	8–12 GHz	58.8 dB
PANI/BF/RGO	–	–	Citrate precursor method	8–12 GHz	31.1 dB
PANI/$Li_{0.5}Fe_{0.5-x}Gd_xO_4$	2 mm	–	In-situ growth	8–12 GHz	42 dB
RGO@polystyrene	–	3.47 vol%	High-pressure solid-phase compression moulding	8–12 GHz	45.1 dB
Segregated RGO/PS	2 mm	10 wt.%	Hot compressed method	0–20 GHz	29.7 dB
Conventional RGO/PS					14.2 dB
Polycarbonate/GNP	1 mm	–	Facile solution method	8–12 GHz	35 dB
	2 mm	–			47 dB

(continued)

Table 6.2 (continued)

Material	Thickness	Loading	Methods	Frequency	Shielding effectiveness
Polyethylene@GNP	–	15.6 vol%	Injection moulding process	18 and 26.5 GHz	16 dB
		19 vol%			31.6 dB
		3 wt.%			12 dB
		10 wt.%			31 dB
GNP/EPDM	0.3 mm	8 wt.%	Ultrasonication technique	8–12 GHz	33 dB
				12.4–18 GHz	35 dB
Hollow Fe_3O_4@GF@PDMS	–	4 wt.%	Solvothermal method	8–12 GHz	45 dB
		8 wt.%			65 dB
		12 wt.%			70.3 dB
3D Graphene Network@ PDMS	0.25 mm	1.2 wt.%	Chemical vapor deposition	8–12 GHz	40 dB
	0.75 mm				90 dB
MXene($Ti_3C_2T_X$)/graphene/ PDMS	1 mm	–	Chemical vapor deposition	8–12 GHz	80 dB
				26.5–40 GHz	77 dB
Graphene flakes@PDMS	–	0.1 wt.%	Mechanical mixing	0.6 THz	6.5 dB
		3 wt.%			12 dB
		10 wt.%			31 dB
Cu@RGOFM@PDMS	0.5 mm	–	Hummers method	8–12 GHz	74.2 dB
GA/PDMS	2.5 mm	–	1. Ultrasonication technique; 2. Hydrothermal method	2–18 GHz	60 dB
Ni@GNS@PVDF	0.5 mm	–	Ultrasonication technique	18–26 GHz	43.3 dB
	0.7 mm				51.4 dB
RGO@CuS@PVDF	1 mm	–	Hydrothermal method	12–18 GHz	Absorber, −25 dB
GNSs-Fe_3O_4/PVDF	0.3 mm	–	Facile layer-by-layer coating	8–12 GHz	52 dB
RGO@Hematite/PVDF	–	5 wt.%	In-situ growth	2–18 GHz	Absorber, −43.97 dB

Gn/SiCnw/PVDF	1.2 mm	–	1. Electrostatic assembly; 2. Solution casting method	8–12 GHz	32.5 dB
PVDF/graphene	20 μm	15 wt.%	Solution casting method	8–12 GHz	47 dB
PVDF/GNP-Ni-CNT	0.6 mm	–	Solvent casting method	12–18 GHz	46.4 dB
PVDF/n-Fe	1.93 mm	–	Hot-moulding process	12–18 GHz	40.21 dB
PVDF/PFC	2 mm	1 wt.%	Solution blending process	12–18 GHz	Absorber, −29.7 dB
CuNWs-TAGA/Epoxy	–	7.2 wt.%	Thermal annealing method	8–12 GHz	47 dB
RGO-CF/EP	3–5 mm	–	1. Electrophoretic deposition; 2. Chemical reduction	8–12 GHz	37.6 dB
GNP/Fe_3O_4/Epoxy	–	7 wt.%	Co-precipitation method	8–12 GHz	37.03 dB
GNP/Fe/Epoxy	5 mm	5 wt.%	Sonication method	1–65 GHz	Absorber, −78 dB
TGO/CI/Epoxy	4 mm	–	Centrifugal mixing method	8–12 GHz	40 dB
GCF/MG_3/EP	–	0.5 wt.%, 9 wt.%	Hummers Method	18–26 GHz	51.1 dB
RGO/PF/Epoxy	3 mm	60 wt.%	Solution mixing method	2–18 GHz	Absorber, −10.26 dB
Polylactic acid/Biochar/Graphite	0.25 mm	–	Hot-pressing method	18–26.5 GHz	30 dB
Polypyrrole/BST/RGO/Fe_3O_4	22.8 × 10.03 × 2.5 mm	–	Chemical oxidative polymerization	8–12 GHz	48 dB
RGO@PEI		2.5 wt.%	Hummers Method	8–12 GHz	26 dB
Fe_3O_4@RGO/TPU	1 mm	–	Solution casting method	8–12 GHz	~15.51 ± 1.6 dB
G/Polyurethane sponge	9 mm	18.7 wt.%	Hydrothermal method	8–12 GHz	35 dB
TPU/TRG	2 mm	5.5 vol%	Solution blending method	12–18 GHz	32 dB
BaFe@TRCO@TPU	0.25 mm	–	Solution casting method	0.1–20 GHz	Absorber, −61 dB
PEDOT/RGO/$SrFe_{12}O_{19}$	2.5 mm	–	In-situ growth	8–12 GHz	42.29 dB
	4.66 mm				62 dB

(continued)

Table 6.2 (continued)

Material	Thickness	Loading	Methods	Frequency	Shielding effectiveness
PEDOT/RGO/$PbTiO_3$	2.5 mm	–	Chemical oxidative polymerization	12.4–18 GHz	51.94 dB
PEDOT:PSS-Fe_3O_4-RGO	3.86 mm	–	Hydrothermal method	2–18 GHz	Absorber, −61.4 dB
Fe_3O_4/C:PPy	0.8 mm	2.8 wt.%	1. Hydrothermal method; 2. Chemical oxidative polymerization	2–8 GHz	>28 dB
FeCo@RGO@PPy	2.5 mm	–	1. Hydrothermal method; 2. In-situ growth	2–18 GHz	Absorber, −40.7 dB
RGO-PANI-$NiFe_2O_4$	2.4 mm	–	1. Hummers method; 2. Solvothermal method	2–18 GHz	Absorber, −49.7 dB
RGO-PPy-$NiFe_2O_4$	1.7 mm				Absorber, −44.8 dB
RGO-PEDOT-$NiFe_2O_4$	2 mm				Absorber, −45.4 dB
Graphene/$Li_{0.35}Zn_{0.3}Fe_{0.35}O_4$/PMMA	4 mm	–	3D printing method	2–18 GHz	Absorber, −46.1 dB
PMMA/RGO	2.9 mm	2.6 vol%	Self-assembly technique	8–12 GHz	63.2 dB
PMMA/graphene	2 mm	20 wt.%	Hot compression method	8–12 GHz	21 dB
PVC/graphene					31 dB
Fe_3O_4@SLGAPC@PVA	0.3 mm	–	Solution casting method	8–12 GHz	20 dB
PVA/Gr/Fe_3O_4	0.2 mm	0.08 wt.%	Hummers method	8–12 GHz	40.7 dB
GNP/CLF/PEEK	–	9 wt.%	Compression moulding method	8–12 GHz	27.1 dB
$NiFe_2O_4$-RCO-Polypropylene	2 mm	5 wt.%	1. Hummers method 2. Hot press method	6–8 GHz	29.4 dB

combinations, where PDMS, PANI, PMMA and PVDF were effective in the X-band range. In the higher frequency range, PDMS, epoxy, PVDF and PANI polymers were astonishing in providing effective shielding. Additionally, intrinsically conducting polymer (ICP) and conductive filler-incorporated conductive polymer-based composites (CPCs) not only provide excellent EMI shielding but also have the advantages of electromagnetic wave absorption rather than reflection (Jiang et al. 2019; Ayub et al. 2021).

Despite considerable advancements, polymer-based EMI shielding and absorber materials face many obvious challenges. The significant limitations include the following: insulating nature of polymers that may lead to the enlargement of skin depth for electromagnetic attenuation (if a sufficient conducting network is not established by the conductive fillers); difficulty in processing and reconstruction of defects; surface functionalization of fillers that harms the physical properties of the composites and processing; poor thermal stability that hinders their applications in high-temperature operation; understanding the thermodynamics of polymer blends and corresponding preferential localization of fillers; tailoring of morphology and physical properties of the fillers, etc. Although multicomponent structures and lightweight foams can be efficient electromagnetic absorbers, the factors that determine their bandwidth of electromagnetic absorption and skin depth need to be determined, and it is not clear how the temperature influences the electromagnetic properties of materials in nanoscale polymer nanocomposites or nanohybrids. However, lightweight multicomponent nanostructures and three-dimensional architectures of polymer-based materials with good mechanical properties and high thermal stability will lead to the realization of highly efficient EMI shielding for various applications, including 5G communication systems (Jiang et al. 2019).

References

Armstrong EIK (1999) Design techniques for EMC part 4: shielding. https://www.emcstandards.co.uk/files/design_techniques_for_emc_1999_part_4_shielding.pdf

Ayub S, Guan BH, Ahmad F, Oluwatobi YA, Nisa ZU, Javed MF, Mosavi A (2021) Graphene and iron reinforced polymer composite electromagnetic shielding applications: a review. Polymers 13:2580

Cohen J (2018) Solving EMI issues in mobile electronic devices. https://blog.parker.com/site/usa/en-US/details-home-page/solving-emi-issues-in-mobile-electronic-devices-us

Dixon P (2019) EMI mitigation at millimeter wave. In: 2019 European microwave conference in Central Europe (EuMCE), pp 610–613

Han Y, Liu Y, Han L, Lin J, Jin P (2017) High-performance hierarchical graphene/metal-mesh film for optically transparent electromagnetic interference shielding. Carbon 115:34–42

Henkel (2017) Component level EMI shielding for semiconductor packages. https://dm.henkel-dam.com/is/content/henkel/webinar-henkel-package-level-emi-shielding-externalpdf

Ji K, Zhao H, Zhang J, Chen J, Dai Z (2014) Fabrication and electromagnetic interference shielding performance of open-cell foam of a Cu-Ni alloy integrated with CNTs. Appl Surf Sci 311:351–356

Jiang D, Murugadoss V, Wang Y et al (2019) Electromagnetic interference shielding polymers and nanocomposites – a review. Polymer Reviews 59(2):280–337

Jin L, Haiyan Z, Ping L, Xijiang Y, Guoxun Z (2014) The electromagnetic shielding effectiveness of a low-cost and transparent stainless steel fiber/silicone resin composite. IEEE Trans Electromagn Compat 56:328–334

Kim WM, Young KD, Lee I, Seo YW, Cheong B, Lee TS et al (2005) The electromagnetic interference shielding effect of indium–zinc oxide/silver alloy multilayered thin films. Thin Solid Films 473:315–320

Kim H-R, Kim B-S, Kim I-S (2012) Fabrication and EMI shielding effectiveness of Ag-decorated highly porous poly(vinyl alcohol)/Fe_2O_3 nanofibrous composites. Mater Chem Phys 135:1024–1029

Li J, Tsai M, Chiu R, He E, Hsieh A et al (2020) EMI shielding technology in 5G RF system in package module. In: 2020 IEEE 70th electronic components and technology conference (ECTC), pp 931–937. https://doi.org/10.1109/ECTC32862.2020.00152

MacArthur D (2019) Let's talk about shielding at the PCB level. In Compliance - Electronic Design, Testing & Standards. https://incompliancemag.com/article/lets-talk-about-shielding-at-the-pcb-level/

Raad SH (2021) Analysis and design of absorbers for electromagnetic compatibility applications. In: Electromagnetic compatibility. IntechOpen. https://www.intechopen.com/chapters/79519

Singh P, Ameri SK, Chao L, Afsar MN, Sonkusale S (2013) Broadband millimeterwave metamaterial absorber based on embedding of dual resonators. Progress In Electromagnetics Research 142:625–638. https://doi.org/10.2528/PIER13070209

Taranovich S (2021) How to shield your 5G systems from EMI. https://www.electronicdesign.com/technologies/analog/article/21181685/electronic-design-how-to-shield-your-5g-systems-from-emi

TDK (2020) Ferrite world – Vol. 11 Radio wave absorbers, anechoic chambers, and ferrite. https://www.tdk.com/en/tech-mag/ferrite02/011

Tong XC (2008) Advanced materials and design for electromagnetic interference shielding. CRC Press, New York

Tsai M, Chiu R, He E, Chen JY et al (2019) Innovative EMI shielding solutions on advanced SiP module for 5G application. In: 2019 IEEE 21st electronics packaging technology conference (EPTC), pp 601–607. https://doi.org/10.1109/EPTC47984.2019.9026564

Vishwanath SK, Kim D-G, Kim J (2014) Electromagnetic interference shielding effectiveness of invisible metal-mesh prepared by electrohydrodynamic jet printing. Jpn J Appl Phys 53(5S3):05HB11

Wanasinghe D, Farhad A (2019) A review on recent advancement of electromagnetic interference shielding novel metallic materials and processes. Composites Part B 176:107207

Wu B et al (2014) Experimental demonstration of a transparent graphene millimeter wave absorber with 28% fractional bandwidth at 140 GHz. Sci. Rep. 4:4130. https://doi.org/10.1038/srep04130

Xu Z, Hao H (2014) Electromagnetic interference shielding effectiveness of aluminum foams with different porosity. J Alloy Comp 617:207–213

Zwenger C (2021) Enabling the 5G RF front-end module evolution with the DSMBGA package. Chip Scale Rev 25(5):26–33. https://chipscalereview.com/wp-content/uploads/2021/09/ChipScale_Sep-Oct_2021-digital.pdf

Chapter 7
Thermal Management Materials and Components for 5G Devices

Abstract 5G devices range from base stations, antenna arrays, edge data centers, and transceivers to handsets. Effective thermal management solutions can help 5G devices maintain their increasingly slim footprint while still maintaining the ability to sustain 5G connections without performance drops. Form factors of different 5G devices constrain the selection of different thermal management solutions, including passive cooling, active cooling and hybrid cooling strategies. Materials with excellent thermal performance are required to improve heat transfer capabilities and properties, ensuring the ability to cope with ever-increasing reliability requirements. For example, thermal interface materials (TIMs) are used to fill the air gaps, establish an effective heat conduction channel between electronic components and the radiator, reduce the heat transfer resistance, and improve the heat dissipation performance. This chapter will provide a brief review of thermal management challenges in 5G devices, advanced thermal management solutions for different 5G devices ranging from base stations and antenna arrays, edge data centers and 5G mobile phones, as well as various thermal management materials and components, including thermal interface materials, polymer-based thermal materials and thermal metamaterials.

7.1 Thermal Management Challenges and Strategies in 5G Devices

5G technology, radio units and active antenna devices contain a greater density of high-power components in smaller spaces. This increased power requires more energy consumption, which generates more heat. Effective thermal management solutions can help 5G devices maintain their increasingly slim footprint while still maintaining the ability to sustain 5G connections without performance drops.

More 5G-capable devices will begin to appear in a variety of form factors, beginning at the base station level and filtering down to the edge. At the handset level, extended battery life is always in demand, which requires running devices at lower

C. Tong, *Advanced Materials and Components for 5G and Beyond*, Springer Series in Materials Science 327, https://doi.org/10.1007/978-3-031-17207-6_7

power than pre-5G devices. Despite the goals of running at lower power, thermal management will remain a top priority in new designs, especially in mmWave handsets. At the base station and data center level, offloaded data processing tasks will consume progressively more power as more data are collected and processed in the cloud. At both levels, successive 5G rollouts will continue to move beyond 6 GHz, requiring greater use of GaN devices to power mmWave 5G devices and networks. Managing the heat and temperature in these systems is critical to extending the useful lifetime, particularly for components. At high temperatures, electromigration also occurs at a faster rate, noise is more intense, embedded antennas radiate less strongly at the desired frequencies, and a host of other problems arise in 5G systems. Solutions to these challenges must be addressed at the board level, circuit design/operation level, and through the use of active thermal management solutions. Any solution will need to work within device form factor requirements without interfering with core functionality (Peterson 2020).

7.1.1 *Form Factor-Constrained Thermal Management Solutions*

Thermal management challenges within the 5G network must be addressed to provide reliability, equipment uptime, and flexibility for different operating environments. Form factors in the variety of devices that make up 5G infrastructure and devices will dictate which thermal management strategies are suitable, as well as their effectiveness. The solutions implemented at the base station will not match those implemented in a handset. Obviously, given the size of current and future handsets, powerful cooling fans cannot be placed in consumer cell phones. Additionally, unwanted EMI from thermal management solutions (fans and heat sinks) can interfere with low-level signals received at a base station. At the data center level, the thermal workload placed on rack-mount servers cannot be overlooked. However, as these units are constantly monitored and are not subject to the same constraints as base stations, more creative active thermal management solutions are feasible in this environment. Figure 7.1 shows a summary of the potential thermal management strategies that can be implemented at the data center, base station, and handset/IoT levels (Peterson 2020).

7.1.2 *5G Mobile Device Level Thermal Management*

Handsets carry such small form factors that active management strategies (cooling fans or liquid cooling) are simply not possible. The same applies to the majority of IoT devices (wearables, smart home systems, etc.), although this category is so broad that it encompasses a huge range of materials and component products. In

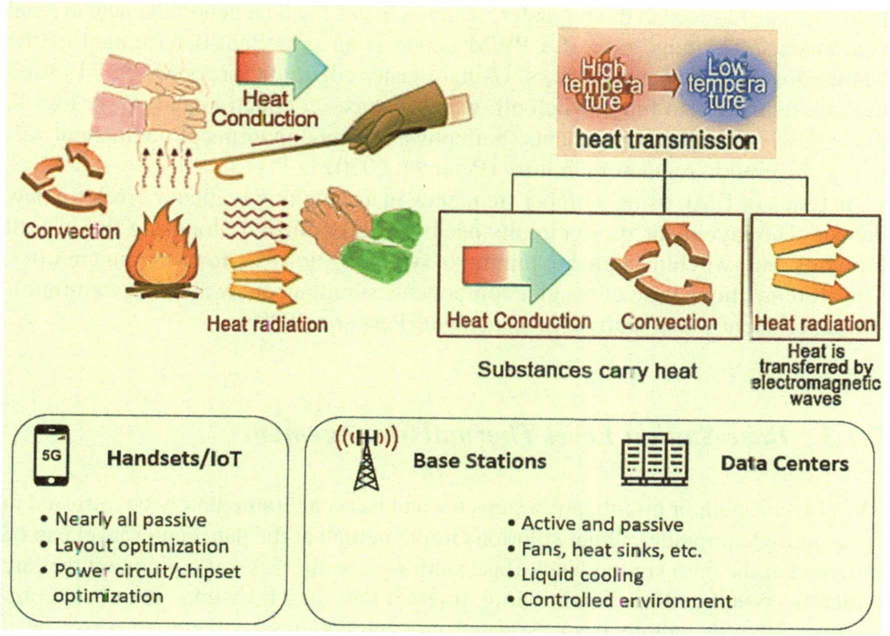

Fig. 7.1 Potential thermal management strategies that can be implemented at various levels in 5G products. (Adapted from Peterson (2020), credit: Northwest Engineering Solutions LLC)

these products, implementing passive thermal management techniques at the board level alongside circuit optimization is key to preventing overheating of critical components. In IoT devices that are large enough to accommodate bulky heat sinks or large cooling fans, combined passive and active cooling measures can bring temperatures low enough to allow persistent 5G operation. Emerging thermal interface materials (TIMs) with very high thermal conductivity can help remove heat from critical components and dissipate heat to the device enclosure. In addition, innovative high-frequency laminates and unique substrate materials may offer promise in equalizing temperatures throughout rigid boards, although this solution is infeasible in boards for handsets as they move to an all-flex architecture. In mmWave 5G handsets, the solution involves continuous optimization of RF chipsets and power electronics, as well as the use of innovative TIMs, heat spreaders (such as CVD diamond) and heat pipes or vapor chambers for heat dissipation. Bulky, flat copper heat sinks, heat pipes and vapor chambers are already present on the enclosure of some smartphones, but more needs to be done to prevent the aforementioned failures in elevated temperature environments (Peterson 2020).

One place to start is in optimizing power regulators for various functional blocks in a handset. In switching regulator circuitry, using smaller discrete components is desirable, as this increases the roll-off frequency at the output. By running power regulators at a higher PWM switching frequency with faster edge rates, thermal losses can be reduced in the regulator, which reduces the heat generated near critical components. The edge rate of a PWM signal is an important determinant of the heating losses in power regulators. Using a faster edge rate forces MOSFETs used in these regulators to fully switch off, which reduces conduction and power losses, thus allowing smaller components (both physically and in terms of component values) to be used in regulator circuitry (Peterson 2020).

In terms of EMI, using a higher frequency in a switching regulator creates a new danger. The layout for these circuits needs to be optimized alongside the overall topology and switching signal to minimize switching noise in downstream circuitry. The aforementioned use of smaller components simultaneously aids layout minimization, efficiency, and thermal management (Peterson 2020).

7.1.3 Base Station Level Thermal Management

The power regulator optimization steps for handsets can immediately be mirrored in base station equipment. Other solutions implemented at the data center level can be mirrored at the base station level. Base stations contain racks of equipment that are typically cooled with air conditioning, internal fans, and heatsinks on critical components. Liquid cooling is investigated and implemented to remove waste heat, effectively copying solutions being brought into data centers. This solution allows equipment to be made smaller and lighter by removing bulky (and electrically noisy) cooling fans. Advanced thermoelectric assemblies are also available for outdoor deployment, including at base stations. These assemblies are an ideal replacement for compressor-based air conditioning systems, as they can force cool air directly into rack-mounted base station equipment. For example, the following different applications will require enhanced thermal management for a structured 5G and beyond network (dmiyares 2017; Peterson 2020):

(a) Central Offices: The future of the traditional telecommunications central office will likely change due to the upgrade in efficiency that the 5G network will present. The central office will look and operate more like a data center than a traditional telecommunications central office, which will present unique thermal management challenges for this part of the 5G network. A few major considerations for equipment installed in these offices are uptime, power usage efficiency, equipment compatibility, and cooling. Today, cooling is usually achieved at the server level with heatsinks, fans or recirculating water chillers and at the rack and room level with very sophisticated climate control systems, but the new framework offers the opportunity for emerging solutions to break through.
(b) Cell Tower Cabinets: Cell tower cabinets have sophisticated electronics, which often require active cooling and climate control in extreme temperature environments because they are located outdoors. Similar to central offices, cooling is traditionally achieved with compressor-based systems, which have proven

inefficiencies in regard to cooling. Because of these inefficiencies, thermal management systems that offer a reduced form factor and are able to fit in space-constrained implementations will prove to be beneficial moving forward. Unique approaches such as geothermal or evaporative cooling may be able to provide ultraefficient cooling for this use case.

(c) Small Cells: Small cells have been added to network infrastructure to increase data capacity, speed, and overall network efficiency. This technology is typically used in very densely populated areas where many people are using data at a given point in time. Traditionally, many cellular equipment sites consist of large cabinets and use noisy fans to provide cooling. When deciding to utilize small cells in densely populated neighborhoods, it will be important to implement a cooling solution that can offer adequate efficiency without the noise and size of legacy systems.

(d) Access Points: 5G networks will also allow WiFi access points to more transparently tie into the network. Depending on how they are deployed, this presents the possibility for temperature-sensitive electronics to be used in extreme temperature environments, where TECs could act as a heating and cooling solution specifically for that temperature-sensitive component. Using this approach allows users to avoid bulky heatsinks or electric heaters while packing more power and performance into devices.

(e) High thermal conductivity PCBs and circuitry substrates: Within base station equipment, thermal demands at the board level motivate the use of high thermal conductivity substrate materials in conjunction with high frequency laminates. For example, composite, ceramic, or metal-core PCB substrates provide thermal conductivities reaching up to a factor two hundred times larger than that of standard FR4 cores and laminates. Ceramic substrates also do not suffer from fiber weave effects, which create some interesting signal integrity and EMI problems at mmWave frequencies.

7.1.4 Emerging Thermal Management Challenges and Strategies

There is still plenty of innovation to come in 5G and beyond, which will bring about new solutions and challenges for thermal management. Innovations in beamforming, hybrid digital/analog base station design, the possibility of radio-over-fiber, and more advanced materials for components are just a few areas that require rethinking thermal management strategies.

(a) From Si to wide-bandgap (WBG) semiconductors

Si power amplifiers used for RF transmission and signal processing at mmWave frequencies are notoriously inefficient, reaching only up to 30%. These components may have been appropriate for earlier incarnations of smartphones, but newer devices in handsets and base stations require a more power-efficient solution. III-V semiconductors are implemented in preparation to satisfy greater demand for sub6 GHz and mmWave 5G components. Power amplifiers and other RF components built from GaAs, GaN on Si, and GaN on SiC can see efficiencies well above 60%. Although these materials are in production and can be purchased at volume, the costs involved are still higher than the cost of equivalent all-Si components. In the mmWave regime, GaN on SiC is the best choice for mmWave components, including power amplifiers, ADCs, and analog front-end arrays, thanks to its broader bandwidth. GaN is already ideal for passively cooled base station electronics and other applications, and this material will inevitably find its way into mobile handsets. Another candidate III-V material that is receiving attention is InGaAs. These WBG-based devices can provide the high performance required in 5G power applications. However, their extremely high power densities should be properly managed, making innovative thermal management techniques a crucial aspect to consider.

(b) Radio-over-fiber

A new paradigm for delivering broadband wireless access services at microwave/mmWave frequencies is radio-over-fiber (ROF). In ROF, a radio wave is upconverted from a baseband signal and used to modulate an infrared light source (laser diode). This modulated signal is then transmitted over a fiber link. This optical communication method enables multigigabit transmission to base stations over long distances with SMF or MMF optical fibers with DWDM. This methodology provides a higher bandwidth and can be used to replace multiple lossy coaxial cables. Despite the promise of this method, conversion between digital and RF analog pulses and back can create a new source of heat that needs to be removed from sensitive optical components. The methods for base station cooling mentioned above are applicable to radio-over-fibers, although the appropriate choice of FET laser diode drivers and other optical components is also critical. Laser diode driver circuitry, PIN detectors, the lasers themselves, and other optical components are very sensitive to changes in temperature. Power output and sensitivity generally drop as temperature increases. FET drivers for long-range links should, at minimum, include a grounded heat sink that is attached to a highly thermally conductive TIM. In many cases, thermoelectric coolers (TECs) have been used to provide active cooling and precise temperature control in the very confined space of the laser diode package. By using thermoelectric devices, organizations can ensure that the laser emits the same wavelength and data rate, so the data are transmitted faithfully to the end point. In a 5G world, the industry will enable a higher data rate and lower latency fiber optic networks. However, it is anticipated that lower power consumption and even smaller form factors will be required for new TECs used in these applications to meet operating requirements (dmiyares 2017; Peterson 2020).

(c) Software solutions

New design software solutions are already allowing innovators to experiment with novel 5G-capable systems and deploy them quickly by taking a fully modular approach. Design tools, such as ARM's DesignStart and Geppetto from Gumstix, allow innovators to develop new systems-on-chip (SoC) and single-board computers for IoT products, respectively. A modular approach is a great design strategy for electronics engineers who may not have design expertise in a particular area, such as single-board computer design or SoC design. These tools will allow designers to move their processing workloads away from data centers and perform them at the edge, which relieves the corresponding processing workload at base stations. These tools also enable new embedded products for applications in machine learning/AI, robotics, smart infrastructure, medical, automotive, and much more—all of which can be made 5G-capable as needed. Different thermal management strategies can be implemented in products created with these design tools, allowing new designers to quickly begin innovating novel 5G-capable products and solutions (Peterson 2020).

7.2 Thermal Management Materials and Components for 5G-Enabled Mobile Devices

The rise of 5G-enabled smartphones makes thermal management more challenging than ever, resulting in significant design alterations and the adoption of advanced materials. Thermal management materials are primarily those that enable the heat to be dissipated strategically and evenly (mitigating hot spots). These can broadly be split into heat spreaders (such as graphitic sheets, heat pipes and vapor cambers), mainly for x-y or in-plane dissipation, thermal interface materials (TIM), mainly for z- or through-plane dissipation, and thermal insulation materials to prevent surface hot spots with minimal RF interference. These materials and components are essential and work in the tangent direction for various application areas. The primary locations for these materials and components in a smartphone are as follows: heat spreader at the back of the phone, heat spreader for the central motherboard, TIM for the lithium-ion battery, TIM for motherboard (e.g., PMICs), heat spreader behind the display, TIM for the antenna, and thermal insulation for 5G mmWave antennas to fit on top of 5G mmWave antenna modules to prevent surface hot spots with minimal RF interference. All of these will have their own requirements and considerations. The key property is the thermal conductivity in either the x-y or z-plane, but it is essential to not overlook many other factors, including adhesiveness, viscosity, coefficient of thermal expansion (CTE), bond line thickness, reworkability, and longevity. At the device level, the main changes to thermal management materials will be around the motherboard and the antenna. This is somewhat obvious for the motherboard, with significantly increased download rates enabled by 5G, and the key antenna changes will be for mmWave applications. There will be the same trends in batteries, displays, and performance for smartphones. One option is to use a vapor chamber or heat pipe instead of a graphitic heat spreader for the motherboard. There are undoubtedly performance advantages but also many tradeoffs (Collins 2020).

7.2.1 Thermal Management Design and Fundamental Solutions for Smartphones

As the performance of smartphones increases rapidly, the power required by major components has increased. To maximize the usage time in the power demand scenario, the number or size of components has been reduced by merging parts to secure maximum battery space within a limited smartphone size. Additionally, a dual stack and triple stack of PCB is introduced to increase the mounting area while reducing the area occupied in the smartphone. This complicated thermal coupling between components on multiple stacks of PCB intensified the hot spot problem. For 5G communication, a 5G dedicated antenna module is additionally needed. Radio frequency (RF) parts related to the 5G modem and additional power-related parts were generated due to increased power consumption. In particular, power parts have to be used with many passive devices, so power-related areas have become an important factor in the PCB configuration. Major parts that consume considerable power include AP, memory, modem, camera, and 5G antenna. Power IC (PMIC: Power management integrated circuit) and major parts are located as close as possible. In most cases, power IC and AP are placed between the PCB top and bottom (Cho et al. 2021). These chips become the major heat source of a smartphone. When the 5G base stations communicate with smartphones at the peak rate, the chip in smartphones must confront dramatic computations that can generate a large amount of heat and raise the temperature of the chip, as shown in Fig. 7.2 (Yang et al. 2019). Therefore, the trend toward smaller form factors combined with higher performance means that overheating, such as hot spots, becomes more prevalent and problematic to resolve.

Smartphone thermal management approaches typically include both software- and hardware-based solutions. On the software side, static or adaptive algorithms are used to control power dissipation by changing both the processor clock frequency and voltage, thereby throttling performance as needed. An adaptive system

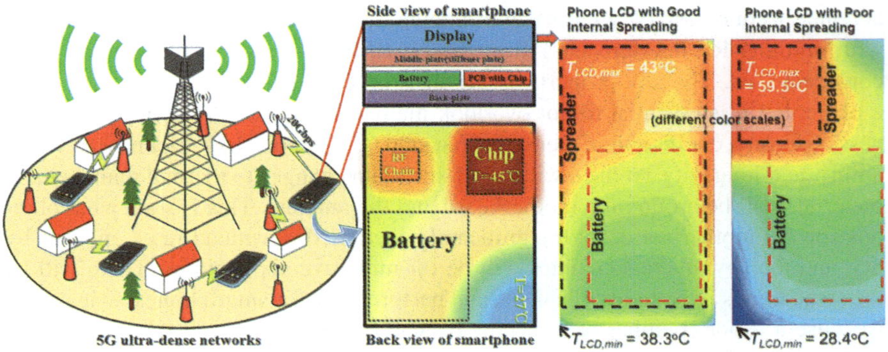

Fig. 7.2 Structure and temperature distribution of smartphones. (Adapted from Chiriac et al. (2016) and Yang et al. (2019) with permission from IEEE)

tries to predict if more power will be needed and adjusts the load accordingly. On the hardware side, mobile phone thermal design requires efficient heat spreading within a highly constrained thickness to mitigate hot spots and ensure surface temperature uniformity on the outer skin of the device. Internal air gaps can be used strategically to limit heat flow to the outer surface and thereby avoid surges in skin temperatures. The thermal inertia of the battery and frame are also used as a buffer (Garimella et al. 2017).

7.2.1.1 Thermal Management Design Guideline

For 5G smartphones, it is becoming more important to balance computing performance and power consumption. Key factors for innovative, active thermal solutions include heat radiating capacity, noiseless and vibration-free operation, scalability to smaller form factors, reliability, and low cost. For wearables, low-power operation is crucial both from the perspective of battery life and ergonomic constraints related to skin temperature. The constraint of greatest interest for smartphone thermal design is not the junction temperature but the skin temperature at the outer surface of the device. Designs target a maximum skin temperature of 40–45 °C and the highest possible temperature uniformity (i.e., an average-to-maximum temperature ratio above 0.8). Portable electronics pose particular challenges for thermal design, given the extremely limited physical space available for inserting thermal solutions and the limits on allowable surface temperature. This currently limits power dissipation in smartphones to 2–3 W. Emerging devices such as smart watches amplify the challenge due to their continuous contact with the human body (Garimella et al. 2017).

In the design of thermal management controls for mobile electronics, complete system models are used to correctly represent the thermal behavior from the transistor level (microsecond response times) to the device level (response times up to minutes). Different devices have different ergonomic limits and usage times, from seconds for phones to minutes for tablets. These thermally aware systems contain several sensors for predictive temperature control, e.g., by throttling performance. Internal heat conduction and capacitance are optimized using TIM layers, but the layout remains a challenge for architectures with memory on top of a CPU, resulting in a restricted thermal pathway to the outside. In mobile smartphones, a system-on-a-chip (SoC) package contains a processor, memory, mixed-signal and radio-frequency functions in a single package. Some challenges include the doubled heat flux in two-die stacks and the need to satisfy different thermal constraints in a single package by stacking memory on top of logic processors. Thermal design for portable electronics requires approaches that are quite different from the traditional paradigms in large-scale electronics cooling. Here, the focus is on customer-centered performance, and thermal targets are driven by ergonomic considerations such as limits on skin temperature. However, with increasing performance and decreasing form factors, this end of the design spectrum also relies on concurrent interaction between electrical and thermal engineering teams. Electrothermal codesign has emerged as a common theme for next-generation systems and devices, whether at large or small scales (Garimella et al. 2017).

Additionally, PCB design also plays an important role in effective heat mitigation. Carefully planned PCBs, configured to optimize thermal management, can lower internal temperatures with little or no cost or performance overhead. Adding extra layers of solid ground or power planes to a PCB connected directly to heat sources with multiple vias (typically hollow and cylindrical copper metal transfers sandwiched between layers) is a technique that is frequently used to increase the effective surface area for heat dissipation.

7.2.1.2 Fundamental Thermal Management Solutions

Various cooling methods have been developed that are generally categorized into passive, active and hybrid cooling. Passive cooling includes techniques that use a heat pipe, vapor chamber, thermoelectric cooler, heat spreader, heat sink, and thermal-interface materials. Active cooling includes forced air and forced liquid cooling techniques. Hybrid cooling is the combination of passive and active cooling. However, the small form factor and light weight requirements of smartphones prevent the adoption of active cooling, and smartphones thus mostly rely on passive cooling. In addition to these cooling techniques, thermal management also uses software-based power control usually at an electronic component level. Modern application processors (APs) support DVFS and CPU hot-plug to manage power consumption and protect devices from overheating. A thermal protection mechanism suppresses temperature increase by reducing power consumption by lowering the operating voltage/frequency of a CPU or turning off CPU cores. For example, thermal management in Android smartphones uses Linux kernel's thermal engine to monitor CPU temperatures from on-die sensors and controls the temperature by mainly using two algorithms: dynamic control and threshold control. The dynamic control algorithm checks whether the core temperatures exceed the threshold values and determines whether to throttle clock frequency or turn off cores. The threshold values are manufacturer-dependent and set by the threshold control algorithm. Android also implements an additional thermal protection mechanism by which a device is shut down when the battery temperature exceeds the threshold (e.g., 68 °C) (Kang et al. 2019).

7.2.1.2.1 Heat Conduction and Spreading

(a) Utilization of high thermal conductivity substrate materials

For high heat flux applications (e.g., high-end servers, transportation, and defense), near-junction high thermal conductivity materials and interfacing techniques have been developed to facilitate targeted heat conduction and spreading. For heat spreading applications, diamond has an exceptionally high conductivity. Carbon-based materials are being considered for the production of solutions with a high conductivity and low density. However, graphene and carbon nanotubes only feature a high conductivity in one or two directions, whereas diamond offers high 3D conductivity (Garimella et al. 2017).

The needs for high thermal conductivity materials are mainly addressed in two areas with extreme limits for electronics cooling: (i) 3D integrated packages with stacked memory and logic chips for high-performance computing (HPC) and (ii) high-electron-mobility wide bandgap transistor electronics (e.g., using GaN, GaAs, AlGaAs, InGaAs) for radar and 5G telecommunication applications. These wide bandgap electronics face extremely high (>10 kW/cm^2) local fluxes or > 1 kW/cm^2 averaged over the chip. Techniques are being developed for the co-fabrication of GaN with a 30 nm-thick intermediate amorphous layer of SiN or SiC bonded onto a diamond substrate. There is an evolution toward substrates of lower thermal resistance, using different material combinations to different extents (e.g., GaN on silicon, GaN on SiC, GaN on diamond) (Cho et al. 2013; Garimella et al. 2017).

(b) Heat spreading and storage in mobile devices

Mobile phone thermal management approaches typically include both software- and hardware-based solutions. On the software side, static or adaptive algorithms are used to control power dissipation by changing both the processor clock frequency and voltage, thereby throttling performance as needed. An adaptive system tries to predict if more power will be needed and adjusts the load accordingly. On the hardware side, efficient heat spreading is required within a highly constrained thickness to mitigate hot spots and ensure surface temperature uniformity on the outer skin of the mobile device. Internal air gaps can be used strategically to limit heat flow to the outer surface and thereby avoid surges in skin temperatures. The thermal inertia of the battery and frame are also used as buffers.

Compact thermal models can be used for design and operational control, but the models must account for the thermal coupling between the battery and processor. Power consumption in a mobile phone processor comprises a static component (due to leakage and standby current) and a dynamic component (due to capacitance current). To handle the dynamic response, energy storage is typically used in the form of thermal mass of the battery and mechanical components, but phase-change materials (PCMs) have also been proposed for use. However, PCMs typically have low thermal diffusivity, making it difficult to remove heat at high fluxes. Graphite sheets used for heat spreading are only 10–100 μm thick; replacement of their volume with PCMs (in small quantities) could only be part of the solution because it would only take a few seconds to heat up and melt (Garimella et al. 2017).

Thermal management solutions for mobile phones have evolved in steps (Garimella et al. 2017): when the Apple iPhone was introduced in the mid-2000s, performance throttling and a TIM on a metal surface were sufficient. Since then, a thermal management unit with multiple sensors and graphite sheet heat spreaders has become necessary. Now, these thermal management units have become more advanced, with multiple embedded sensors. While it is unlikely that active liquid cooling will appear in mobile phones, heat pipes are now being adopted to provide more efficient spreading (e.g., SonyXperiaZ5, Samsung Galaxy S7, and others). These solutions typically place a heat pipe near the edge of the phone, likely due to severe thickness limitations in the middle of the frame (less than ~0.4 mm); much thinner vapor chamber technologies are required to place these solutions closer to the heat source. While vapor chambers may offer a viable solution for passive spreading within mobile devices, conventional design approaches and performance metrics have focused on increasing the maximum achievable heat load and power density. The requirements for mobile thermal management are starkly different and call for heat spreading efficiency to meet ergonomic constraints at ultrathin form factors. The key heat transfer mechanisms governing the performance of vapor chambers with shrinking thickness have been identified to delineate the performance thresholds within which the effectiveness of a vapor chamber, as a function of geometry and heat input, is greater than that of a comparable solid heat spreader. For ultrathin vapor chambers operating at a low power, the heat spreading resistance is dictated by the vapor-phase behavior; design approaches and working fluid selection criteria must be revised with these considerations.

For even more powerful and effective cooling, it is possible to dissipate heat through a graphene heat spreader to instantly transform heat into an active cold plate via thermoelectric cooling. Thermoelectric coolers (TECs) are thin and compact devices typically inserted between a heat source and a heat sink to enable rapid heat dissipation. When voltage is applied to a TEC, a temperature differential is formed between each side of the device, allowing heat transfer to take place via conduction.

7.2.1.2.2 Convective Air Cooling

Many smaller platforms currently rely on radiation and natural convection as the ultimate heat sink, yet the higher-end devices may soon resort to forced air convection. Fan-based air cooling has proven very resilient in the face of growing thermal demands. Advances in alternative air-moving techniques as described below may gain importance for the following situations: (i) to enhance natural convection without resorting to rotary fans for small-scale platforms or (ii) to supplement fans in the cooling of server components with moderate heat fluxes, e.g., alongside liquid cooled CPUs in hybrid-cooled servers (Garimella et al. 2017).

(a) Piezoelectric fans have proven to be a viable cooling solution in controlled environments. Their main advantages are low cost, low power consumption, low noise and good reliability. However, in harsh environments, dust fouling and accumulation on the blade changes the resonance frequency of the oscillating fan. In this case, vibration frequency tracking can provide a partial solution, changing the actuation frequency based on a measurement of the free resonance frequency. Good thermal performance is observed in the moderate heat flux range of 0.1–1 W/cm^2, making these most suitable for low-power platforms (below 40 W) and thereby bridging the overlap zone (15–40 W) between natural and forced convection heat sinks (Johnson and Go 2015; Garimella et al. 2017).

(b) Synthetic jets could also provide solutions in the moderate heat flux range between natural and forced convection, although with properly integrated design of the actuator and ducting, these devices can achieve significant local cooling comparable to steady impinging jets in the heat flux range 0.2–5 W/cm^2. Since both synthetic jets and piezo fans are essentially agitators, the overall air path should be carefully designed to avoid recirculating hot air. A synthetic jet can be a quieter than an axial fan for the same or better thermal performance. Furthermore, a pair of phase-controlled adjacent jets can be used to control the angle of the jet flow, thus providing opportunities for active cooling control. Synthetic jets (and piezo fans) can be used in combination with low-speed fans to decouple local heat transfer from thermal advection due to bulk air flow. With the introduction of compact liquid cooled heat sinks on server CPUs instead of high pressure drop air cooled heatsinks, the overall pumping power budget may shift dramatically to enable synthetic jets and piezo fans to obtain a foothold in server cooling. In data centers, this could reduce maintenance costs and increase energy efficiency (Fanning et al. 2015).

(c) Other alternative air movers have also been studied, from bladeless fans to avoid blade passage-related tonal noise to electroaerodynamics for small platforms (< 3 mm) for which bearings are unsuitable. The latter could use low-voltage piezoelectric transformers to generate a corona discharge in air (Johnson and Go 2015).

Other opportunities to enhance air cooling include (i) advanced flow simulation and optimization tools and (ii) combining numerical shape optimization techniques with additive manufacturing of validation prototypes such as selective laser sintering of aluminum alloys (e.g., AlSi12) (Dede et al. 2015).

With high-end portable electronics such as tablet computers and 5G smartphones pushing the limits of natural convection and radiation as ultimate heat spreaders and heat sinks, some of these novel air-moving techniques could find acceptance in a range of emerging platforms.

7.2.1.2.3 Convective Liquid Cooling

Since liquids (and water in particular) have a much greater volumetric heat capacity than air, component-level liquid cooling remains an active research area, focusing on targeted liquid chip cooling and heat transfer enhancement. In single-phase cooling systems, an increase in the heat transfer coefficients would require a significant reduction in the hydraulic diameter of the cooling channels. The use of such small structures would drive up the cost significantly and increase the risk of clogging. Two-phase operation can provide a solution without the need for such small-scale flow channels, although the overall system cost could be higher because of the added complexity of a two-phase flow loop (Garimella et al. 2017).

7.2.2 *Material Selection for Heat Spreaders and Heat Sinks*

Air cooling usually dissipates heat through heat spreaders and heat sinks (Fig. 7.3) and works by making the object to be cooled by a relatively larger surface area, an increased flow of air over its surface, or both. For high heat flux, a push-pull air flow arrangement with air moving devices at both the bottom and top of the PCBs is used to provide high pressure drop capability. Low-power electronic systems are conveniently cooled by natural convection and radiation. When natural convection is not adequate, forced convection is adopted by a fan or blower to blow the air through the enclosure that houses the electronic components.

A heat spreader is a primary heat exchanger most often in the form of simply a highly thermal conductive plate, which moves heat between a heat source and a secondary heat exchanger such as a heat sink (Fig. 7.3a). The secondary heat exchanger is always larger in cross-sectional area, surface area and volume. The heat flow is the same in both heat exchangers, but the heat flux density is less in the secondary, so it can be made of a less expensive material such as aluminum and is a better match to an air heat exchanger, since the low heat transfer coefficient for air convection is adequate for a low heat flux. A heat spreader is generally used if and only if the heat source tends to have a high heat-flux density, and for whatever reason, heat cannot be conducted away effectively by the secondary heat exchanger. A sufficiently high heat exchanger transfer coefficient is often sufficient to avoid the need for a heat spreader. The ever-increasing processing speeds in high-speed digital systems and high-frequency 5G systems make the presence of heat spreaders much more complex in terms of their impact on the electrical behaviors of the system. For example, mitigation of electromagnetic interference (EMI) by grounding the heat spreader or by other solutions may need to be considered during the design of the heat spreader system. Various heat spreading methods have been widely used in electronic packaging. Heat spreaders are usually used in die level packaging to spread heat from the microprocessor chip into the package. For instance, multichip modules have been built to incorporate a copper heat spreader. On a component or

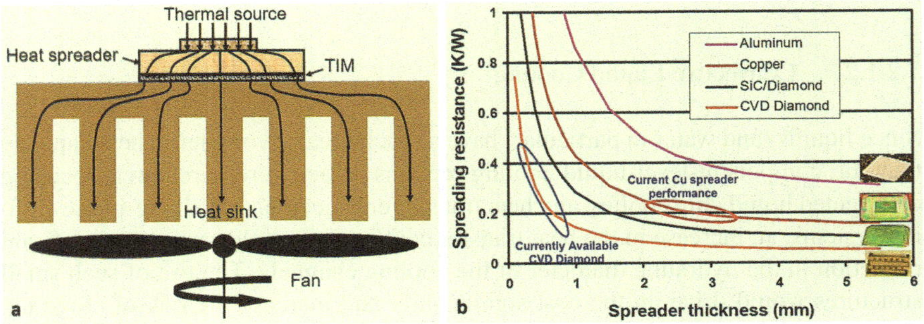

Fig. 7.3 Typical heat spreading and heat sink system (**a**) and spreading model with different materials that can be used to get the same spreading resistance with different spreading thickness (**b**). (Adapted from Tong (2011), credit: Springer Nature)

board level, active spreaders such as heat pipes are used to move heat from chip packages to heat sinks. Heat sinks themselves can incorporate spreaders such as a copper base or heat pipes embedded in an aluminum base. Passive heat spreaders have been incorporated into plasma display panels. Spreaders are commonly used within electronic enclosures (such as smartphones) to move heat from discrete components to the walls of the enclosure. As shown in Table 7.1, materials and process technologies to make heat spreaders have been involved in metallic heat spreaders made of solid copper, aluminum, copper-tungsten or copper-moly with thermal conductivity of less than 400 W/m-K, composite spreaders including ceramic, metallic

Table 7.1 Common materials used for fabricating heat spreaders (Tong 2011)

Materials type	Composition	Thermal conductivity (W/mK)	Coefficient of thermal expansion (ppm/K)	Density (g/cm^3)
Dielectric heat spreading materials (Electrical inslulators)	Alumina (Al_2O_3)	30	7.3	3.9
	AIN	170–200	4.3	3.3
	BeO	250	8–8.5	2.9
	SiC	250	2.4	3.2
	Si	150	2.6	2.3
	CVD diamond	1200–1700 (Z); 1000–1300 (X&Y)	0.8–1.5	3.5
	SiC/Diamond	600	5–10	3.3
Metallic materials (Electrical conductors)	Al	237	24	2.7
	Cu	401	17	8.9
	Metal-diamond composites	~600	3.0–7.5	3.3–5.8
	Copper tungsten			
	90wt%W + 10wt%Cu	200	6.4	17.3
	85wt%W + 15wt%Cu	216	7.2	16.6
	80wt%W + 20wt%Cu	219	16.2	7.6
	75wt%W + 25wt%Cu	228	8.1	15.7
	Molybdenum copper			
	85wt%Mo + 15wt%Cu	155	6.9	10
	80wt%Mo + 20wt%Cu	165	7.5	9.9
	75wt%Mo + 25wt%Cu	175	8.0	9.8
	Al/SiC			
	37vol%Al + 63vol%SiC	180	8.8	3
	45vol%Al + 55vol%SiC	180	10.6	2.96
	63vol%Al + 37vol%SiC	170	11.5	2.9
	Kovar			
	29wt%Ni + 17wt%Co + 53wt%Fe	17.3	5.1	8
Carbon materials (Electrical conductors)	Natural graphite	140–500 (a, b) 3–10 (c)	50(L); −0.5 (T)	1.1–1.7
	Annealed pyrolytic graphite (APG)	1700 (x, y) 8 (Z)	20 (L); −1 (T)	2.24–2.25

and carbon composites (< 1200 W/m-K); and advanced heat spreader technologies including high conductivity plates (500–5000 W/mK), vapor chambers (>5000 W/mK), and oscillating flow heat spreaders (up to 120,000 W/mK). For some applications, such as laser diodes and many semiconductor integrated circuits, which require both electrical isolation and thermal conduction, dielectric heat spreaders made with nonmetallic materials are ideal. For example, CVD diamond is the ultimate heat spreader material for 5G mobile devices. In addition, graphene or few-layer graphene (FLG) has been explored for use in lateral heat spreaders in silicon-on-insulator (SOI)-based chips and 5G electronics (Tong 2011).

Heat sinks can usually be divided into two types: active and passive. Active heat sinks utilize power and are usually a fan type, or liquid cooling or thermoelectric cooling. Passive heat sinks are fully reliable, as they have no mechanical components. Passive heat sinks are usually made of a metal such as an aluminum-finned radiator that dissipates heat through convection. Next-generation heat sinks and leading-edge materials for air-cooled heat removal have been developed, such as sintered heat sinks, metal injection molding, high-performance fans, aluminum extrusions at a 25:1 ratio and self-fanning piezoelectric fin heat sinks. High thermal conductivity polymers, carbon-based composites, sintered metallic powders, phase-change compounds, synthetic diamonds and strand-oriented graphite materials are all among the leading edge materials for heat sinks. The most desirable properties include controlled levels of conductivity, high machinability, low weight, low coefficient of thermal expansion, low toxicity and, above all, the cost will be expected to be lower than that of aluminum. Composite heat sinks are emerging and have received increasing attention since nanotechnology can be applied (Tong 2011).

7.2.3 Flat Plate Heat Pipes and Vapor Chambers for Mobile Electronic Devices

Mobile electronic devices such as smartphones and tablets are trending toward lower thickness and increased functionality, leading to higher heat generation density from active components. Active air cooling or large heat sinks are not suitable to remove the generated thermal energy due to consumer-driven size requirements. Natural convection cooling is a passive cooling strategy that is primarily dependent on the mobile device surface area and the temperature difference between the outer casing and surrounding ambient conditions, the latter of which is predominantly fixed by user comfort. Thus, to optimize device performance and skin temperature, the generated thermal energy from the CPU, battery, and other active components must be spread across the entire device surface, enhancing natural convection cooling. Flat plate heat pipes or ultrathin vapor chambers may offer a viable solution to passively spread the generated thermal energy within mobile devices and increase thermal capacity for transient conditions. A vapor chamber or a flat plate heat pipe is a highly efficient, two-phase cooling solution that passively transports heat from a localized source to a much larger heat rejection surface, as shown in Fig. 7.4. With particular advantages in temperature uniformity and local hot-spot elimination, they enable efficient transport of thermal energy by utilizing the latent heat associated with phase change, with effective thermal conductivities that are an order of magnitude larger than their solid material counterparts (Gibbons et al. 2021).

Heat pipe performance is strongly dependent on the wick's effectiveness to recirculate the working fluid from the condenser to the evaporator. A wick structure's performance is characterized by two critical parameters, permeability and capillarity. Heat pipe wicks are traditionally constructed using woven wire meshes, metal foams, sintered powders, or advanced etching and coating techniques to the inner surfaces of the heat pipe chamber. Wire meshes have been prominently employed as a wicking structure in ultrathin flexible heat pipe applications due to their low cost, ductility, and ease of manufacturing (Lee and Byon 2018).

Three metrics are typically used to characterize heat pipe performance: (i) thermal conductivity, $k = \frac{qL}{\Delta TA}$ [W/(mK)], where q is the applied power, L is the length between temperature measurements, ΔT is the temperature difference across the characteristic length, and A is the cross-sectional area perpendicular to the direction of heat flow. (ii) thermal resistance, R = ΔT/q [K/W], which characterizes the temperature drop across the device for a given applied thermal power. It does not account for the geometry of the device. (iii) dimensionless Coefficient of Thermal

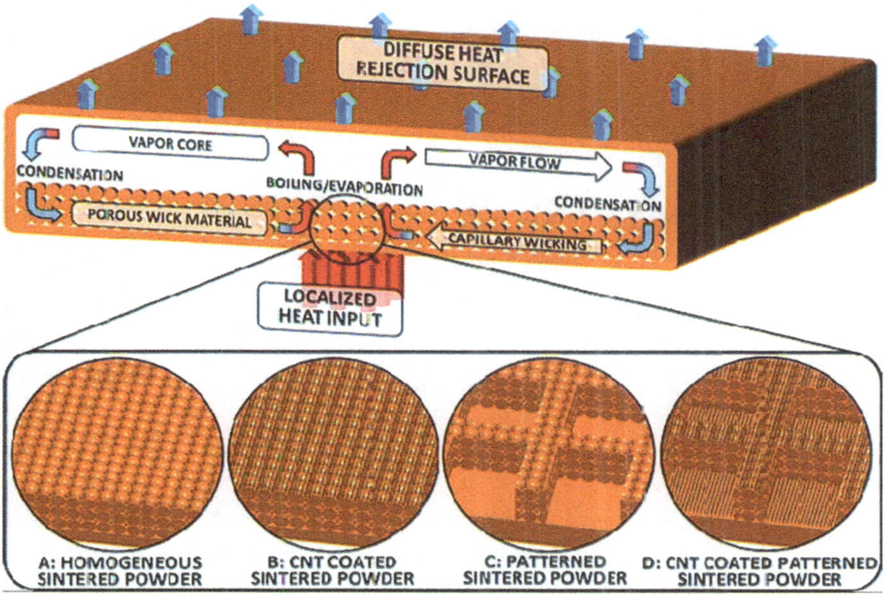

Fig. 7.4 Schematic diagram of internal transport processes for a vapor chamber heat spreader with micropatterned and nanostructured surfaces. (Adapted from Weibel et al. (2012) with permission from IEEE)

Spreading (CTS = $(T_{ave} - T_{\infty})/(T_{max} - T_{\infty})$), where T_{ave} is the average surface temperature, T_{∞} is the ambient far field temperature, and T_{max} is the maximum surface temperature. The CTS defines the effectiveness of heat spreading by accounting for the uniformity of the surface temperature. It can be calculated utilizing IR thermography or multiple local temperature measurements. It is an important metric for mobile electronics, which are typically cooled due to natural convection. A low temperature uniformity indicates poor thermal spreading. Considering the negative impact of a 1 °C temperature difference across a device for screen and battery performance and using a thermal design power (~5 W), thermal management solutions for mobile electronics should target a thermal resistance of <0.2 K/W. Higher power devices require even lower thermal resistances. For context, a basic calculation using the above thermal design power and dimensions of a Galaxy Z Fold 2, assuming the max dimension as the characteristic length, shows that an ~16 mm thick solid copper spreader (k = 390 W/m-K) would be required to maintain a 1 °C temperature difference across the device. This is 2.3 times thicker than the actual Galaxy Z Fold 2. To achieve the desired temperature difference across the device, for a small form factor (thickness, t < 1 mm), a thermal spreader with a thermal conductivity of >6250 W/m-K is needed. Heat pipes are an ideal technology to achieve this (Gibbons et al. 2021).

Figure 7.5 compares the thermal characteristics of different vapor chambers or flat heat pipes. The filled markers denote when the entire top surface is implemented as a condenser. In this configuration, heat is typically applied centrally, and heat transfer paths are facilitated laterally and vertically through the heat pipe. For the empty markers, there exists an adiabatic section where heat is transported in one

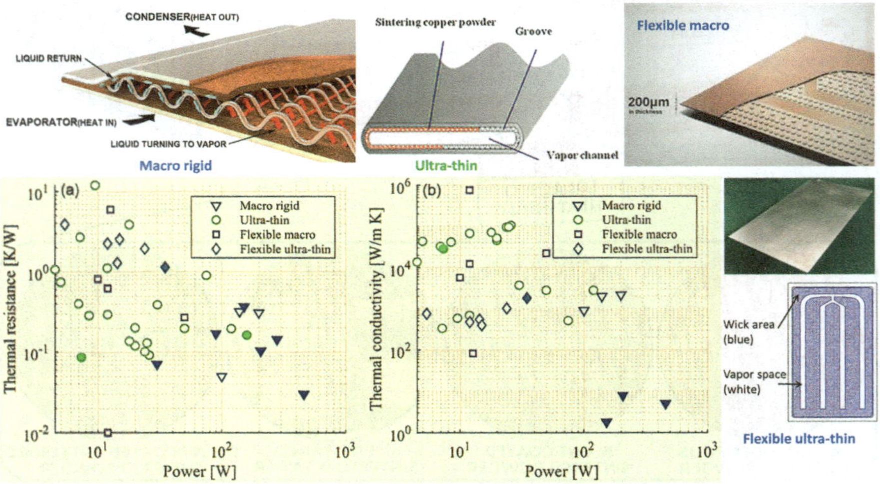

Fig. 7.5 Thermal characteristics of different vapor chambers or flat heat pipes, (Adapted from Gibbons et al. (2021) with open access (Elsevier)): (**a**) thermal resistance and (**b**) thermal conductivity. Filled markers denote when the entire top surface is implemented as a condenser

direction along the heat pipe. Figure 7.5a contrasts the applied thermal power transferred by the heat pipe and heat pipe thermal resistance. Generally, the macroscale rigid and rigid ultrathin heat pipes transfer larger thermal powers for lower thermal resistances (temperature drop between evaporator and condenser). In contrast, the flexible macroscale and ultrathin heat pipes transfer lower thermal powers for higher thermal resistance. Figure 7.5b contrasts the applied power with the heat pipe thermal conductivity. Ultrathin rigid heat pipes exhibit the greatest thermal conductivities. The macroscale rigid heat pipes show low thermal conductivities but the largest thermal power transferred. Both the macroscale and ultrathin flexible cases demonstrate intermediate thermal power transfer and thermal conductivities. The flexible heat pipes' lower thermal performance is sensible, given the performance penalties: (i) the altered vapor space and wicking structure when the heat pipe is deformed, limiting working fluid recirculation; (ii) the typically lower thermal conductivity of the flexible materials employed in the heat pipe casing; and (iii) the difficulty in maintaining high vacuum due to these more permeable casing materials than their rigid counterparts (Gibbons et al. 2021).

Ultrathin heat pipe technology is still underdeveloped. At normal operating temperature, the heat transfer performance of a heat pipe is limited by the wick capillary pressure. The pressure balance inside a heat pipe at equilibrium is given by (Chen et al. 2019):

$$\Delta P_{cap} \geq \Delta P_v + \Delta P_l + \Delta P_g \tag{7.1}$$

where ΔP_{cap} is the capillary force provided by the wick, ΔP_v is the vapor pressure from the evaporation section to the condensation section, ΔP_l is the liquid pressure drop from the condensation section to the evaporation section, and ΔP_g is the pressure drop due to gravity. The following are potentially impactful factors for future research in the field of ultrathin flexible heat pipes:

1. Shell material

The starting point in the development of flexible, ultrathin heat pipes is investigating a suitable shell material. Potentially promising materials are polymer-based casings. The significant advantages of using polymers in the development of an ultrathin flat plate heat pipe are their light weight, flexibility, chemical resistance, electrical insulating characteristics, and compatibility with most electronics fabrication processes. The main challenges and possible solutions to the adoption of polymer heat pipes are as follows (Liu et al. 2019; Weibel et al. 2012):

(a) Polymers generally possess lower thermal conductivities than metals. This can be overcome using thermal vias, high thermal conductivity composite materials, or hybrid structures.
(b) Metal wick structures, such as sintered copper powder and meshes, cannot be sintered together with the polymer casing, increasing their thermal resistance.
(c) Polymer-based heat pipes can yield outgassing problems, as well as the generation of noncondensable gases limiting long-term reliability. Solutions include

incorporating a flexible gas barrier, such as aluminum foil, or a coating, such as indium or copper.

(d) Polymers generally have lower wettability with working fluids. This may be advantageous for promoting dropwise condensation in the condenser section. This can be overcome by utilizing a sufficiently hydrophilic wicking structure or augmented evaporator section.

(e) Polyimide, ultrathin copper, and hybrid materials incorporating nonpermeable layers provide promising flexible casing materials. Integration and joining of the flexible material to a rigid section (with higher wall conductivity) are also worth considering. Challenges here include maintaining minimal thickness while not sacrificing internal vapor core thickness. Close integration with material sciences will be crucial to the development of a future solution.

(f) An evident challenge for flexible heat pipe technology is the disruption of normal vapor flow and fluid recirculation during articulation, which leads to increased liquid-vapor interfacial thermal resistance in the evaporator section. Articulation can lead to the formation of a liquid plug in the internal cavity due to the reduced effective channel area of the wick for fluid flow as the heat pipe curvature increases, resulting in a reduced allowable heat dissipation rate. This can be reduced by implementing a support structure at the articulation joint to prevent vapor space collapse. In tandem, the heat pipe can incorporate design for articulation (a bend line), reduced performance while folded, and then normal performance when open. Full performance might only be needed when the foldable device is open. An updated pressure balance for a flexible heat pipe:

$$\Delta P_{\mathrm{cap}} \geq \Delta P_{\mathrm{v}} + \Delta P_{\mathrm{l}} + \Delta P_{\mathrm{g}} + \Delta P_{\mathrm{bend}} \tag{7.2}$$

2. Porous wick structure

As seen in Eq. 7.2, the capillary wick inside a heat pipe should be able to provide sufficient capillary force as well as space for vapor flow, which is required to achieve the desired heat transfer performance. According to the Young-Laplace formula, ΔP_{cap} can be calculated by:

$$P_{cap} = \frac{2\sigma cos\theta}{r_{eff}} = \frac{2\sigma}{r_{cap}} \tag{7.3}$$

where σ is the surface tension of the working fluid, θ is the working fluid-wick contact angle, r_{eff} is the effective capillary radius of the wick and r_{cap} is the capillary radius. For a mesh screen, the effective capillary radius is equal to the distance between the wires in the mesh ($r_{eff} = (w + d_w)/2$), where w is the spacing between wires and d_w is the wire diameter. For channels between micropillars, the capillary radius is equal to half the channel width. The employed wick should maximize hydrophilicity and permeability; some strategies include the following (Feng et al. 2018; Weibel et al. 2012):

(a) Surface oxidation, chemical corrosion and deposition to form a micro/nano-structured morphology on the surface.
(b) Maximizing the wicking potential using nanoscale wicks, such as carbon nanotubes.
(c) Hybrid structures, such as grooves and fibers, and pillars/grooves and meshes/fibers. Multiscale wicks can meet the requirements for sufficient capillary force and low vapor flow resistance.
(d) Spiral woven meshes consist of braided fibers enabling excellent flexibility and wicking potential, and they are well suited for manufacturing ultrathin heat pipes.
(e) The minimum possible wick thickness should be chosen such that the capillary pressure in the wick is twice the total pressure drop in the wick and the vapor core. The greater the wettability of the wick is, the thinner it can be, maximizing the vapor space. K_l is the permeability of the wick. For the commonly applied screen mesh wick, this can be calculated using (Marcus 1972):

$$K_l = \frac{d_\omega^2 \varepsilon^3}{122(1-\varepsilon)^2} \tag{7.4}$$

$$\varepsilon = 1 - \frac{1.05\pi N d_\omega}{4} \tag{7.5}$$

where N is the mesh number of the screen mesh and ε is the porosity.

For a grooved wick, the permeability is given by (Faghri 1995):

$$K_l = \frac{d_h^2 \varepsilon}{2\left(fRe_{L,h}\right)} \tag{7.6}$$

where d_h is the hydraulic diameter:

$$d_h = \frac{4A_l}{C_l} \tag{7.7}$$

where C_l is the wetted perimeter of the liquid channel.

For a monoporous structure, such as a sintered wick, the permeability can be defined by (Weibel and Garimella 2012):

$$K_l = \frac{d_p^2 \varepsilon}{32} \tag{7.8}$$

where d_p is the pore diameter.

(f) Wick design for nonpulsating flexible heat pipes centered on single and multiple monoporous meshes. Challenges include facilitating a flexible wicking structure that enables a strong thermal connection to the shell material, high thermal conductivity, capillary pressure, and low permeability. Future solutions will incorporate a strong material science and surface wettability collaboration. Potential solutions include copper chemical-plating modified polyacrylonitrile-based carbon fiber wick, flexible graphene coated polyurethane, and laser-induced graphene on polyimide substrate, among others.

3. Vapor-fluid space design

The maximum heat transfer decreases with a decrease in the vapor chamber thickness. The liquid and vapor spaces of the internal cavity require careful design optimization through numerical and analytical models. The vapor-fluid space design should focus on optimizing the vapor and wick domains.

(a) The thermal resistance of vapor chambers becomes dominated by the temperature gradient in the vapor core as the thickness is reduced. An ultrathin vapor core induces a high-pressure gradient and hence a high saturation temperature gradient. Fluid thermophysical properties govern this thermal resistance. Such a high-temperature gradient along the vapor core leads to a high-temperature variation along the condenser surface (Gibbons et al. 2021)
(b) Intelligent geometry configurations should be considered to minimize the pressure drop (radial channels, leaf/blood capillary design). This can be done through analytical and numerical modeling (multiobject genetic algorithms or topology optimization).

The design of the vapor-fluid space is intrinsically linked to the wick design and the outer shell layer thickness. The heat and mass transfer in an ultrathin heat pipe is realized by steam flow and condensate reflux. The pressure drop due to the vapor flow (ΔP_v) can be calculated using the Hagen-Poiseuille equation (Chen et al. 2019):

$$\Delta P_v = \frac{8\mu_v q_{\max} L_{eff}}{\pi \rho_v A_v d_h^2 h_{fg}} \tag{7.9}$$

where μ_v is the dynamic viscosity of the vapor, q_{max} is the maximum power at the capillary limit, and L_{eff} is the effective length for the heat pipe ($L_{eff} = \frac{L_e + L_c}{2} + L_a$).

ρ_v is the vapor density; A_v is the cross-sectional area of the vapor channel; d_h is the hydraulic diameter of the wick ($d_h = 2A_l/C_l$), where A_l is the cross-sectional area of the liquid channel and h_{fg} is the specific enthalpy of vaporization. The permeability of the vapor space (K_v) is defined as (Lee and Byon 2018):

$$K_v = \frac{t_v^2}{12} \tag{7.10}$$

where t_v is the thickness of the vapor space. ΔP_l can be calculated using Darcy's law (Chen et al. 2019):

$$\Delta P_l = \frac{\mu_l q_{\max} L_{eff}}{K \rho_l A_l h_{fg}} \tag{7.11}$$

where ρ_l is the density of the working fluid. According to the Clausius-Clapeyron relation, the vapor pressure in the condenser section (P_{cond}) and the evaporator section (P_{evap}) can be calculated by (Liu et al. 2019):

$$P_{cond} = P_{\infty} e^{\left(\frac{h_{fg}}{R_g} \frac{T_{cond} - T_{\infty}}{T_{cond} T_0}\right)} \tag{7.12}$$

$$P_{evap} = P_{\infty} e^{\left(\frac{h_{fg}}{R_g} \frac{T_{evap} - T_{cond}}{T_{evap} T_{cond}}\right)} \tag{7.13}$$

where P∞ and T∞ are the ambient pressure and temperature, respectively, and R_g is the gas constant. The pressure drop due to gravity is given by (Gibbons et al. 2021):

$$\Delta P_g = \rho_l g L_{eff} \sin \phi_{incl} \tag{7.14}$$

where g is the acceleration constant due to gravity and ϕ_{incl} is the inclination angle of the heat pipe. Combining the above equations, assuming a horizontal heat-pipe orientation, the maximum power at the capillary limit is given by (Chen et al. 2019):

$$q_{\max} = \frac{2\sigma cos\theta}{r_{eff} L_{eff}} \left[\frac{8\mu_v}{\rho_v A_v r_h^2 h_{fg}} + \frac{\mu_l}{K_l \rho_l A_l h_{fg}} \right]^{-1} \tag{7.15}$$

4. Evaporator

Evaporator design should focus on the integration of a high thermal conductivity structure layer. The combination of a rigid, high thermal conductivity material at the evaporator and condenser section with a flexible articulating adiabatic section is a potential direction. A strong thermal connection between the shell material and porous wick structure in the evaporator is essential to minimize thermal resistance. A composite polymer and copper design have been shown using thermal vias to reduce through-wall thermal conductivity. For a sufficiently thin low thermal conductivity casing material, the temperature drop through the casing may be insignificant compared to the temperature drop across the entire system. In the confined geometry of the flat plate heat pipe, no nucleated boiling occurs despite a superheat of 8 K due to the large latent heat and surface tension. During the optimum performance in the nuclear boiling regime, film boiling is subdued due to the mesh structure and reflow of the condensed work fluid, aiding in bubble detachment. The bubble size is limited to that of a single mesh pore and the periodic movement of flat bubbles in the vapor space (Gibbons et al. 2021; Zu et al. 2021).

5. Condenser

Mobile devices are limited by natural convection from the outside shell. To mitigate hot spots on these devices, effective thermal spreading must be achieved. This is tied to the liquid-vapor space ratio.

(a) To improve the condenser uniformity, it has been suggested to (i) increase the thermal conduction resistance in the central region of the condenser-sidewall and wick and (ii) reduce the vapor core thermal resistance. One way to achieve both effects is to eliminate parts of the wicking layer of the central condenser region. This would locally reduce the wick's effective thermal conductivity in this region by replacing the porous sintered copper with a layer of lower-conductivity liquid (water). The effective permeability (of the sintered copper wick and the grooves) would increase, allowing for a thinner wick over the entire condenser side inner wall while maintaining the same pressure drop, hence reducing the vapor-core thermal resistance (Patankar et al. 2016).
(b) The combination of a superhydrophobic condenser and superhydrophilic evaporator can significantly improve the vapor chamber's performance. This superhydrophobic surface has several advantages: (i) it can effectively prevent the generation of a liquid film that may significantly degrade the condensation efficiency; (ii) the vapor can easily condense to a droplet on this surface, which then drops down directly into the wick. This can shorten the water feeding route and prevent dry-out in the center wick (Sun and Qiu 2014).

6. Working fluid

A suitable approximation of the optimum filling ratio was when the working medium was enough to immerse the structure of the capillary mesh. The choice of a working fluid is crucial in the design of such vapor chambers. Given the principle of operation of a vapor chamber (a two-phase thermodynamic cycle), the fluid's thermophysical properties significantly impact its performance. A separate figure of merit for the vapor (M_v) and liquid (M_l) phases has been defined for the best working fluid for a given geometry (Patankar et al. 2016):

$$M_l = \frac{\sigma \rho_l h_{fg}}{\mu_l} \tag{7.16}$$

A higher value for this figure of merit indicates that the vapor chamber can operate at a larger power prior to reaching the capillary limit.

$$M_v = \frac{P_v \rho_v h_{fg}^2}{\mu_v R_g T_v^2} \tag{7.17}$$

where P_v is the vapor pressure and T_v is the vapor temperature. A higher value for this figure of merit corresponds to a lower thermal resistance in the vapor core. To achieve maximum thermal transfer at minimal thicknesses, a working fluid with a higher M_l is required to minimize the required wick thickness, thus maximizing the available vapor space. Design for ultrathin heat pipes should maximize both measures to optimize performance. The unique temperature dependence of thermophysical properties for each fluid governs fluid selection; caution must be exercised to ensure a reasonable vapor pressure at which the vapor chamber's structural integrity is not compromised.

Therefore, the realization of ultrathin flexible heat pipe technology is a complex challenge incorporating evaporation, condensation, two-phase flow, flexible porous wicking structure and flexible impermeable shell material design, all for an ultrathin package size. Prominent challenges to overcome include (i) the development of a shell material with suitable flexibility, thermal conductivity, impermeability and the capacity to thermally bond to the implemented porous wicking structure. (ii) Multiscale flexible porous wicking structure to enable high capillary pressure, permeability and flexion and (iii) optimization of the fluid-vapor space design to allow efficient two-phase flow within the flexible heat pipe. These three components are all intrinsically linked, providing an exciting opportunity for future research. The proposed research is interdisciplinary in nature, incorporating experimental heat transfer, fluid visualization, analytical and numerical modeling, and material sciences (Gibbons et al. 2021).

7.2.4 Thermal Interface Materials

Thermal interface materials (TIMs) play a key role in thermally connecting various components of electronic assemblies in computing, information, communication, energy harvesting, energy storage and lighting technologies. Any engineering surface of the electronic components is rough on a microscopic level. When two such rough surfaces come in contact during electronic assembly, the actual contact occurs only at a few discrete spots, approximately 3% of the apparent contact area (Fig. 7.6). A thermal interface material is commonly in the form of paste, adhesive, solder or a resilient sheet that serves to fill a gap between the two adjoining surfaces. The heat transfer of this complex joint is governed by contact pressure; number, size and shape of contact spots and voids; types of fluid in voids; pressure of fluid in voids; hardness and flatness of contact surfaces; modules of elasticity of contact surfaces; surface cleanliness; and property of the thermal interface material used in the joint. The development of thermal interface materials requires a sophisticated understanding of material and surface sciences, heat transport at submicron scales, and the manufacturing processes used in assembly of microelectronics and other target applications.

Proper selection and application of thermal interface materials are crucial for electronic device efficiency and reliability. Technically selected criteria include (a) low thermal impedance; (b) high thermal conductivity; (c) good conformability with minimum thermal expansion stress when joining two contact surfaces; (d) mechanically sound joint, great adhesion, and high reliability with long service life;

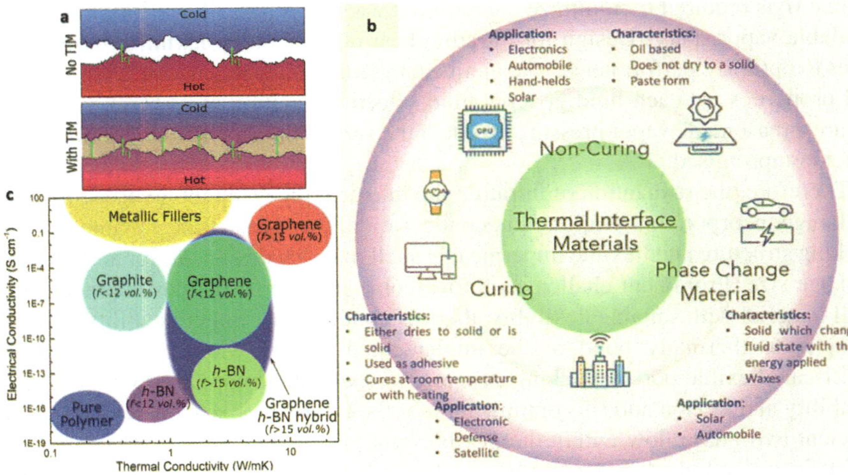

Fig. 7.6 Sketch for the real area of contact of a joint formed by two rough surfaces and the TIM in the real application scenario. (Adapted from Lewis et al. (2021) with permission from IOP Publishing): (**a**) primarily air-gapped exaggerated physical interface without (top) and with (bottom) a TIM; (**b**) uses for different types of TIMs; (**c**) Typical composite properties for different TIMs with un-oriented fillers

(e) lack of sensitivity to moisture and temperature changes; and (e) good manufacturability with ease of handling, application and use. It is often better to invest in the interface material instead of the sophisticated cooling technique. Without good thermal contact between the connecting components, the use of expensive thermally conducting materials for them is a waste.

As shown in Fig. 7.6 and Table 7.2, thermal interface materials can be classified as thermal greases, gels and pastes, adhesives, pads, tapes and films, phase change materials, graphite, solders, liquid metals and compressible metals, all of which have different capabilities (Gordon 2015). However, advances in all these and novel emerging technologies are blurring the distinction, enabling next-generation thermal interface materials to be designed with new combinations of properties expanding the traditional boundaries of performance.

Thermal greases are typically silicones loaded with thermally conductive fillers, which were initially used to fill gaps between power devices and heat sinks. They do not need curing, and they can flow and conform to interfaces. They also offer reworkable thermal interface layers. Manufacturing with greases is messy and imprecise. On extended operation and over time, some greases can degrade, pump out, or dry out, which affects thermal transfer performance. A more reliable grease to use against the pump-out phenomenon is to use a nonsilicon-based and preferably binder liquid that has higher surface tension liquid polymers. Thermal greases are popular in industrial computing and consumer electronics (Tong 2011; Gordon 2015).

Table 7.2 Advantages and disadvantages of conventional TIMs (Bahru et al. 2020)

TIMs	Materials	Advantages	Disadvantages
Carbon structure	CNTs, graphene, graphite	High thermal conductivity (>10 W/mK)	High internal resistance
			Very stable, inert (poor compatibility with other materials)
		Strong mechanical	Fabrication in harsh condition (high temperature, longer reaction time)
		Well absorbed stress	
Greases	Silicon-based matrix loaded with particles, AIN, Ag, ZnO, Silicon oil	High bulk thermal conductivity (3–5 W/mK)	Susceptible to grease pump-out and phase separation
		Thin border line thickness (BLT) with minimal attach pressure	Manufacturing process is complicated and imprecise due to a tendency for migration
		Low viscosity enables matrix material to fill the surface crevices easily	
		Curing is not required	
PCM	Polyolefin, epoxy, low-molecular-weight polyesters, acrylics typically with BN, or Al_2O_3 fillers	High viscosity (stability and less susceptibility to pump out)	Lower thermal conductivity than grease
			Higher thermal resistance than greases; suggest reducing by thermal pre-treatment
		Easy to handle compared with grease	
		Curing is not required	Applied pressure to increase thermal effectiveness for increasing mechanical stresses
		Thermal conductivity (0.5–5 W/mK)	
Gels	Al, Al_2O_3, and Ag particle in silicone, olefin matrices	Good bulk thermal conductivity (3–4 W/mK)	Curing is needed
			Lower thermal conductivity than grease
		Conforms to surface irregularity before curing	

(continued)

Table 7.2 (continued)

TIMs	Materials	Advantages	Disadvantages
			Lower adhesion than adhesive; delamination can be a concern
		No pump-out nor migration concerns	
Adhesive	Ag particles in a cured epoxy	Conforms to surface irregularity before curing	Curing is needed
			Delamination post reliability testing is a concern
		No pump-out	
		No migration	Considering that cured epoxies have high post-cure modulus, the coefficient of thermal expansion (CTE) mismatch-induced stress is a concern.
Metal/ solder	Pure In, In/Ag Sn/Ag/ Cu In/Sn/Bi	High thermal conductivity (30–86 W/mK)	Reflow is needed
			Stress crack
		Easy handling	Delamination
		No pump-out	Voiding possible, not reusable

Thermal gels are low modulus, paste-like materials that are lightly cross-linked. They can be made out of silicone resins, epoxy resin or other chemically cross-linkable resins. Thermal gels perform like grease with respect to their ability to conform to surfaces while displaying reduced material pump-out.

Thermal gap filling pads are gel-like compounds that form sheets of 0.05 mm to 0.5 mm or thicker that can be placed and compressed to shape with the components. They are usually fabricated by molding silicone elastomer reinforced electrically conductive fillers. Reinforcements for thermal pads can include woven glass, metal foils, polymer films, and nanoscale conductive fillers. Thermal pads are typically precut in sizes to accommodate different size components. Therefore, they are cleaner, easier to use and more precise than thermal greases (Tong 2011). These pads fill air voids between mating surfaces when subjected to assembly pressures from clips or screws and exhibit continually improving thermal performance over time as further flowing of the material occurs. Pads of this kind are used in LED lighting and telecommunications equipment. Elastomeric pads will become the most widely used TIM (Gordon 2015).

Phase change materials undergo a transition from a solid to a semisolid phase with the application of heat from the power devices. The semisolid PCM readily conforms to both contact surfaces. This ability to completely fill the interfacial air gaps and surface voids, usually under light clamping pressure, allows performance comparable to thermal grease. While less 'runny' than grease, PCMs contain wax, and once the melt-on temperature is reached, they may flow out of tight areas.

Advanced phase change type materials are not wax based and will not drip. At room temperature, these materials are firm and easy to handle. This allows more control when applying the solid pads to a heat sink surface. After installation, some phase-change pads create a strong adhesive bond between the power device and the heat sink (Tong 2011). Compressible phase change pads that not only phase-change to a liquid form to eliminate trapped air along the interface, it is also compressible and conforms to the slight or large curvature of electronic devices. The performance is similar to that of the best thermal greases and is particularly good for defense and aerospace applications (Gordon 2015).

As one special form of phase change material, polymer solder hybrids provide superior long-term reliability performance and exhibit the lowest thermal impedance of the phase change family. For optimum performance, the material must be exposed to temperatures above the melting point of the contained solder(s) during operation or by a burn-in cycle to achieve the lowest thermal impedance and highest thermal performance. Upon reaching the required burn-in temperature, the material will fully change phase and attain a minimum bond-line thickness of less than 0.001 inches or 0.0254 mm and maximum surface wetting. They are typically used in microprocessors, graphics processors, chipsets, memory modules, power modules, and power semiconductors. Material may flow when oriented vertically, especially at higher temperatures. This does not affect thermal performance but should be considered if appearance is important (Tong 2011).

Thermal adhesives are one- or two-component systems containing conductive fillers. They are typically applied by dispensing or stencil printing. These adhesives are cured to allow for cross-linking of the polymer, which provides adhesive properties. The major advantage of thermal adhesives is that they provide structural support, therefore eliminating the need for mechanical clamping. Most small heat sinks are attached to other components with double-sided pressure-sensitive adhesive thermally conductive tape. Factors for tapes include peel strength, lap- and die-shear strength, holding power, and thermal resistance. Thermally conductive adhesive tapes are considered to be convenient for heat sink attachment with mid-range thermal performance. While they replace mounting hardware, thermal tapes often have problems with the lack of flatness on component surfaces. Plastic ICs are usually concave in the center, and heat sink surfaces vary as well. This can result in air gaps in the interface. One thermal adhesive tape consists of a finely woven nickel-coated copper mesh that conforms closely to irregular mounting surfaces varying up to 50% of its thickness (Tong 2011). It is also common to attach the circuit board to the metal chassis of the product itself with thermally conductive tape. These double-sided adhesive materials combine high structural bond strength and good heat transfer characteristics. Products are available for different thicknesses and surface areas of the mating parts (Gordon 2015).

Graphite is electrically conductive, low cost and has been used as a TIM for a long time. It does not change with time or temperature cycling by releasing volatiles or creeping. Therefore, there is no fogging in LED lighting applications; its high stability is good for chip-testing equipment. It can work at higher temperatures than polymeric TIMs, enabling the performance of higher power density lighting displays. For some applications, a dielectric coating must be used to make it electrically insulating and prevent short circuiting. The orient fibers can be laminated in a horizontal plane, resulting in very different thermal conductivity measurements. For example, pyrolytic graphite sheet (PGS) is a thermal interface material made from a highly oriented graphite polymer film. It is synthetically made and has high thermal conductivity (700–1950 W/mK). It is very thin, making it suitable for providing thermal management in limited spaces, and very lightweight. This material is flexible and can be cut into customizable shapes (Tong 2011).

In addition to these polymeric and graphite TIMs, there are also a wide range of metallic TIMs available. Although the use of lead solders is limited to a few industries, such as defense and aerospace, indium and gallium alloys are increasingly popular. The variety and breadth of thermal materials available today are huge. Metallic thermal interface materials can be made in many different forms and are no longer limited to solder applications. In some applications, metallic TIMs are totally reworkable and recyclable. The need for better performing TIMs in devices such as power amplifiers and IGBT modules has prompted the exploration of other types of metallic TIMs, such as liquid metals, phase change metals, and soft or compressible metal alloys. The soft or compressible metal thermal interface material is the most easily adopted metallic TIM because it does not need to be reflowed or contained in a gasket such as a solder or liquid metal. Metallic TIMs are very thermally conductive, reliable, and in the case of compressible metals, easily adopted. A new hybrid material consists of a thermally conductive silicone film on one side bonded to a copper film. The advantage of this material is that it can be used to manufacture flex circuits as well as provide EMI and RFI noise protection (Tong 2011).

Traditionally, TIMs have been microstructured, but advanced thermal interface materials are increasingly focused on nanostructuring. This can involve using nanomaterials such as carbon nanotubes or graphene, using metallic nanoparticles as fillers for polymeric TIMs. However, the TIM industry is extremely conservative. Change is only considered when the current technology has reached its absolute limit, and the substitute must be mature. Nevertheless, the consumer demand for increasingly smaller and faster electronics will drive new innovations and result in the market for thermal interface materials growing (Gordon 2015).

7.2.5 Thermal Insulation Materials

Thermal insulation materials are specifically designed to reduce or inhibit the heat flow by limiting heat conduction, convection, radiation or all three while performing one or more of the following functions: (1) conserving energy by reducing heat loss or gain; (2) controlling surface temperatures for personnel protection and comfort; (3) facilitating vapor flow and water condensation of a process; (4) increasing the operating efficiency of heating/ventilating/cooling, plumbing, steam, process and power systems found in commercial and industrial installations; (4) assisting electrical or mechanical systems in meeting standard criteria in related products; (5)

energy storage and temperature regulation for energy facilities, textiles/clothing, buildings, vehicles, warships, etc

Thermal insulation materials come in a range of forms; primarily, they are low-density foams or fibrous materials that restrict conductive and convective heat transfer. Depending on the thermal environment, appropriate materials may be polymers, glasses, ceramics or carbons. Conventional organic insulation materials include cork, cotton, felt, carbonated cork, and foam rubber and are suitable for use at temperatures below 150 °C. Asbestos such as asbestos, glass wool, quartz sand, diatomaceous earth, magnesium carbonate powder, magnesia powder, calcium silicate, pearlite, etc., are used, and most of them can withstand high temperature use. These are the limits of softening and decomposition temperature of each material. In addition, a supercooler of approximately −200 °C is alternately covered with aluminum foil and glass wool, plastic is packed, and the inside air is removed. On the other hand, refractory bricks made by bonding refractory materials in a porous form are used in most of the heat insulating materials used at 1000 °C or higher. In this case, in addition to the thermal conductivity, a thermal expansion coefficient (thermal expansion coefficient) and shrinkage ratio are needed. These insulation materials are used in many places, such as the outer wall of the furnace, the reaction tower, the oil storage tank, the steam conduit (conduit), the outer wall of the water pipe, and the outside of the refrigerator.

The most commonly used conventional insulation materials include calcium silicate heat insulators and heat insulation ceramics. Calcium silicate insulation is mainly composed of calcium silicate hydrates such as tobermorite ($5CaO \cdot 6SiO_2 \cdot 5H_2O$) or zeolite ($6CaO \cdot 6SiO_2 \cdot H_2O$) and asbestos, which are generated by hydrothermal reactions of quartzite and silica. They are light and have good thermal insulation, so they are used for building materials. Ceramic insulation is usually used to prevent the conduction of heat, has a high heat resistance property and is not easily deformed. Typically, it is made of diatomaceous earth, pearlite, etc., or a bubble glass heated by mixing a blowing agent with a glass powder. Ceramics are solid materials that have been heat treated at a high temperature after molding inorganic nonmetal raw materials. Among these ceramics, a common feature of ceramics used as insulation is that many small holes are pierced. These bubbles interfere with heat transfer and therefore have the effect of adiabatic effects. Typical ceramic thermal insulation materials include diatomaceous earth and pearlite, which are natural raw materials, and bubble glass, which is heated by mixing a blowing agent with glass powder. Compared to corks and foamed polystyrenes, they have higher heat resistance and are not easily deformed.

A range of advanced insulation materials, including aerogels, have been used in military, automobile, aerospace, and 5G device applications to shield and safeguard electronic components; for protection of control equipment, these materials tend to be bespoke designs. With the introduction of more electronics into demanding environments, such as all-electric vehicles, aircraft and jet engines, there will be requirements for even higher performance thermal insulation and multiple functions, which may require new materials and approaches. For instance, as a space application, the reflective insulation/heat radiator combination is a key approach to ensure that the electronics within satellites are maintained within their operating temperature range. For the Mars Pathfinder Rover, NASA used a solid silica aerogel to insulate electronics and restrict heat loss. Silica aerogel is the lowest density solid material and is a remarkable insulator because silica is a poor conductor and the material structure restricts convective transfer; the typical conductivity is 0.02 W/m-K. Aerogels can be manufactured from a number of materials, and in the early 1990s, carbon aerogels were developed. The major factor preventing the more extensive use of aerogels has been their high cost of production and health issues. However, the take-up of the materials has been significant in thermal insulation applications such as buildings and pipelines, and they have been used for automotive 'under-the-underthe-hood' thermal management. With the ongoing cost reductions and the development of new materials and composites, this class of super insulator should become very attractive for electronics thermal shielding.

Nanomaterials and their coatings may play a role, particularly in hybrid constructions. An oxide nanocomposite material may be applied as a liquid coating, exhibiting a total conductivity of 0.017 W/m-K because the nanosized structure presents tortuous thermal paths and consequently poor thermal transfer. The creation of a high interface density within a nanolaminate material has produced a very effective barrier to heat transfer and a material with a thermal conductivity three times smaller than that of a conventional insulator. The heat flow is limited at the interface, as the differences in lattice vibration between the materials restrict the transfer of energy across the interface (Tong 2011).

7.2.6 Thermal Metamaterials

Thermal metamaterials have amazing properties in heat transfer beyond naturally occurring materials owing to their well-designed artificial structures. The idea of thermal metamaterials has completely subverted the design of thermal functional devices and makes it possible to manipulate heat flow at will, such as thermal cloaks, concentrators, rotators/inverters, and camouflages, as shown in Fig. 7.7. These advances imply new opportunities to guide heat transfer in complex systems and new packaging approaches related to the thermal management of electronics. Such aspects are important, as trends of electronics packaging toward higher power, higher density, and 2.5D/3D integration are making thermal management even more challenging (Narayana and Sato 2012; Han et al. 2014).

While conventional cooling solutions based on large thermal-conductivity materials as well as heat pipes and heat exchangers may dissipate the heat from a source to a sink in a uniform manner, thermal metamaterials could help dissipate the heat in a deterministic manner and avoid thermal crosstalk and local hot spots, and associated packaging designs may benefit from the new capabilities of controlling heat transfer paths, underlying the scope and aim of thermal metamaterial design. Possible thermal metamaterial approaches include (1) anisotropic heat spreaders and diffusers, (2) heat cloaking and isolation, and (3) heat guiding and bending.

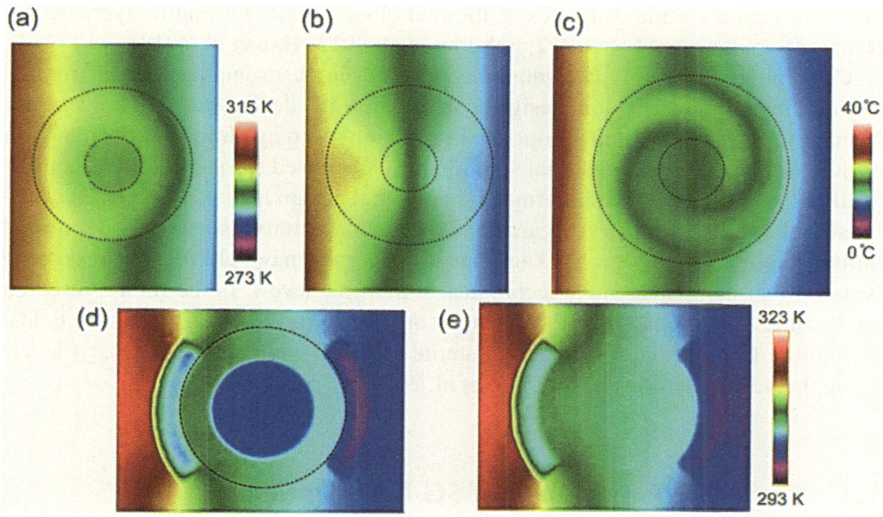

Fig. 7.7 Experimental demonstration of different thermal metamaterial devices. (Adapted from Han et al. (2014) with permission from John Wiley and Sons): (**a**) Cloak. (**b**) Concentrator. (**c**) Rotator/inverter. (**d**) Camouflage: single conducting cylinder appears like two insulating segments. (**e**) Equivalent temperature profile of camouflage illusion, with two insulating segments actually present and no conducting cylinder

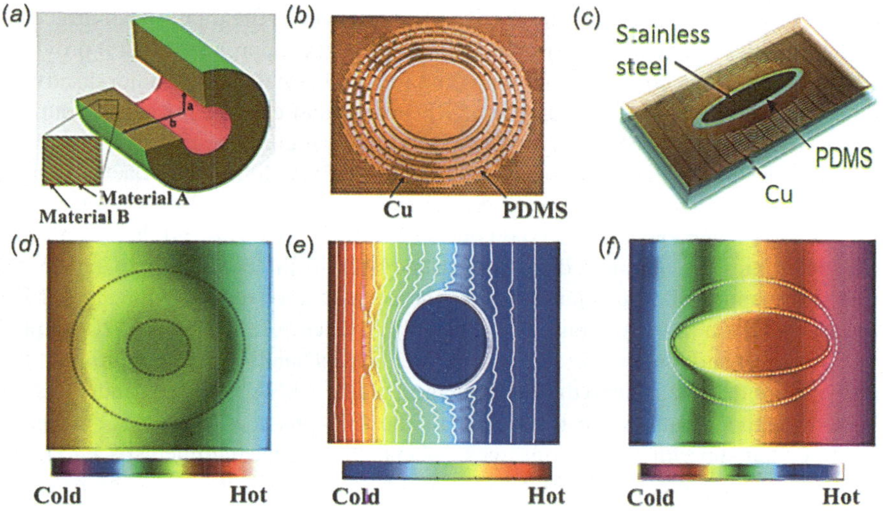

Fig. 7.8 Thermal cloaks made with multilayered composites. (Adapted from Han et al. (2018) with permission from John Wiley and Sons): (**a**) The steady-state thermal cloak composed of latex rubber film (material A) and silicone elastomers (material B) with agar-water as the background material. a = 50.8 cm and b = 52.7 cm. (**b**) The transient thermal cloak composed of Cu and PDMS. (**c**) The omnidirectional elliptical thermal cloak composed of a stainless steel object, a Cu shell, and a PDMS insulating layer. (**d**) The measured temperature profile of (**a**). (**e**) The measured temperature profile of (**b**) at t 5120 s. (**f**) The measured temperature profile of (**c**) at t 55 min

Figure 7.8 shows some examples of thermal cloaks made with multilayered composites (Narayana and Sato 2012; Schittny et al. 2013; Han et al. 2018).

Conventional thermal management approaches in electronic packaging are usually confined to an individual design scale, such as the device, package, or system. Many prior solution techniques operate in isolation and may ultimately conflict with each other. Thermal metamaterial solutions could be used for thermal management at all of these scales. While a thermal metamaterial design may be a good heat transfer solution in itself, further research and development into associated and required multidisciplinary electronic packaging methodologies may hold the key to synergy between thermal management techniques at multiple levels. In the future, research on thermal management may logically combine efforts across technical fields, including ECAD design tools, base material systems, and active devices, by leveraging the metamaterial concept (Kim et al. 2021).

7.3 Thermal Management of 5G Base Station Antenna Arrays

5G base station antenna arrays usually include two types of active electronically scanned arrays (AESA) that are presented, namely, traditional AESA and planar AESA. An example of the traditional structure of AESA is given in Fig. 7.9a. In such constructions, a number of circuit cards are placed orthogonal to the antenna array. Each card feeds a row of antenna elements with individual transmit/receive (T/R) modules mounted on them. The advantage of this approach is that it provides a large surface area for the T/R modules and for the thermal load. A major disadvantage is the need for a large number of RF boards and cabling for signal routing. Moreover, in the case of having large T/R modules, it could be difficult to make a compact design. Figure 7.9b visualizes a planar AESA. In this approach, a single multilayer RF board is used to integrate the antenna elements and RF beamformers. The radiating elements are placed on one side of the board while all the beamforming electronics are mounted directly onto the back of the board. Compared to the traditional approach, this alternative has the advantage of reducing the area of RF boards and the number of connectors and cables. Therefore, planar AESA can provide low-cost and low-profile apertures, which makes handling and mounting of the arrays more flexible. Due to these advantages, planar AESA are good candidates for high-throughput but low-cost antennas in the first phase of digital 5G systems. Despite their superiority in compactness, planar AESA's have limited space to fit all the beamforming electronics. With increased frequency, since the spacing between the radiating elements becomes even smaller, the problem of excessive heat dissipation becomes more problematic. A possible solution might be adding more functionality into a single IC, which is quite challenging. A more straightforward option is to create a sparsity in the radiating elements and thus to increase the spacing between the T/R modules for passive heat removal (Aslan et al. 2018).

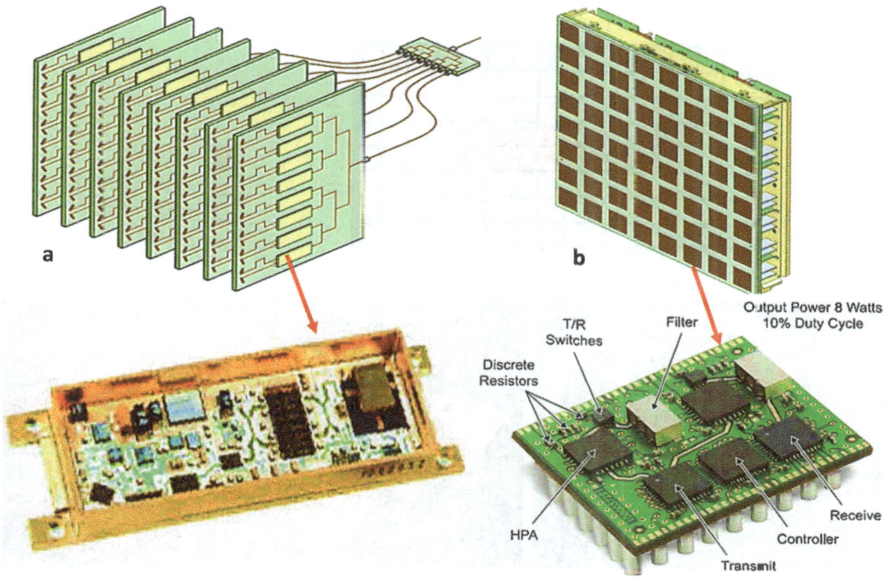

Fig. 7.9 Illustration of AESA construction. (Adapted from Herd and Conway (2016), with permission from IEEE): (**a**) traditional approach and its T/R module, (**b**) planar approach and its T/R module

7.3.1 Cooling in Traditional AESA's

Figure 7.10a shows the temperature distribution of a 16x8 traditional AESA array, assumed around which an ideal heat sink with a constant temperature of 30 °C is maintained. It clearly shows that the heat becomes trapped in the center of the array, resulting in an unacceptable temperature rise. Therefore, for such arrays, it is necessary to transfer the generated heat toward the edges via cold plates or fluid channels. As an example, an X-band airborne phased array antenna is given here in Fig. 7.10b, c, which shows many microwave modules having transmit and receive circuitry with GaAs power amplifiers that are attached to liquid cooled slats (Wilson 2003; McCune 2017).

7.3.2 Cooling in Planar AESA's

Thermal management in planar AESA's is different than the traditional counterparts due to the placement of the beamforming and amplification chips. Since the chips are now located at the opposite site of the substrate compared to the radiating elements, external heat sinks can be attached to the chips without affecting the radiation performance. As seen in Fig. 7.11a, thermal interface materials (TIMs) or some flat plates with large thermal conductivities are used to enhance the thermal

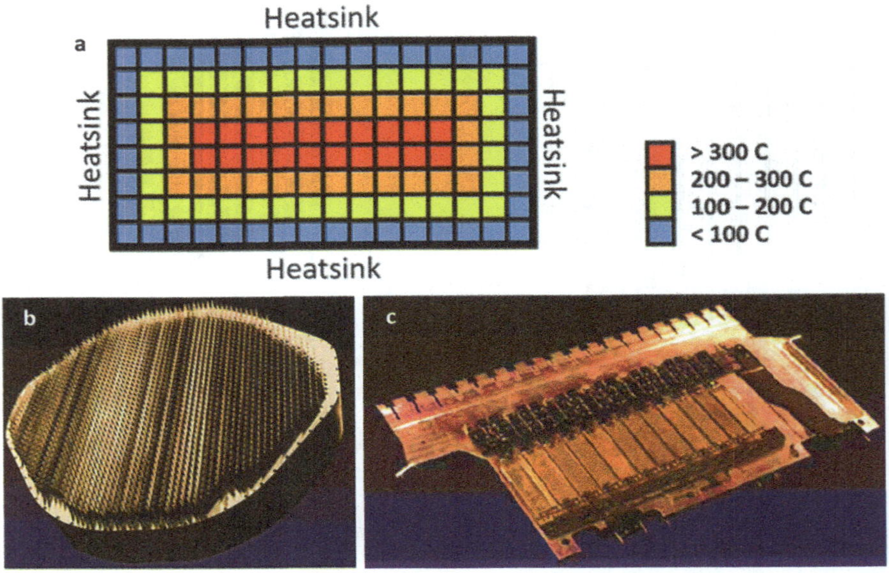

Fig. 7.10 Cooling in traditional AESA's. (Adapted from Wilson (2003), credit: Electronics Cooling; McCune (2017) with permission from IEEE): (**a**) Temperature distribution across a traditional 8 × 16 AESA; (**b**) array aperture; and (**c**) a row of microwave modules on a slat of an airborne phased array radar antenna

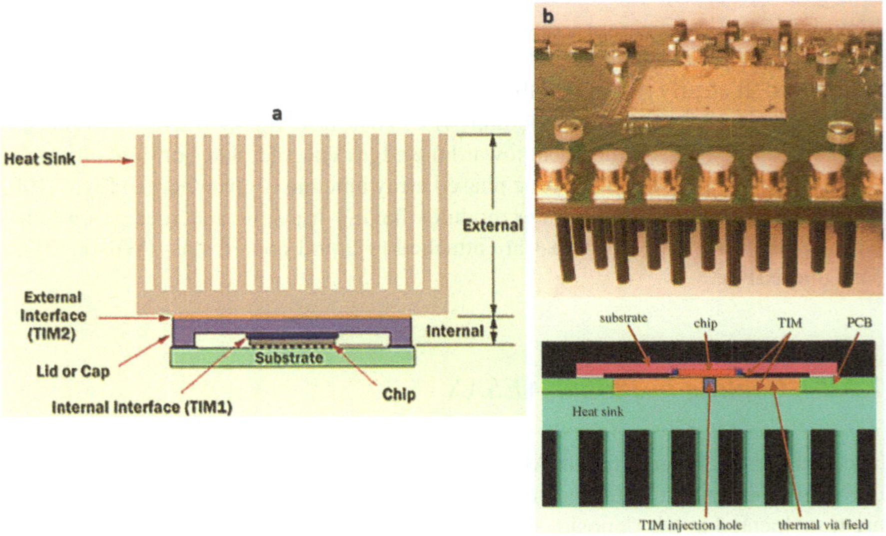

Fig. 7.11 Cooling in planar AESA's. (Adapted from Lasance and Simons (2005), credit: Electronics Cooling; Kam et al. (2011) with permission from IEEE): (**a**) Chip package attached to a heat sink, (**b**) AiP package mounted on evaluation board with heat sink

conduction and heat spreading from the chip to the lid and heat sink. A sample design with a package mounted on an evaluation board with a heat sink is shown in Fig. 7.11b, which presents a low-cost 60 GHz antenna-in-package (AiP) phased array (Lasance and Simons 2005; Kam et al. 2011).

7.3.3 Antenna Array Cooling at Millimeter Waves

Existing cooling strategies for mm-wave antennas are mainly active cooling methods exploiting either forced air or liquid cooling. Despite being effective in removing heat, active cooling systems with fans or pumps require the use of electricity and make the system more complex and hard to maintain. On the other hand, passive thermal management is a relatively cheaper and more energy-efficient solution. However, since the heat is removed only passively via natural convection by utilizing only heat spreaders or heat sinks, it is not easy to achieve thermal performance similar to that of the active counterparts unless there is a sufficient surface area that is in contact with the outside.

To achieve fully passive cooling in 5G base station arrays, CPU coolers (without fans) can potentially be used in planar AESA, for which the heat sink can be attached to the chips at the opposite side of the radiating elements. Although they are bulky and heavy, the heat sink modules for CPU coolers are able to provide competitive results in heat removal when compared to the active systems. A sample CPU cooler is illustrated in Fig. 7.12, which is the CPU processor package of Fujitsu Primepower 2500. Figure 7.13 shows how the heat can be absorbed from the chips by the heat pipes (which are totally passive with low thermal resistance and efficient heat transfer) and transported to the fin stack, which is cooled via natural convection, and a sample CPU cooler fin stack from FSP Windale (Wei 2007; Aslan et al. 2018).

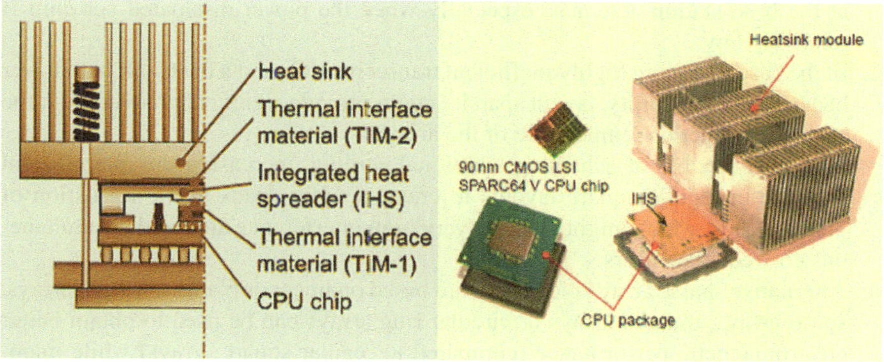

Fig. 7.12 CPU processor package of Fujitsu Primepower 2500. (Adapted from Wei (2007), credit: https://www.fujitsu.com/global/documents/about/resources/publications/fstj/archives/vol43-1/paper14.pdf)

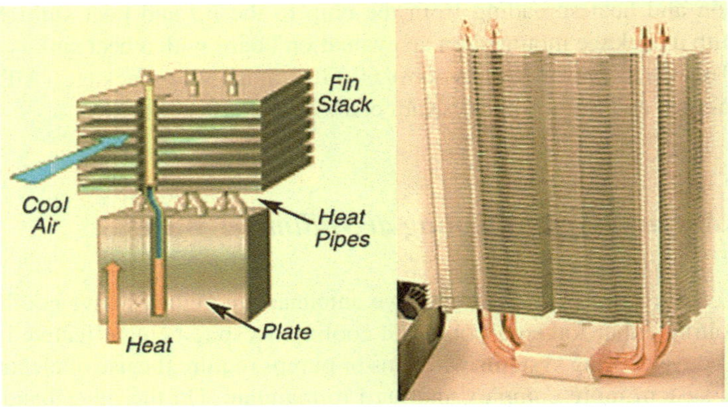

Fig. 7.13 CPU cooling via heat pipes and a remote fin stack. (Adapted from Aslan et al. (2018) with open access (IEEE))

There are also some techniques that focus on cooling on the antenna element level instead of the array (or system) level. One idea is to create a heat spreader connecting the transistor to the radiating patch. Such an antenna is able to provide desirable radiation and thermal performance at the same time. Therefore, it could be a good alternative to exploit heat spreaders between the chips, ground plane and the radiating elements in planar AESA for cooling enhancement. Other techniques concentrate more on finned or fractal element structures for dual (electromagnetic and thermal) functionality. Although their structure can be optimized for a certain desired radiation pattern, they are not suitable for flexible beam forming as required by AESA (Aslan et al. 2018).

In addition, the following approaches may enhance the thermal management performance of the antenna arrays (Aslan et al. 2018):

1. Fanless CPU coolers with relatively large h.t.c.'s can provide sufficient cooling at the base station antennas, especially when the power dissipated per chip is relatively low.
2. In the case of having highly inefficient transceiver ICs and a heat sink with lower heat transfer capability, layout sparsity could provide significant reduction in the maximum junction temperature of the array.
3. Although capable of achieving additional cooling, increasing the interelement distance in regular square layouts to create sparsity leads to the formation of grating lobes, which might result in very high interference among the simultaneous co-frequency users.
4. Alternative sparse array layouts that are based on linear or planar irregular arrays, spiral arrays, thinned arrays or circular ring arrays can be used to obtain better electromagnetic performance (compared to regular square arrays) while maintaining the cooling capacity.
5. For different sparse topologies, the major factor to reduce the temperature is the average distance between the elements. The sunflower topology is the best one

from this point of view since the interelement spacing in this topology can be the largest among the sparsest toplogies while maintaining low sidelobe levels.

Furthermore, creating a conduction path between the chips, ground plane and radiators by exploiting heat spreaders located in the middle of the patches provides additional cooling without affecting the radiation performance. Cooling with spreaders increases with increasing power dissipation per chip and decreasing h.t.c. of the heat sink attached to the chips.

7.4 Thermal Management of 5G Edge Computing

Edge computing, or mobile edge computing (MEC), is an alternative to traditional cloud computing. Rather than handling the bulk of the data processing at a central location, MEC moves that data processing closer to the consumer or to the "edge" of the network. The ultimate goal of this shift is not necessarily to replace cloud computing entirely; more accurately, the goal is to improve the idea of cloud computing by merging its technology with the geographic distribution of processing centers to solve many of traditional cloud computing's limitations. As shown in Fig. 7.14, Edge computing pushes intelligence, processing power and communication capabilities from a network core to the network edge and from an edge gateway or appliance directly into devices. The benefits include improved response times

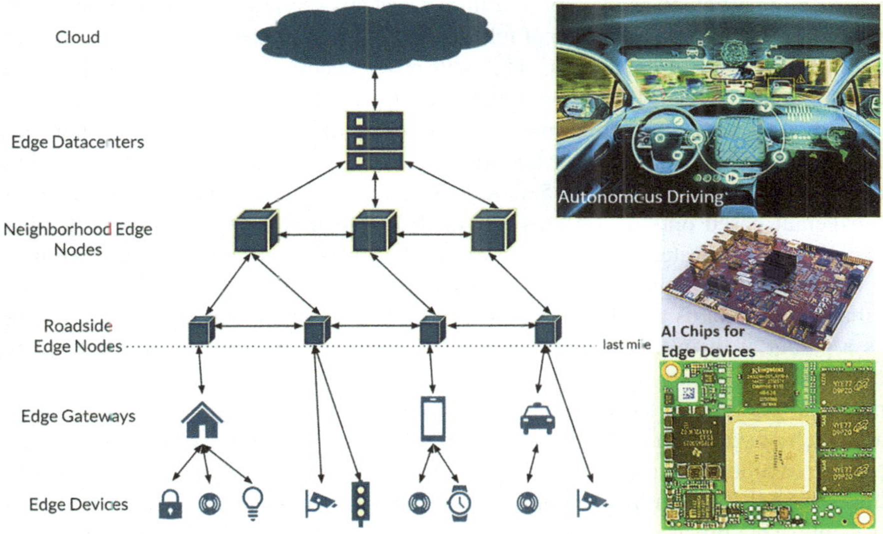

Fig. 7.14 Edge computing architecture scheme with both the computing power and latency decreasing downwards. (Adapted from Scheuermann and Bischoff (2019); credit: https://www.inovex.de/de/blog/edge-computing-introduction/)

and better user experiences. In the edge computing architecture scheme, the layers are included (O'Day and Quesnel 2020):

(a) Cloud: In this layer, compute power and storage are virtually limitless. However, latencies and the cost of data transport to this layer can be very high. In an edge computing application, the cloud can provide long-term storage and manage the immediate lower levels.
(b) Edge Node: These nodes are located before the last mile of the network, also known as downstream. Edge nodes are devices capable of routing network traffic and usually possess high computing power. The devices range from base stations, routers and switches to small-scale data centers.
(c) Edge Gateway: Edge gateways are similar to edge nodes but are less powerful. They can speak the most common protocols and manage computations that do not require specialized hardware, such as GPUs. Devices on this layer are often used to translate for devices on lower layers. Alternatively, they can provide a platform for lower-level devices such as mobile phones, cars, and various sensing systems, including cameras and motion detectors.
(d) Edge Devices: This layer is home to small devices with very limited resources. Examples include single sensors and embedded systems. These devices are usually purpose-built for a single type of computation and are often limited in their communication capabilities. Devices on this layer can include smart watches, traffic lights and environmental sensors.

Edge computing is becoming essential where time-to-result must be minimized, such as in smart cars. Bandwidth costs and latency make crunching data near its source more efficient, especially in complex systems such as smart and autonomous vehicles that generate terabytes of telemetry data. MEC allows for a more efficient and intuitive flow of data traffic. This will prove necessary with the growth of 5G capabilities, as MEC is the key component behind the success of 5G technology.

Although edge computing involves fewer racks, the data need vital cooling protection. For edge computers located in remote locations, the availability of cooling resources may vary. Figure 7.15 shows some air cooling options. Both water- and refrigerant-based options are also provided. Refrigerant cooling provides flexible installation, and water-based cooling brings the advantage of ambient air assistance for free cooling. Naturally, thermal management solutions must meet the power and reliability standards that MEC demands. Passive two-phase cooling is widely seen as a potential fit in this regard. Many MEC hardware designs call for cold plate evaporators, thermosyphons, and heat pipes to dissipate the heat generated by the powerful electronics to more active cooling approaches at the room level. This approach is seen as a cost-effective, reliable, high-performing solution. Similarly, passive two-phase immersion cooling could be the key to unlocking the full potential of the MEC, optimizing its innumerable data centers to be as energy-efficient as possible. This process involves submerging servers in specially designed, nonconductive fluid; as the liquid passes over the electronic components, it removes heat and evaporates, rising through the system before condensing and falling back over the components. This creates a passive, closed-loop system that reduces energy

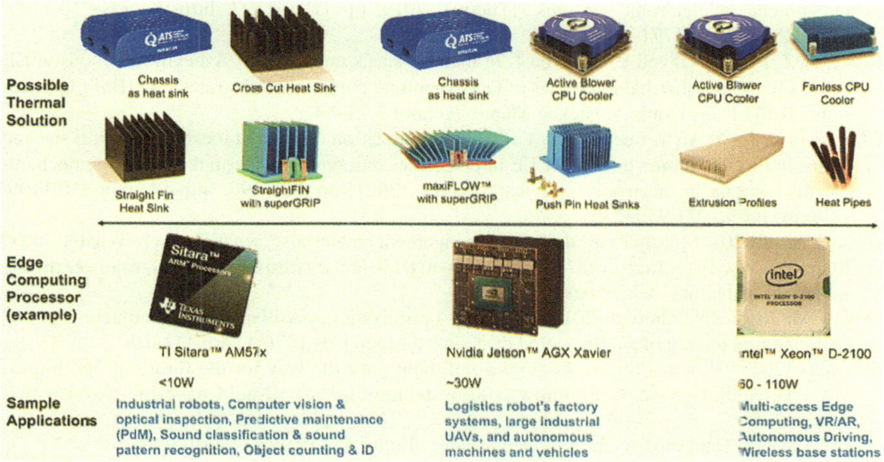

Fig. 7.15 Edge computing Air Cooling Options (Adapted from O'Day and Quesnel (2020); credit: https://www.qats.com/cms/2020/01/14/edge-computing-and-thermal-management/)

costs by an average of 22.8% compared to air-cooled systems. With the emergence of MEC as a more efficient means of data processing and transfer, the cooling solutions that support its related technologies must be equally proficient in reducing energy costs. In comparison to in-row air cooling, passive two-phase cooling for electronics is expected to be a major improvement in terms of system and energy usage efficiencies. In fact, air cooling is gradually switched to cooling to passive two-phase—and immersion—cooling techniques. With the industry building toward a future dependent upon countless edge data centers, these energy savings will add up quickly. Two-phase cooling, then, is the key to making MEC a sustainable option for the future (Amalfi et al. 2021).

References

Amalfi RL, Enright R, Kafantaris V (2021) Emerging 5G networks: potential economic benefits of two-phase thermal management. Electron Cool 2021:19–23

Aslan Y, Puskely J, Janssen JHJ, Geurts M, Roederer A, Yarovoy A (2018) Thermal-aware synthesis of 5G base station antenna arrays: an overview and a sparsity-based approach. IEEE Access 6:58868–58882. https://doi.org/10.1109/ACCESS.2018.2873977

Bahru R, Zamri MFMA, Shamsuddin AH, Shaari N, Mohamed MA (2020) A review of thermal interface material fabrication method toward enhancing heat dissipation. Int J Energy Res 2020:1–21. https://doi.org/10.1002/er.6078

Chen Z, Li Y, Zhou W, Deng L, Yan Y (2019) Design, fabrication and thermal performance of a novel ultrathin vapor chamber for cooling electronic devices. Energy Convers Manag 187:221–231. https://doi.org/10.1016/j.enconman.2019.03.038

Chiriac V, Molloy S, Anderson J, Goodson K (2016) A figure of merit for mobile device thermal management. 15th IEEE intersociety conference on thermal and thermomechanical phenomena in electronic systems (ITherm), 2016, pp 1393–1397. https://doi.org/10.1109/ITHERM.2016.7517712

Cho J, Li Z, Bozorg-Gayeli E, Kodama T, Francis D, Ejeckam F, Faili F, Asheghi M, Goodson KE (2013) Improved thermal interfaces of GaN-diamond composite substrates for HEMT applications. IEEE Trans Compon Packag Manuf Technol 3:79–84

Cho Y, Yoo J, Im Y, An T, Lee H, Heo Y (2021) Optimization of hotspot location on multi stacked board of mobile phone. In: 20th IEEE intersociety conference on thermal and thermomechanical phenomena in electronic systems (iTherm), 2021, pp 484–490. https://doi.org/10.1109/ITherm51669.2021.9503236

Collins R (2020) Smartphone thermal management materials: what's new? What's next? https://www.idtechex.com/en/research-article/smartphone-thermal-management-materials-whats-new-whats-next/22181

Dede EM, Joshi SN, Zhou F (2015) Topology optimization, additive layer manufacturing, and experimental testing of an air-cooled heat sink. J Mech Des 137(11):art. 111403

dmiyares (2017) How cooling technologies will help pave the way for the future of 5G. https://www.5gtechnologyworld.com/how-cooling-technologies-will-help-pave-the-way-for-the-future-of-5g/

Faghri A (1995) Heat pipe science and technology. Taylor and Francis, Washington.

Fanning E, Persoons T, Murray DB (2015) Heat transfer and flow characteristics of a pair of adjacent impinging synthetic jets. Int J Heat Fluid Flow 54:153–166

Feng C, Yugeswaran S, Chandra S (2018) Capillary rise of liquids in thermally sprayed porous copper wicks. Exp Thermal Fluid Sci 98:206–216. https://doi.org/10.1016/j.expthermflusci.2018.05.031

Garimella SV, Persoons T, Weibel JA, Gektin V (2017) Electronics thermal Management in Information and Communications Technologies: challenges and future directions. IEEE Trans Compon Packag Manuf Technol 7(8):1191–1205. https://doi.org/10.1109/TCPMT.2016.2603600

Gibbons MJ, Marengo M, Persoons T (2021) A review of heat pipe technology for foldable electronic devices. Appl Therm Eng 194:117087

Gordon R (2015) Opportunities in the $1.3 billion thermal interface material market. http://www.idtechex.com/research/articles/opportunities-in-the-1-3-billion-thermal-interface-material-market-00008238.asp

Han T, Bai X, Thong JT et al (2014) Full control and manipulation of heat signatures: cloaking, camouflage and thermal metamaterials. Adv Mater 26:1731–1734

Han T, Yang P, Li Y, Lei D, Li B, Hippalgaonkar K, Qiu CW (2018) Full-parameter omnidirectional thermal Metadevices of anisotropic geometry. Adv Mater 30(49):1804019

Herd JS, Conway MD (2016) The evolution to modern phased array architectures. Proc IEEE 104(3):519–529

Johnson MJ, Go DB (2015) Piezoelectric transformers for low-voltage generation of gas discharges and ionic winds in atmospheric air. J Appl Phys 118:art. 243304

Kam DG, Liu D, Natarajan A, Reynolds SK, Floyd BA (2011) Organic packages with embedded phased-array antennas for 60-GHz wireless chipsets. IEEE Trans Compon Packag Manuf Technol 1(11):1806–1814

Kang S, Choi H, Park S, Park C, Lee J, Lee U, Lee S (2019) Fire in your hands: understanding thermal behavior of smartphones. In: The 25th annual international conference on mobile computing and networking. Article No.: 13, pp 1–16. https://doi.org/10.1145/3300061.3300128

Kim JC, Ren Z, Yuksel A, Dede EM, Bandaru PR, Oh D, Lee J (2021) Recent advances in thermal metamaterials and their future applications for electronics packaging. ASME J Electron Packag 143(1):010801. https://doi.org/10.1115/1.4047414

Lasance CJM, Simons RE (2005) Advances in high-performance cooling for electronics. Electron Cool 11:22–39

Lee D, Byon C (2018) Fabrication and characterization of pure-metal-based submillimeter-thick flexible flat heat pipe with innovative wick structures. Int J Heat Mass Transf 122:306–314. https://doi.org/10.1016/j.ijheatmasstransfer.2018.01.135

Lewis JS, Perrier T, Barani Z, Kargar F, Balandin AA (2021) Thermal interface materials with graphene fillers: review of the state of the art and outlook for future applications. Nanotechnology 32:142003

Liu C, Li Q, Fan D (2019) Fabrication and performance evaluation of flexible flat heat pipes for the thermal control of deployable structure. Int J Heat Mass Transf 144:118661. https://doi.org/10.1016/j.ijheatmasstransfer.2019.118661

Marcus BD (1972) Theory and design of variable conductance heat pipes. https://ntrs.nasa.gov/api/citations/19720016303/downloads/19720016303.pdf

McCune E (2017) Energy efficiency maxima for wireless communications: 5G, IoT, and massive MIMO. In 2017 IEEE custom integrated circuits conference (CICC) 2017, pp 1–8. https://doi.org/10.1109/CICC.2017.7993614

Narayana S, Sato Y (2012) Heat flux manipulation with engineered thermal materials. Phys Rev Lett 2012(108):214303

O'Day R, Quesnel N (2020) Edge computing and thermal management. https://www.qats.com/cms/2020/01/14/edge-computing-and-thermal-management/

Patankar G, Weibel JA, Garimella SV (2016) Patterning the condenser-side wick in ultrathin vapor chamber heat spreaders to improve skin temperature uniformity of mobile devices. Int J Heat Mass Transf 101:927–936. https://doi.org/10.1016/j.ijheatmasstransfer.2016.05.093

Peterson ZM (2020) Thermal management challenges in the 5G Era. https://www.nwengineering-llc.com/article/thermal-management-challenges-in-the-5g-era.php

Scheuermann JM, Bischoff M (2019) The edge is near: an introduction to edge computing! https://www.inovex.de/de/blog/edge-computing-introduction/

Schittny R, Kadic M, Guenneau S, Wegener M (2013) Experiments on transformation thermodynamics: molding the flow of heat. Phys Rev Lett 110(19):195901

Sun Z, Qiu H (2014) An asymmetrical vapor chamber with multiscale micro/nanostructured surfaces. Int Commun Heat Mass Transf 58:40–44 https://doi.org/10.1016/j.icheatmasstransfer.2014.08.027

Tong XC (2011) Advanced materials for thermal management of electronic packaging. Springer, New York

Wei J (2007) Thermal management of Fujitsu's high-performance servers. Fujitsu Sci Tech J 43(1):122–129

Weibel JA, Garimella SV (2012) Visualization of vapor formation regimes during capillary-fed boiling in sintered-powder heat pipe wicks. Int J Heat Mass Transf 55:3498–3510. https://doi.org/10.1016/j.ijheatmasstransfer.2012.03.021

Weibel JA, Kousalya AS, Fisher TS, Garimella SV (2012) Characterization and nanostructured enhancement of boiling incipience in capillary-fed, ultrathin sintered powder wicks. In The 13th intersociety Conference on Thermal and Thermomechanical Phenomenon in Electronic System. San Diego, CA, 2012, pp 119-129

Wilson J (2003) Challenges in thermal control of military electronics systems. Electron Cool 9:18–24

Yang J, Ge X, Zhong Y (2019) How much of wireless rates can smartphones support in 5g networks? IEEE Netw 33(3):122–129. https://doi.org/10.1109/MNET.2018.1800025

Zu S, Liao X, Huang Z, Li D, Jian Q (2021) Visualization study on boiling heat transfer of ultrathin flat heat pipe with single layer wire mesh wick. Int J Heat Mass Transf 173:121239. https://doi.org/10.1016/j.ijheatmasstransfer.2021.121239

Chapter 8
Protective Packaging and Sealing Materials for 5G Mobile Devices

Abstract Apart from semiconductor, conductor, dielectric, magnetic and other functional materials, 5G electronics require protection against RF/EM interference, the environment, in-operation stress, moisture, and corrosion. Therefore, protective materials that seal the devices and secure the enclosures are important members of the 5G material family. Taking the shell/case material as an example, to minimize the effect on avoiding radio/electromagnetic wave communication, the mobile phone case needs to be changed from metal to ceramic/glass/polymer. Compared with metal materials, ceramic/glass/polymer allows better RF/EM penetration and can improve the wireless charging efficiency. The material candidates for mobile phone front covers should be lightweight and have high transparency, an antiglare property, strong fingerprint resistance and scratch resistance. For the rear cover, wear resistance and efficient heat dissipation are usually considered. Tempered glass, ceramic or polymer composites (e.g., polycarbonate (PC)/poly(methyl methacrylate) (PMMA)) can be good competitors in the mobile phone shell market. For the PC/PMMA composite, PMMA ensures the scratch resistance of the product, while PC with sufficient toughness can ensure the overall impact strength. The middle frame of a 5G mobile phone is still mainly based on metal, and aluminum alloy/stainless steel with high strength is the future development trend. This chapter will provide a brief review of the current status and future trends of protective packaging and sealing materials for 5G mobile devices.

8.1 Design of 5G Millimeter Wave Compatible Covers for High-End Mobile Devices

5G connectivity will make use of millimeter wave technology (at frequencies above 20 GHz) to allocate more channels and alleviate the ever-increasing demand of wireless capacity. The smaller wavelengths at mm-wave frequencies result in smaller antennas, arrays of which can be installed in compact end user devices such as mobile phones. Beam steering of such small arrays is a key enabler of the 5G

C. Tong, *Advanced Materials and Components for 5G and Beyond*, Springer Series in Materials Science 327, https://doi.org/10.1007/978-3-031-17207-6_8

vision, since it allows the device to select the best direction to establish communication with another device or base station while improving the link budget by making a more focal usage of the radiated power. The mm-wave antenna module can be manufactured in a package that integrates both the small antenna array and the radio frequency front end. This antenna-in-package design (AiP) approach mitigates the high losses in the transmission lines at high frequencies. AiP modules can be produced at scale and can in principle be installed in any device, including mobile phones. In the millimeter frequency range, the enclosure found in most consumer electronic devices is too thick to be ignored. Figure 8.1 shows the transmission characteristic of a dielectric slab as a function of its thickness in wavelengths. Considering the back of a modern mobile phone to be made of glass with a permittivity of 6.84, as an example, at 28 GHz, the optimum thickness for maximum transmission – approximately half a wavelength, depending on the scanning angle – would be 2.05 mm. If the total thickness of the device is approximately 8.0 mm, this optimum design thickness would certainly be too bulky to be of practical use. Therefore, optimal design is crucial for the cover of a mobile device (Enjiu and Rütschlin 2019).

8.1.1 Dielectric Cover Design

The dielectric cover can be divided into two categories: thin radomes and thick radomes. Thin radomes are characterized by a dielectric slab not thicker than 10% of the wavelength of interest. For the glass backed phone, being thin would require

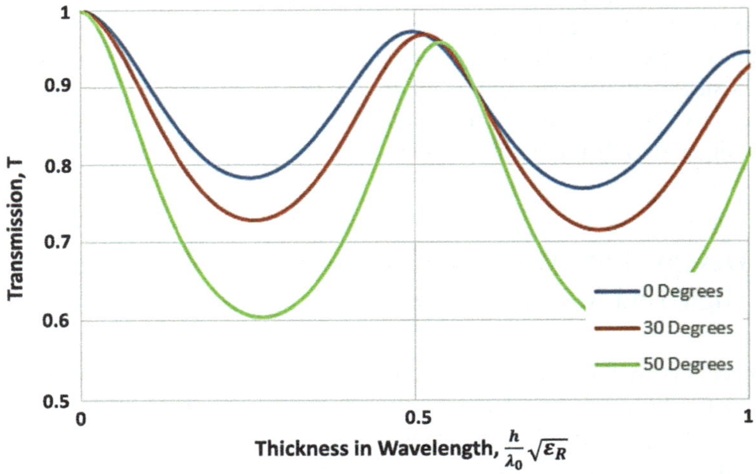

Fig. 8.1 Transmission characteristic of a dielectric slab with thickness, h, and permittivity ε_R, for different angles of incidence, as a function of the slab thickness normalized by the wavelength inside the material. (Adapted from Enjiu and Rütschlin 2019, credit: Dassault Systèmes/the 3DEXPERIENCE Company)

the radome to have a maximum thickness of 0.65 mm. A thin cover may lack the mechanical toughness to be used in the phone. Thick radomes, on the other hand, are designed in which the dielectric slab has multiples of half of the wavelength of interest. Therefore, the optimum thickness h of the thick radomes can be determined by (Volakis 2007):

$$h = \frac{n\lambda_0}{2}\frac{1}{\sqrt{\varepsilon_R - \sin^2\theta_i}}; n = 1, 2, 3, \ldots \tag{8.1}$$

where n is the order of the radome and θ_i is the angle of incidence. In general, first-order thick radomes are preferred. Equation 8.1 can also be rewritten to find the optimum ε_R values if the thickness h is given (Enjiu and Rütschlin 2019):

$$\varepsilon_R = \frac{\lambda_0^2}{4h^2} + \sin^2\theta_i \tag{8.2}$$

Therefore, for the cover material with a constant permittivity, the higher the permittivity is, the smaller the optimum thickness dependency to the incident angle; for the cover with a constant thickness, the lower the design thickness is, the higher the required permittivity. To better visualize these results, Fig. 8.2 shows the thickness

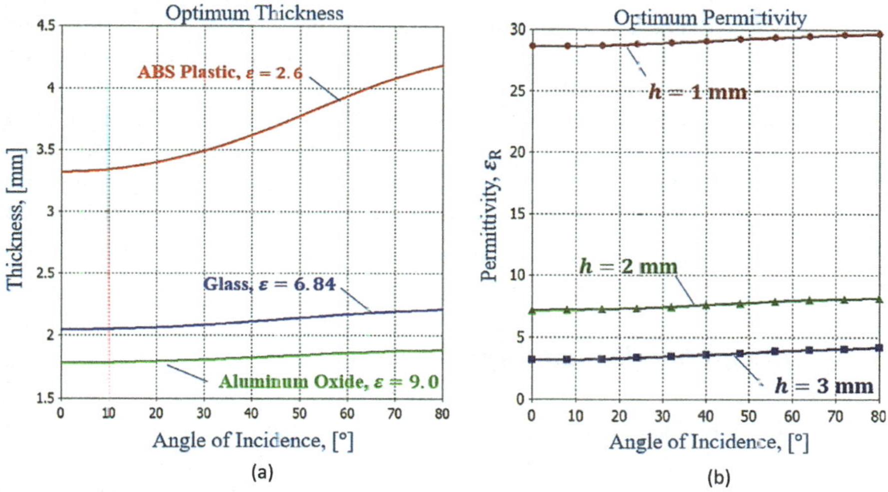

Fig. 8.2 The selection of thickness and permittivity for optimum transmission properties given three different materials and three different design thickness. (Adapted from Enjiu and Rütschlin 2019, credit: Dassault Systèmes/the 3DEXPERIENCE Company): (**a**) Optimum transmission thickness as a function of the incident angle for ABS plastic (red curve), glass (blue curve) and aluminum oxide ceramic (green curve). (**b**) Optimum transmission material permittivity as a function of the incident angle for a material with 1 mm (dark red), 2 mm (dark green) and 3 mm (dark blue) of thickness

and permittivity for optimum transmission properties given three different materials and three different design thicknesses, respectively (Enjiu and Rütschlin 2019).

8.1.2 Metallic Cover Design with Inserted Dielectric Slots

Ceramic monolithic radome designs might present a possible solution, but mechanical requirements may hinder their application. In older generation high-end mobile phones, metal back covers were widely used. Wireless communication was enabled by inserting dielectric slots to act as electromagnetic windows in the enclosure. The slots could also separate different regions of the metallic housing, allowing the enclosure to be used as an integrant part of the antenna design. At mm-wave frequencies, the electromagnetic window can be realized by a frequency selective surface (FSS). The FSS is a 2D periodic structure that can be designed to be transparent in a given frequency band and frequency range. As shown in Fig. 8.3a, b, the FSS-loaded back cover is designed for an enclosure of 1 mm thickness. The FSS layer has a thickness of 0.1 mm and is sandwiched between two dielectric layers with a relative permittivity of 9. The same material composes the nonconductive areas of the FSS layer. The chosen geometry for the FSS is a circular loop design, since it offers a more compact design and stability over frequency and angle of incidence. For a ring dielectric width of 0.2 mm, the goal function was as follows: return <-10 dB, 5G band – 26.5 GHz $< f <$ 29.5 GHz, $0° < \theta < 30°$. The geometry dimension had a ring diameter of 1.18 mm and interelement spacing of 1.29 mm. The performance results for the return loss for both polarizations are shown in Fig. 8.3 c, d (Enjiu and Rütschlin 2019).

As a result, the metal cover loaded with an FSS is a promising solution, at least in terms of electromagnetic performance, since antenna matching and beam steering requirements can be met even for small gaps between the enclosure and the antenna module. This performance comes at the cost of design and manufacturing complexity.

8.1.3 Integration Design Consideration

5G networks provide reliable high-speed data links between base stations and mobile devices. Base stations featuring antenna arrays with a large number of elements operating at high millimeter wave frequencies (28 GHz, for example) use beamforming to allow efficient, targeted communication with mobile phone handsets. In the handsets themselves, the use of several small chip-integrated arrays in each device becomes feasible due to the small physical size of antennas at these high frequencies. However, spatial and material constraints make the integration of such antenna arrays into mobile phone handsets challenging. Chip-based arrays may be accommodated below the back cover of the device, which, for high-end

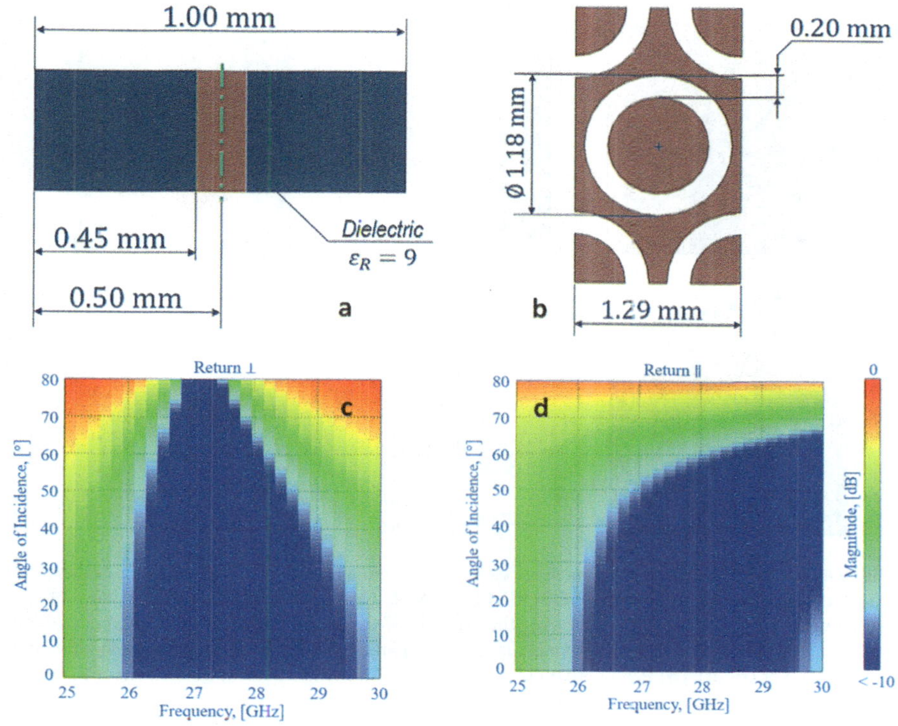

Fig. 8.3 The inserting dielectric slot design example. (Adapted from Enjiu and Rütschlin 2019, credit: Dassault Systèmes/the 3DEXPERIENCE Company): (**a**) View of the layers of the back cover; (**b**) view of the unit cell of the FSS design; (**c**) and (**d**) The reflection coefficient of FSS unit cell analysis. In both graphs the vertical axis represents the angle of incidence and the horizontal axis represents the frequency, while the magnitude of the return loss is depicted by the color ramp. The (**c**) map shows the result for the perpendicular polarization, and the (**d**) map shows the result for the parallel polarization

phones, may be made of metal or glass. A metal cover would act as a very effective shield, preventing communication entirely. Glass may allow electromagnetic energy to propagate through it, but its electrical thickness at high mm-wave frequencies may substantially influence the array performance. Therefore, it is crucial to design the back cover of a mobile phone when integrating a chip-based antenna array with adequate scanning behavior across the frequency of interest, providing the efficient high data-rate connections that users require without sacrificing aesthetics and the tactile experience while handling the device.

For example, the AiP used consists of a 2 × 2 array of stacked patch antennas designed to operate in the 26.5–29.5 GHz frequency range and to steer the beam to +/− 30 degrees in both axes with dual polarization. It is square with a length of 10.71 mm along one edge. The module is installed in the upper right corner quadrant of the back of the phone, as shown in Fig. 8.4a, with a gap of 1 mm between the

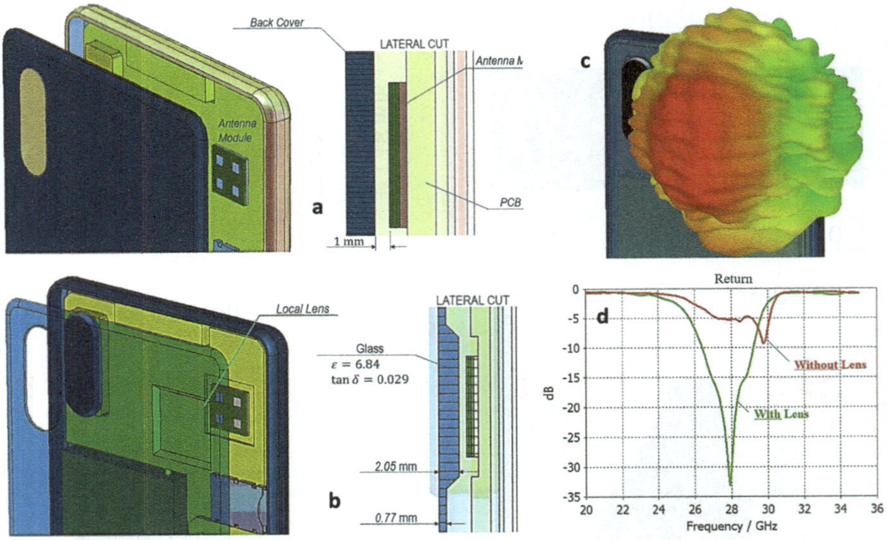

Fig. 8.4 Integration design of phone back cover and antennas. (Adapted from Enjiu and Rütschlin 2019, credit: Dassault Systèmes/the 3DEXPERIENCE Company): (**a**) Perspective view of the phone with the back cover lifted showing the position of the antenna module (left); and lateral cut at the antenna module showing the stack up arrangement of the phone PCB, antenna module and the back cover (right). (**b**) Perspective view of the phone showing the local thickening of the glass back cover (left); and lateral cut at the antenna module showing the stack up arrangement of the design (right). (**c**) Directivity pattern of the phone at the boresight, the rippled look of the pattern is due to part of the energy being propagated along the PCB surface and part being guided inside the dielectric. (**d**) Return loss response of the module with and without the local dielectric lens for a back cover with 0.77 mm thick glass

module and the cover. In the monolithic radome design, glass is used as the cover material. Instead of thickening the entire back cover, such as an electromagnetic window, local thickening is applied to the region of the radome immediately above the antenna to achieve the required 2.05 mm thickness, as shown in Fig. 8.4b. The electromagnetic performance of this enclosure is suboptimal. Although an improvement in the antenna matching and efficiency is observed, the radiation pattern is quite distorted by ripples (Fig. 8.4c, d), as part of the energy is guided inside the dielectric cover or propagates along the surface of the PCB before being radiated (Enjiu and Rütschlin 2019).

For the metallic back, the FSS must be at least slightly larger than the antenna footprint to perform adequately. The integration of the FSS in the phone's back cover provides an excellent electromagnetic window that allows the antenna module to radiate efficiently with the desired – almost ripple-free – far field pattern, as shown in Fig. 8.5 (Enjiu and Rütschlin 2019).

Therefore, the monolithic radome design seems to be a suboptimal solution for this example application, since it produced a rippled radiation pattern. However, the applicability of dielectric enclosures has the potential to deliver the desired

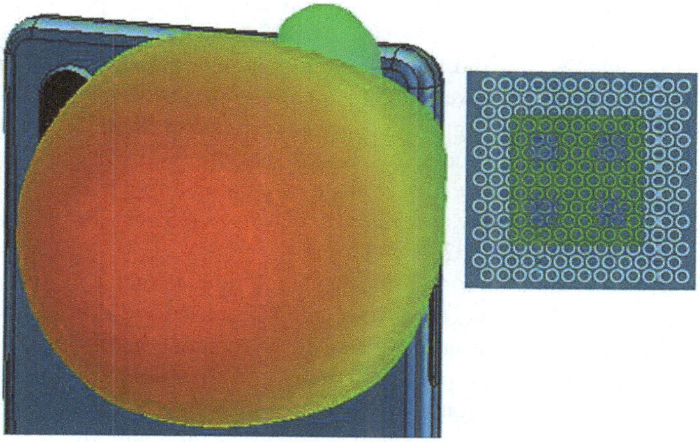

Fig. 8.5 Directivity pattern in 3D for the phone with FSS with 13 × 15 elements loaded metallic back. The radiation pattern shown is for the boresight radiation, minimum distortion of the 3D pattern is observed, and the same holds when the beam is steered. (Adapted from Enjiu and Rütschlin 2019, credit: Dassault Systèmes/the 3DEXPERIENCE Company)

performance if the gap between the antenna module and the enclosure can be increased. Comparably, the metal cover loaded with an FSS is a promising solution, since antenna matching and beam steering requirements can be met even for small gaps between the enclosure and the antenna module.

In addition, the material candidates for mobile phone front covers should be lightweight and have high transparency, an antiglare property, strong fingerprint resistance and scratch resistance. For the rear cover, wear resistance and efficient heat dissipation are usually considered. Tempered glass, ceramic or polymer composites (e.g., polycarbonate (PC)/poly(methyl methacrylate) (PMMA)) can be good competitors in the mobile phone shell market. For the PC/PMMA composite, PMMA ensures the scratch resistance of the product, while PC with sufficient toughness can ensure the overall impact strength. The middle frame of a 5G mobile phone is still mainly based on metal, and aluminum alloy/stainless steel with high strength is the future development trend.

8.2 Thin Film Encapsulation in 5G Electronic Packaging

Radio-frequency microelectromechanical systems (RF MEMS) technology is emerging as a key enabling solution to address the demanding requirements of 5G passive devices and networks, such as high operating frequencies, large tunability, reduced hardware redundancy, and low power consumption. These performance benefits, however, are offset by a lack of low-cost packaging available for RF MEMS. The package must protect the device from structural damage and contaminants, introduce minimal RF losses, and not degrade the performance of the switch, circuit or complex passive device. MEMS switches can be packaged using wafer-level techniques to avoid high costs and possible damage due to individual handling and release. Wafer-level bonding and thin film encapsulation are the most common wafer-level packaging methods. Thin film encapsulation typically uses a sacrificial layer to cover the structures to be sealed, followed by cap film deposition. The sacrificial layer is subsequently removed by etching through access holes or by thermal decomposition. Finally, a sealing film is deposited over the cap film to seal the access holes (Persano et al. 2020).

The choice of the material for the encapsulating film is crucial. Different materials (organic, metals, and dielectrics) have been proposed with thicknesses in the range of 1–20μm. Among the various materials, aluminum nitride, amorphous silicon, and silicon nitride are the most commonly used for thin encapsulating layers. In fact, they satisfy some essential requirements to make the packaging process suitable for industrial production, including good structural integrity and selectivity with respect to the sacrificial material during chemical etching, excellent insulation during device operation, and an optimized deposition rate in relation to the thickness to achieve. In particular, mixed frequency plasma-enhanced chemical vapor deposition (PECVD) provides low stress silicon nitride layers that are stable if processed at high temperature. Another crucial issue of thin film encapsulation is that the etch times required to remove the sacrificial material can be excessively long, especially as the size of encapsulation becomes larger, limiting the manufacturability of the package. The long release time may limit the choices of cap and sacrificial layer materials due to the possible damage of the cap layer and/or encapsulated device during the release process. Etch holes distributed over the cap layer or fabricated in its sidewall with standard surface micromachining techniques can alleviate this problem. Anisotropic plasma-assisted etching processes, based on fluorine chemistry, are generally used in reactive ion etching (RIE) or inductively coupled plasma (ICP) configurations to pattern encapsulation caps with etch holes. The drawback of the RIE and ICP methods is that the substrate bias and physical ion bombardment can modify the stress distribution in encapsulated suspended beams due to device heating with consequent beam deformation (Persano et al. 2019).

Inorganic materials, such as silicon oxide and amorphous silicon, are generally used as sacrificial materials, but their etching requires F-based vapors that could be a serious challenge in combination with micromachining technologies for structures containing aluminum or silicon oxide. Organic materials, which can be easily removed with oxygen plasma etching, are interesting candidates for sacrificial layers in fabrication processes at low temperature (<200 °C) (Persano et al. 2020).

In addition, the application of a protective cap to RF passives may introduce various parasitic effects that can detune the electromagnetic performance of the devices and cause additional loss. That is, applying a protective package to RF MEMS means building a housing for the device with a physical part (i.e., a dome) in the vicinity of the RF passive itself. Such a dome, therefore, interacts with the electromagnetic field around the device, thus introducing additional and unwanted capacitive couplings. Moreover, very often the packaging solution requires managing the redistribution of the electrical signals from the RF MEMS device to the external world. This can be done, for instance, by means of through-silicon vertical vias filled with conductive materials or by means of signal underpasses. Regardless of the specific signal redistribution solution, additional resistive losses, discontinuities of the RF paths, series inductive, and shunt capacitive contributions are added, thus jeopardizing the characteristics of the initial RF MEMS device. Therefore, the electromagnetic design of the package should be optimized with at least the same care spent optimizing RF MEMS devices themselves (Iannacci 2013).

Figure 8.6 shows a process flow for the fabrication of thin film encapsulating caps on conductive coplanar waveguides (CPWs), as an example application for packaging of RF MEMS in 5G and modern telecommunication systems. Capping

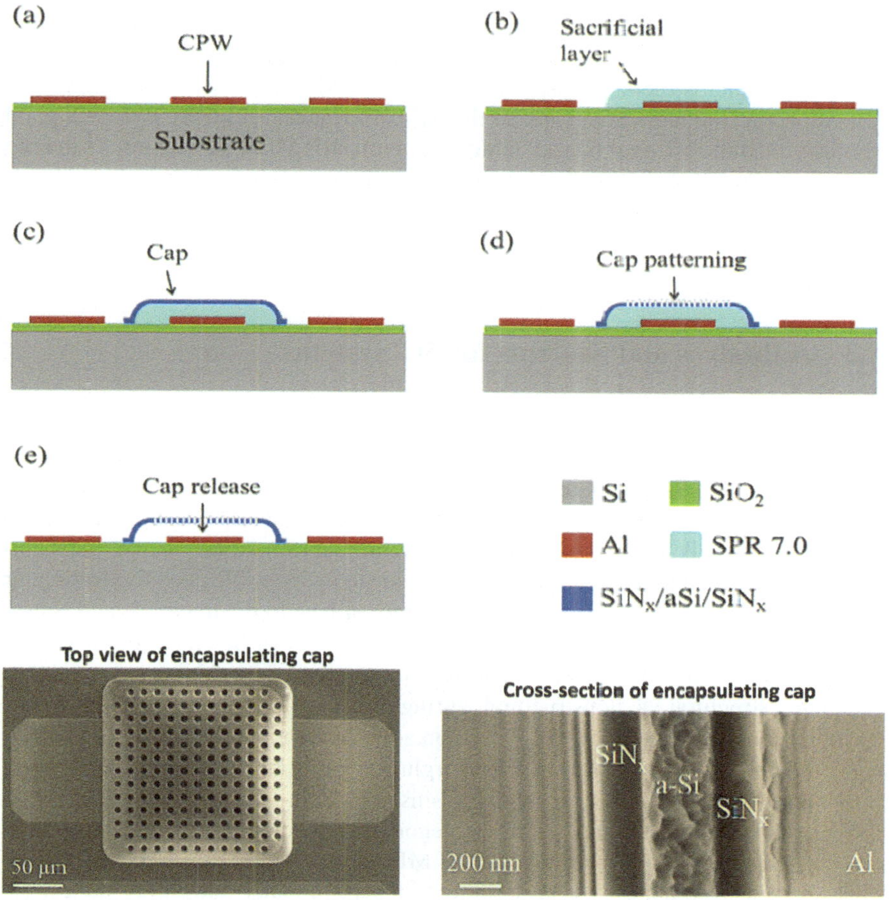

Fig. 8.6 Sketch of the process flow for the fabrication of thin film encapsulating caps (from **a** to **e**), and top view and cross-section of the fabricated encapsulating cap. (Adapted from Persano et al. 2019 with permission from Springer Nature)

layers in thin multilayer SiNx/aSi/SiNx were deposited by mixed frequency PECVD. Barrel etching was used to pattern the capping layers as an alternative to RIE and ICP processes. This allows an entirely chemical process with no ion bombardment to be applied to the device. Moreover, the use of the barrel etcher for both the pattern and release of the capping layer offers the advantage of reducing fabrication costs; since the encapsulating layer is sequentially etched and released in the same chamber, the plasma chemistry only needs to be changed from fluorine to oxygen-based chemistry. The removal of the sacrificial layer was investigated as a function of the percentage of cap perforated area as well as the time (t) and the power (P) of the barrel etching. Moreover, a full three-dimensional (3D) finite element method-based (FEM) simulation model could be used to predict the RF performance of uncapped and capped CPWs (Persano et al. 2020).

The oxygen plasma-based release of caps and the reasonably low temperatures (<200 °C) of all involved processes make the fabricated thin film encapsulation caps compatible with a wide range of microsystem technologies, from surface to high aspect ratio microstructures, also integrated with CMOS devices. In particular, the low impact of the fabricated caps on the RF performance of gold CPWs is a good premise for transferring this packaging solution to RF MEMS switches, switching units, and other complex devices, such as reconfigurable phase shifters, tunable filters, and digital step attenuators, for 5G and modern telecommunication systems applications (Persano et al. 2020).

8.3 Adhesives and Sealants for 5G Systems

Bonding technologies such as adhesives and sealants play a key role in 5G systems. The use of adhesive technologies is considered to be an efficient process in the bonding of materials in various elements and components, owing to their numerous benefits over alternative technologies. For instance, sealants and adhesives help overcome the challenges associated with conventional techniques, such as mechanical bonding, which may weaken the material due to the possible need for drilling, or thermal bonding, which involves the application of heat to the materials and subsequently alters their properties. Adhesives and sealants can be classified in several ways: by chemical composition—adhesive binder, natural vs. synthetic, organic vs. inorganic, structural vs. nonstructural, curing or setting mechanism, etc. However, the most common classification is based on adhesive binders, such as epoxy, acrylics, polyurethanes, cyanoacrylates (superglues), anaerobics, silicones, phenolics, polyimides, biismaleimides, and amino resins (e.g., urea-formaldehyde). Moreover, different adhesive technologies can be categorized by increasing order of load bearing capability, typically from 0.01 to 40 MPa of overlap shear strength, such as pressure sensitive adhesives, reclosable fasteners, contact and spray adhesives, acrylic foam tapes, hot melt adhesives, adhesive sealants based on polyurethane and hybrids, polyurethane adhesives (PUR), epoxy, and acrylic and urethane structural adhesives. Structural adhesives provide several advantages, such as strong bonds, design flexibility and process efficiency. Well-established industrial adhesives encompass different chemistries, curing methods, open times and final bond capabilities, including low-odor and nonflammable versions of certain chemistries to meet specific regulatory and safety requirements (Marques et al. 2020).

The use of adhesives and sealants in 5G electronics contributes directly not only to the manufacture of electronics products but also to their long-term operation and longevity. Major uses of adhesives in electronics include bonding of surface-mount components (SMCs), wire tacking and potting or encapsulating components. The days when solder was the exclusive joining material in electronics assembly have long gone. More and more adhesives have been used as solder alternatives. There is still some way to go for full implementation, as the adhesive will need to take over the mechanical properties of the solder and be electrically and thermally conductive to a very high degree. However, the use and development of adhesives in electronics will certainly continue to grow. Some general applications of adhesives and sealants in electronic assemblies are included (Gould 2004):

1. Surface-mount adhesives can serve two functions: they act as a processing aid (e.g., holding a part temporarily until it can be permanently attached by soldering), and they can also provide stress relief to solder connections during service to prevent premature failure of the electrical connections. After soldering, the now redundant adhesive must not affect the circuitry in any way or must be washable/removable. Depending on the exact function of the surface-mount adhesive, it can be electrically conductive, electrically nonconductive and/or thermally conductive or insulative. Three chemical types of adhesive are commonly used for surface-mount component bonding: acrylics, epoxies and urethane acrylates. Ideally, surface-mount adhesives need to be of the single-component type. This avoids the problems of pot life, air entrapment and changes in viscosity normally associated with dual or multipart systems. Usually, the adhesives will cure within 2 min at elevated temperatures safe for use with PCBs and associated components.
2. Wired connections are common on PCBs. The use of wires not only allows replacement but also augmentation of printed wiring on the PCB. Usually, an insulated wire is stripped and soldered at each end to pads and component leads and then bonded or staked to the board substrate. The bonding prevents the wire from moving around and acts as a structural adhesive in environments where vibration could occur. The types of adhesive that can be selected for wire tacking include cyanoacrylates (so-called 'superglues'), epoxies and light-curable acrylics. In practice, cyanoacrylate is widely used, often in conjunction with a suitable activator to speed the cure time to approximately 20–30 s.
3. Potting is a method of filling small spaces or surfaces with a material that will protect components from physical and environmental damage. Potting components also provide additional insulation capability. Potting compounds usually exhibit good chemical properties and high adhesion to plastics and metals, which are materials used in the construction of containers and components. Typical resins used for potting are epoxies, polyurethanes, silicones and acrylics, the lat-

ter usually being UV-cured formulations. In addition to potting, there are other methods of encapsulating electronic components, namely, casting and molding. Casting uses the same type of adhesive resins as potting, although the container (outer casing) is usually removed after the resin has cured, unlike the potting process where the container becomes an integral part of the component. Molding usually involves the injection of premelted thermoplastic resins into a mold containing electronic components or circuitry.

4. Conformal coatings, while not strictly the result of an adhesive bonding process, utilize the same chemical types of resin (e.g., epoxies, polyurethanes, acrylics and silicones). The aim is to produce a coating that strongly adheres to and covers all the board and the components on it. A conformal coating is typically used to protect the board from environmental factors such as corrosion (e.g., of solder joints), moisture and mildew, changes in service temperature (typically from −40 °C to +200 °C), and short circuits. Conformally coated boards are protected from environmental damage and mechanical and electrical interference (EMI).

Three different processing phases must be considered when selecting an adhesive for electronics applications: the uncured or liquid-resin phase, the curing (transitional) phase and the cured or solid-material phase. The performance of the cured adhesive is ultimately the most important because it affects reliability. The method of applying the adhesive is also of great importance, in particular because of the need to ensure that the correct amount is applied in the correct place. Major methods of applying adhesives in electronics applications are screen printing (squeezing the adhesive through patterns in a screen), pin transfer (using multipin grids that convey patterns of adhesive drops to the board), syringe application (in which shots of adhesive are delivered by a pressure-regulated syringe), and various 3-D printing technologies. Syringe application is still probably the most popular method, usually by way of electropneumatically controlled syringes for moderate production of many different types of PCB. Some types of adhesive are included:

1. Electrically conductive adhesives

   By their nature, most adhesives, both organic and inorganic, are not electrically conductive. This applies to the main types used in electronic applications, such as epoxies, acrylics, cyanoacrylates, silicones, urethane acrylates and cyanoacrylates. However, in many applications, including integrated circuits and surface-mount devices, electrically conductive adhesives are needed. The usual way of converting nonconductive adhesives to electrically conductive materials is to add suitable filler to the base material; usually the latter is an epoxy resin. Typical fillers used to confer electrical conductivity are silver, nickel, carbon, and related nanomaterials. Silver is the most widely used. The conductive adhesives themselves are either in a liquid or preform (reinforced adhesive films die-cut before bonding to the shape required). There are two types of electrically conductive adhesives – isotropic and anisotropic. Anisotropic adhesives conduct in all directions, but an isotropic adhesive conducts in the vertical (z-axis) direction only and is thus unidirectional. The isotropic adhesives lend themselves to fine-line interconnections.

2. Thermally conductive adhesives

   Thermally conductive adhesive is a crucial solution for joining components and removing the extra heat generated during the device's operation. These adhesives are used for joining chips and substrates, metal components, polymer composites, and concrete structures, which are challenging for other techniques. The need for effective heat conduction within electronic equipment has been highlighted more with the development of 5G mobile networks. As a result, thermally conductive adhesives, as a kind of thermal interface material, are essential in modern electronics to ensure efficiency and reliability. Metallic (electrically conductive) or nonmetallic (insulating) powders are blended into the adhesive formulation to make high-viscosity (paste) adhesives, which are highly thermally conductive (in comparison with unfilled adhesives). The most common thermally conductive systems are formulated with epoxy, silicone and acrylics.

3. Thermal insulation adhesives

   Thermal insulation adhesive sealing materials can be used to prevent heat generated from electronics and moving machinery parts. Inhibiting heat transfer is of crucial importance in 5G system bonding, coating and sealing applications. These insulation materials are required to offer high bond strength, low outgassing, thermal stability, excellent corrosion resistance, durability and dimensional stability.

4. Ultraviolet-cured adhesives

   Light-curing adhesives, coatings and encapsulants are being used in the electronics manufacturing industry with increasing frequency because they meet the requirements for materials and processing within this industry. These factors include environmental demands (environmentally damaging solvents and additives are not required), manufacturing-yield improvement and product cost. Light-curing adhesives are simple to use and are quickly cured without the need for elevated temperature curing. The adhesives are normally acrylic-based formulations and contain photoinitiators that, when activated by ultraviolet radiation, form free radicals to initiate the polymer-forming (curing) process. Ultraviolet light must be able to penetrate into the uncured resin – a drawback of light curing adhesives. Deposits of resin that are dark-colored, inaccessible or very thick are difficult to cure.

There are more adhesives that have found use in electronics applications, and doubtless more will follow. For many applications, such as structural adhesive bonds, debonding or removable capabilities of the adhesives are needed. The majority of current debonding technologies involve the use of functional additives that are mixed into the adhesive (or primer layer) in quantities of up to 20 wt%. In all these cases, the mechanical properties of the adhesive are compromised. New debonding materials and methods have been developed for specific applications.

Moreover, replacing hazardous chemicals and manufacturing processes with safer chemicals and greener technologies has been a key investment of adhesive companies to avoid regrettable roadblocks in product commercialization in the near future. In addition to the benefits to companies, the environment, and the health of workers and consumers, this can also have a significant positive impact on the implementation of a circular economy. Due to the growing demand for environmentally friendly, green and sustainable materials, as well as economic and availability issues related to petroleum resources, there has been a growing interest in the development of environmentally friendly adhesives (Marques et al. 2020).

References

Enjiu R, Rütschlin M (2019) Design of 5G mm-wave compatible covers for high end mobile phones. NAFEMS world congress 2019, Quebec City. https://www.3ds.com/fileadmin/PRODUCTS/SIMULIA/PDF/conference-papers/5G-mm-wave-mobile-phone-cover-design-SIMULIA.pdf

Gould J (2004) Adhesives in electronics. https://www.lboro.ac.uk/microsites/mechman/research/ipm-ktn/pdf/Technology_review/adhesives-in-electronics.pdf

Iannacci J (2013) Practical guide to RF-MEMS, 1st edn. Wiley-VCH, Weinheim, pp 1–344

Marques AC, Mocanu A, Tomić NZ, Balos S, Stammen E, Lundevall A, Abrahami ST, Günther R, de Kok JMM, Teixeira de Freitas S (2020) Review on adhesives and surface treatments for structural applications: recent developments on sustainability and implementation for metal and composite substrates. Materials 13:5590. https://doi.org/10.3390/ma13245590

Persano A, Iannacci J, Siciliano P, Quaranta F (2019) Out-of-plane deformation and pull-in voltage of cantilevers with residual stress gradient: experiment and modeling. Microsyst Technol 25:3581–3588

Persano A, Quaranta F, Taurino A, Siciliano PA, Iannacci J (2020) Thin film encapsulation for RF MEMS in 5G and modern telecommunication systems. Sensors 20:2133. https://doi.org/10.3390/s20072133

Volakis J (2007) Antenna engineering handbook. McGraw-Hill Education Ltd, New York

Chapter 9
Perspectives on 5G and Beyond Applications and Related Technologies

Abstract Since its first commercialization in early 2019, 5G has been making progress around the world and penetrating daily lives. Now, research interest in wireless communication is quickly shifting to the next generation mobile system, 6G. 5G is not just an evolutionary upgrade of the previous generation of cell phones, but it is a revolutionary technology envisioned that will eliminate the bounds of access, bandwidth, performance, and latency limitations on connectivity worldwide. 5G and beyond have the potential to enable fundamentally new applications, industries, and business models and dramatically improve quality of life around the world via unprecedented use cases that require high data-rate instantaneous communications, low latency, and massive connectivity for new applications for mobile, eHealth, autonomous vehicles, smart cities, smart homes, and the IoT. Therefore, 5G has become a novel driving force for leading innovation and stimulating new types of information consumption, as well as an emerging engine for promoting industrial upgrading and driving sustained economic growth. Through the construction of 5G network systems and the popularization of 5G terminals, components such as base station antennas, filters and communication PCB boards are increasingly demanded. From a technical perspective, ultralow-loss high reliability copper-clad laminate materials employed in 5G high-speed communications, 5G metamaterials and low-loss high-performance RF technology, and 5G low-loss magnetoelectric functional materials have been developed. For materials and devices, multimodule integrated printed circuits for 5G communication base stations, manufacturing of 5G communication optoelectronic integrated cables, and multichannel high linearity and large dynamic range RF optical transceiver integrated modules in terms of 5G mobile communication will show critical significance in various applications. This chapter will give a brief review of perspectives on 5G and beyond applications and related technologies.

C. Tong, *Advanced Materials and Components for 5G and Beyond*, Springer Series in Materials Science 327, https://doi.org/10.1007/978-3-031-17207-6_9

9.1 Applications in Industry Verticals and Their Needs

5G networks – with the capabilities of network slicing (NS), latency, and ultrabroad bandwidth – can help service providers create new application platforms to enable the next generation of applications and develop new business models. Video streaming and IoT-based applications are the current "killer applications," and these applications, mixed with the virtual/augmented reality capabilities, will create opportunities across several verticals. Table 9.1 presents some emerging applications and services for which 5G will be a pivotal enabler. Possible future extensions and verticals might include the following: Aerospace; Ocean; Threat response; Mobile platform; Terrestrial and distributed computing (cloud/IoT-fog) (IEEE 2017).

The bandwidth and latency requirements of applications based on mobility are mapped as shown in Fig. 9.1 (Warren and Dewar 2014).

9.1.1 5G in Automotive

Connected, cooperative and automated mobility (CCAM) is one of the key enablers to enhance road safety and traffic efficiency and to improve environmental and information flows. In the automotive industry, the fifth generation (5G) of wireless networks is expected to significantly enhance wireless connectivity, support automation and provide a broad range of digital services to vehicles, which will support various advanced vehicular use cases and pave the way to fully connected mobility and autonomous driving. As such, the automotive vertical industry has been one of the main targets of 5G deployments. 5G technology is expected to be a game changer for the automotive industry. The possibility for vehicles to be connected to other vehicles, pedestrians, roadside infrastructure, or application servers enables the development of multiple revolutionary services, including (Chiasserini and Magnan 2017; Bouali et al. 2021):

(a) Vehicle platooning: vehicles dynamically forming a group, driving together, and proceeding at a very short distance from each other, as shown Fig. 9.2;
(b) Advanced driving (autonomous/assisted driving): sharing driving intentions, sensor data, and videos gathered through onboard cameras with roadside infrastructure, other vehicles, pedestrians and network servers for safer traveling, collision avoidance and improved traffic efficiency, as well as semi or fully automated driving;
(c) Support for remote/cloud computing driving: a remote driver or a V2X application that operates a remote vehicle traveling in dangerous environments, with impaired passengers onboard, or public transportation vehicles.
(d) Vehicle data services: focus on collecting actionable information from all the relevant entities (e.g., vehicles and road users) to provide various services via the available 5G infrastructure.

Table 9.1 Emerging applications and services enabled by 5G (IEEE 2017)

Verticals	Drivers	Enablers	5G requirement
Education	Remote delivery	Video streaming	Large bandwidth
	Immersive experiences	Augmented reality/ Virtual reality	Low latency
Manufacturing	Industrial automation	Massive IoT networks	High connection density
			Ultra reliability
			Low power consumption
Healthcare	Remote diagnosis and intervention	Video streaming	Low power
	Long term monitoring	Augmented reality/ virtual reality	High throughput
		Embedded devices, advanced robotics	Low latency
Smart grid	Intelligent demand/ supply control	IoT sensors and networks	High reliability
	Powerline communication		Broad coverage of network
			Low latency
Entertainment	Immersive gaming and media industry	Video streaming	Large bandwidth
	Multimedia experience at 4 K, 8 K resolution	Augmented reality/ virtual reality	Low latency
Automotive/ autonomous cars	Collision avoidance	Vehicle-to-vehicle (V2V),	Large bandwidth and low latencies (<5 ms) and high connection reliability (99.999%)
	Intelligent navigation and transportation systems	Vehicle-to-infrastructure (V2I) and other intelligent transport systems (ITS)	
Smart cities	Connected utilities, transportation, healthcare, education and all amenities	Massive IoT networks	Large bandwidth; High throughput; High connection density; Low latencies
	Automation	Automation	
	Cloud infrastructure	Cloud infrastructure	
	Artificial intelligence	Artificial intelligence	

Each of these use cases entails a few scenarios, where the focus is on a specific service or functionality. Among the key innovations of 5G, it is expected that the following will represent major enabling technologies for automotive services (Chiasserini and Magnan 2017): network slicing; multiaccess edge computing (MEC); eV2X communications; and mmWave communications. Network slices, namely, virtual function graphs and the corresponding resources needed to implement them, should be easy to configure, reusable for one or more network services, with agile lifecycle management, and possibly isolated from other slices for security

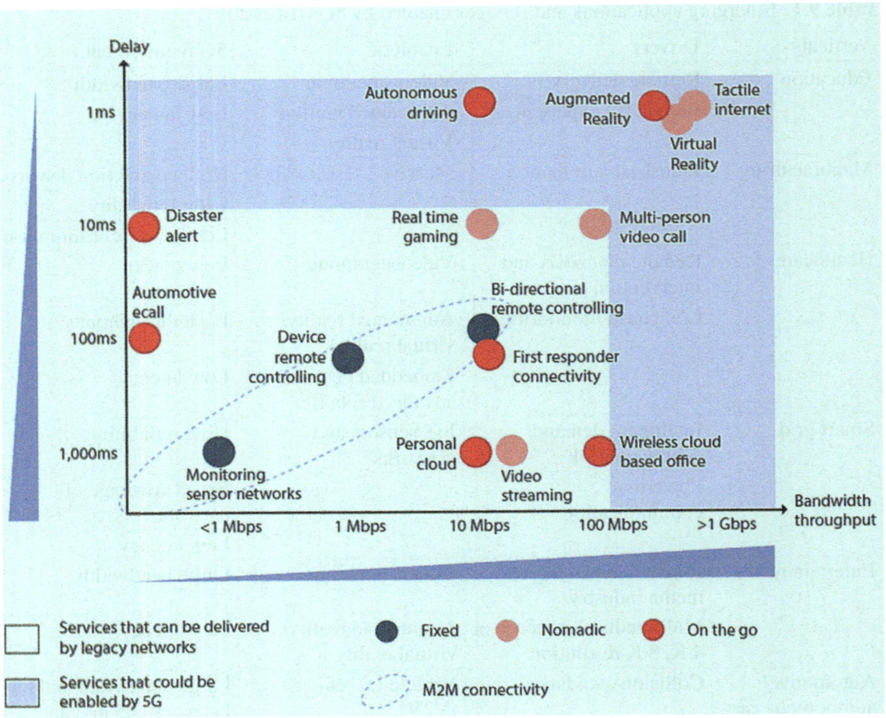

Fig. 9.1 Bandwidth and latency requirements of 5G applications. (Adapted from Warren and Dewar (2014); Credit: https://www.gsma.com/futurenetworks/wp-content/uploads/2015/01/2014-12-08-c88a32b3c59a11944a9c4e544fee7770.pdf)

reasons. Network slicing can involve resources at the edge of the network, leading to the implementation of the MEC concept. MEC is indeed an effective approach capable of guaranteeing low end-to-end delays, low bandwidth consumption, low energy consumption and high resilience, which are of critical importance to automotive services. The idea of exploiting resources at the edge of the network can then be pushed to the extreme case where resources offered by pedestrian and vehicular user equipment are exploited. It is thus fundamental to realize efficient V2X communications to enable user equipment to share computing, network and storage resources.

9.1.2 Big Data Analytics in 5G

Big Data refers to large datasets whose size is growing at enormous speed, making it difficult to handle and manage them using traditional techniques and software tools. It is a step forward from traditional data analysis, considering the following

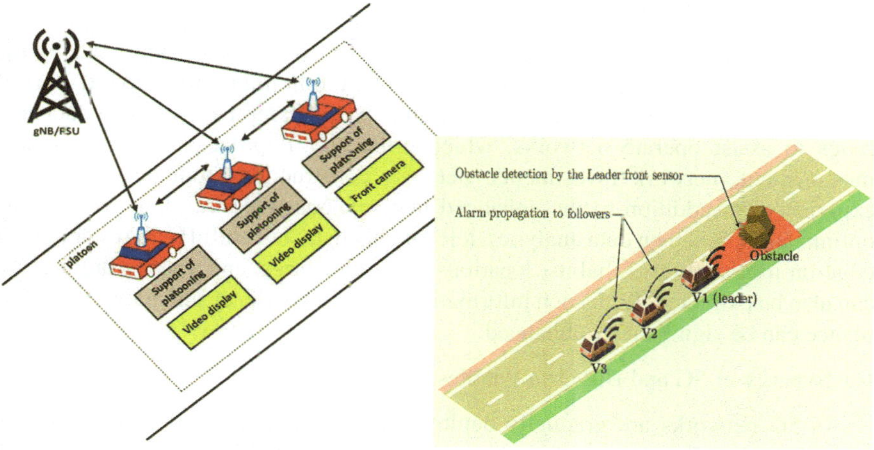

Fig. 9.2 See-through for platooning. (Adapted from Bouali et al. (2021) with open access (IEEE))

aspects (so-called "five Vs"): quantity of data (volume), different types of semistructured and unstructured data (variety), the rate with which data are changing or how often they are created (velocity), the importance of results extracted from data (value), and data quality, including trust, credibility and integrity (veracity). Taking into account the prediction that the number of connected devices will increase 10–100 when 5G is commercially used. Therefore, big data techniques will play an important role, as all the considered usage scenarios are based on extracting knowledge from the enormous amount of heterogeneous data generated by connected devices to support decision making and other mechanisms in future 5G networks.

To accommodate the tremendously expanding communication demands while given limited resources and costs, future communication systems are expected to be deployed with the development of big data technology (Zhu et al. 2019):

(a) 5G Networks Facilitate Big Data

In the big data processing chain, 5G networks, providing the basic infrastructure of transmission, play the role of channels between data sources and data centers. There are tremendous numbers of connected devices in 5G networks, such as mobile user equipment, numerous sensors, and closed circuit television, which act as data sources in the big data chain. Therefore, the volume and variety of data sources would further increase. In addition to the data acquisition, communication networks are required to store and preprocess before transmission.

By means of data compression and data aggregation, data redundancy is implemented for further transmission. In addition, transporting data to the cloud or data centers would depend on wireless networks, consisting of radio access networks and the core network. As a consequence, 5G networks should consider the requirements of big data applications, including reliability guarantees, end-to-end delays, and data integrity.

(b) Big Data Assist 5G Networks

5G wireless networks can facilitate the big data processing chain, and by the same token, telecom big data also has great potential in improving network performance and user experience. Telecom operators have attempted to employ data analytics to assist operate networks, which efficiently improves network resource management, network investments, network construction strategies and users' experience. In addition to the enhanced network construction, management, and optimization, based on data analytics, telecom big data, insightful features or knowledge on usage patterns, habits, locations and trajectories, and service preferences can also be extracted. With such information, users' experience and network performance can be significantly enhanced.

(c) Synergy of 5G and Big Data Enables Vertical Industries

As 5G networks are gradually deployed and commercialized and cooperated with systemized data operations, a wide range of intelligent applications will emerge, enabling vertical industries. Figure 9.3 illustrates a conceptual relationship among 5G, big data and vertical industries. 5G, providing high-speed, robust and low-latency mobile communication networks, ensures that individuals always have

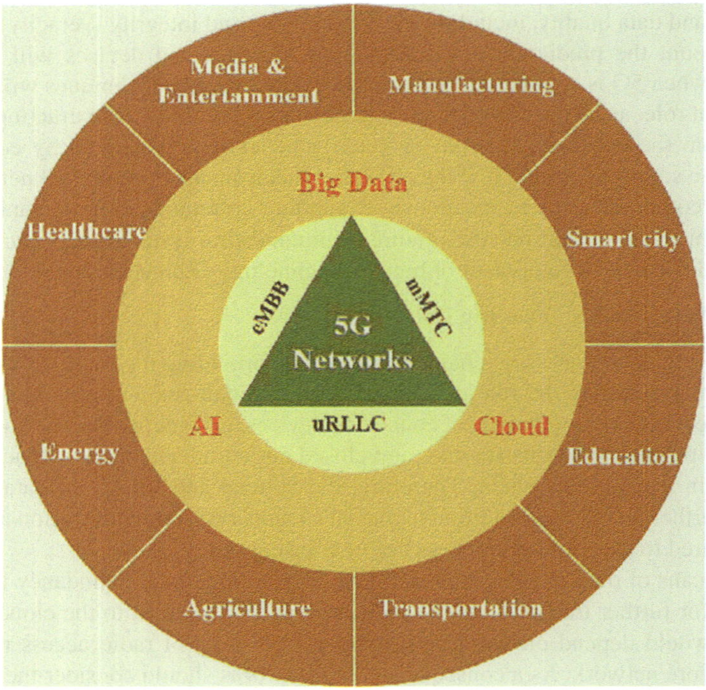

Fig. 9.3 "5G + Big Data" Enables Vertical Industries. (Adapted from Zhu et al. (2019) with permission from IEEE)

online services. The IoT networks, sensing and collecting data for business users, continually deliver high-value information to assist business decisions. Based on distributed computing infrastructure, real-time intelligent services can be supplied. The combination of big data, cloud and artificial intelligence (AI), acting as the brain and engine, builds the intelligence-based platform of computing power, big data, and service-based connection, creating innovative services and boosting vertical industries.

In Fig. 9.3, eight primary industries are shown; nearly all of them are facing challenges on the road of digital transformation. 5G and big data technologies enable novel use cases, such as automatic driving, drones, and virtual reality (VR)/augmented reality (AR), and can help vertical industries address connectivity and data platform demands with customized or generalized solutions. The design schemes of communications and data platforms should comprehensively consider objective applications because the demands of each scenario would be different in terms of transmission bandwidth, end-to-end latency, robustness, data security, real-time/on time processing, and data volumes (Zhu et al. 2019).

However, the challenges for the development of telecom big data in next-generation networks may include the following (Zhu et al. 2019):

(a) The complexity of the networks and complicated usage patterns is inevitably increasing, which makes manual configuration for network planning, maintenance and optimization impossible. Moreover, conventional reactive maintenance is inefficient and cannot ensure the continuity and quality of services all the time. The operational system is supposed to be smart, self-aware, and self-adaptive. Consequently, autonomous and intelligent analytics and tools that help network operators manage networks economically and efficiently in real time are in demand.
(b) Data sharing and data fusion will be increasingly performed. The combined data are merged from different sources with different formats and semantics, while the data can be accessed by different parties so that the insightful benefits can be maximized. However, the data security and privacy preservation should be taken into serious consideration, since a large part of data are related to individuals or commercial actives. How to make the best use of data under data protection needs to be addressed.
(c) The future communication network will facilitate service and application in a flexible manner, including numerous time-sensitive applications (e.g., smart and connected vehicles) that need real-time or near-real-time analytics of the stream of data. As a consequence, a new analytic framework in conjunction with fast/streaming data analytics is required to satisfy these types of applications to provide service-oriented end-to-end services.

However, 5G and big data have become some of the most promising and prevailing technologies for the future. 5G is supposed to meet growing demands from various industries for higher speeds, lower latency, more devices, larger data, and improved operational efficiency. Meanwhile, utilizing the large amounts of data generated from the network, subscribers and vertical industries would enable diversified services and dramatically improve society (Zhu et al. 2019).

9.1.3 5G Emergency Communications

The migration of telecom operators to broadband IP infrastructures enforces emergency systems to follow this path and adapt their emergency communication platforms to fulfil next-generation emergency services regulatory requirements. A key factor pushing toward this end is the support of diverse applications and services with heterogeneous performance requirements, including mission-critical IoT communication, massive machine-type communication and gigabit mobile connectivity. Eventually, emergency systems will have to be upgraded to fulfil the next-generation network (NGN) regulatory requirements. Legacy systems, protocols and the services provided by emergency communications organizations and the public safety sector in general are affected particularly by the increased performance of next-generation wireless and mobile networks, enhanced security and improved device-to-device communications (Markakis and Politis 2017).

The envisioned 5G features will need to be tightly coupled with the stringent operational and management requirements of emergency services, which need to be maintained. Depending on the actual design of the 5G network, emergency communication is expected to support real-time, high priority total conversation services (voice, video, real-time text) and improved immunity to data pollution and security threats that could potentially disrupt the reaction time of the first responders. Additionally, device-to-device communications will increase the availability of the communication channels and the uplink capacity, creating an "always connected" experience for the emergency workers and rendering them capable of utilizing high-quality multimedia content for improved awareness. Toward this end, the network-slicing feature of the 5G network will enable network operators to dynamically adapt the transmission speed and latency of the network, ensuring high priority for first responders' communications (Markakis and Politis 2017).

To satisfy the strict requirements imposed by emergency communications and public protection and disaster relief (PPDR) services in general, the network operators are expected to ensure ultralow latency, ultrahigh availability and reliability for these services. Due to the low latency that will be provided by the underlying access network, next-generation emergency services based on massive IoT and device-to-device communications will be characterized by higher throughput, higher Quality of Service (QoS) and Quality of Experience (QoE) and low buffer requirements for the user devices. Mobile edge computing (MEC) has been proposed to fill this niche not well served by existing cloud architectures. MEC is regarded as a key enabling technology that provides the capability to have a high-performance virtual environment residing at the network's edge. By being adjacent to the IoT environment, MEC can support applications and services with increased bandwidth, low latency and improved QoS (Markakis and Politis 2017).

As shown in Fig. 9.4, potential medical emergencies that would require remote intervention (e.g., emergency calls back to the user, remote calibration of pharmaceutical doses, etc.) is detected at the MEC level. In the case where the emergency requires the intervention of first responders to the area, the health care provider initiates an emergency call to the emergency service operator, augmented with the current health sensor data, along with the patient's location information, medical history and insurance data (Markakis and Politis 2017).

In the case of a fire detected by the building management system, the security service provider who owns the MEC platform can initially verify that it is not a false positive alert from video footage available from the site. Alternatively, an emergency call to the fire brigade is initiated, including a building floor plan for the first responders. Finally, the MEC's remote provisioning system can be utilized to decrease the sensor reporting interval to provide near real-time data delivery. In contrast to what is the traditional case for emergency service networks based on GSM and LTE, where each one of the physical networks is reserved for one use case (e.g., GSM for voice, LTE for mobile data), the proposed network architecture for

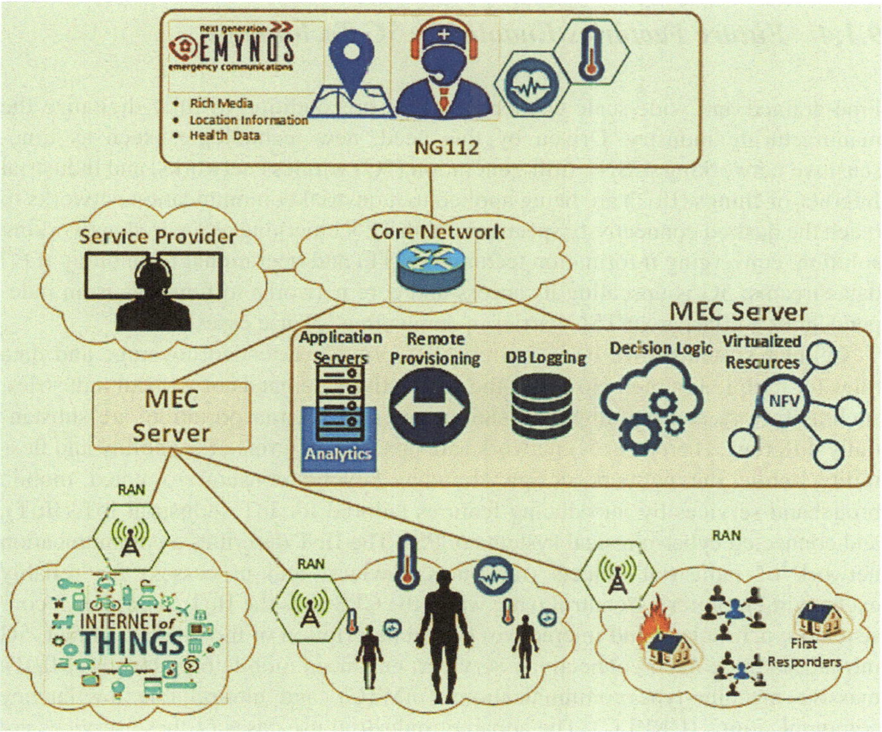

Fig. 9.4 Illustration of 5G Emergency Communications. (Adapted from Markakis and Politis (2017), Credit: https://futurenetworks.ieee.org/images/files/pdf/applications/Emergency-Communications030518.pdf)

5G emergency services can create and manage virtual instances of access networks, hence providing customized network resources to each emergency service agency (i.e., police, ambulance, fire-brigade) to the area of the event. Network slicing will eliminate communication interference between the different networks, ensuring extremely high throughput and ultralow latency. This flexible orchestration of network slices is realized using software-defined functions and programmable infrastructures. RAN's backhaul are governed by the NFV infrastructure, the control of which relies on the MEC. In this case, the bandwidth allocated to each wearable for the health monitoring, or the first responders' communication devices, the management of the traffic including delay, loss, active bearers, etc., are synchronized by the NFV controller located in the MEC (Markakis and Politis 2017).

Therefore, the transition to 5G for emergency service providers is a chance for emergency service providers to migrate their systems to support an amalgam of diverse applications and services with heterogeneous performance requirements, including mission critical IoT communication, massive machine-type communication and gigabit mobile connectivity.

9.1.4 Future Factories Enabled by 5G Technology

Fine-grained and wide-scale connectivity is a precondition to fully digitalize the manufacturing industry. Driven by this need, new technologies such as time-sensitive networking (TSN), fifth-generation (5G) wireless networks, and industrial Internet-of-things (IIoT) are being applied to industrial communication networks to reach the desired connectivity spectrum. With TSN emerging as a wired networking solution, converging information technology (IT) and operational technology (OT) data streams, 5G is upscaling its access and core networks to function as an independent or a transparent TSN carrier in demanding OT use cases.

Cellular systems have been optimized to provide extended coverage and data rates for mobile Internet. However, the connectivity demands of vertical industries, such as factory and building automation, energy, and transportation, are substantially different. Therefore, 5G networks are upscaled in terms of capability and flexibility, while the network design stretches beyond consumer-oriented mobile broadband services by introducing features tailored for IoT, industrial IoT (IIoT), and connected cyber-physical systems (CPS). The IIoT describes a communication network of industrial objects (machines, devices, and processes) for reliably exchanging monitoring/control data, while the CPS uses the IIoT to provide a consistent (synchronized and interactive) digital description of the objects. In fact, 5G introduces three main connectivity services: enhanced mobile broadband (eMBB), massive machine-type communications (mMTC), and ultrareliable low-latency communications (URLLC). The apparent industrial use cases of these services are shown in Fig. 9.5, yet by combining these services, 5G can support extreme variations in IIoT applications (Mahmood et al. 2021).

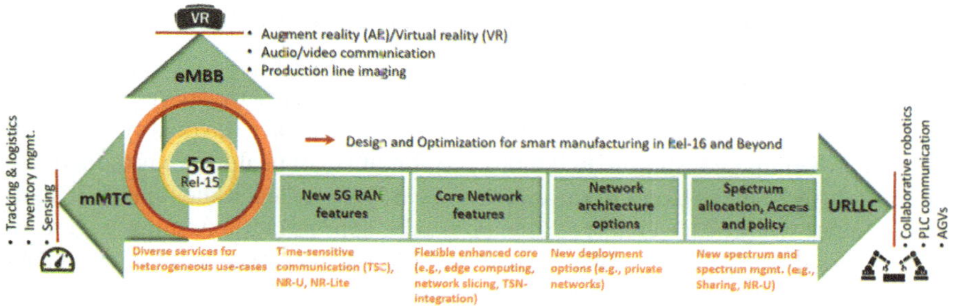

Fig. 9.5 5G's design and optimization for diverse industrial applications. 3GPP Release 15 (Rel-15) defined eMBB, mMTC, and URLLC services for 5G New Radio (NR); Rel-16 focused on IIoT-related enhancements, including URLLC for TSC, 5G-TSN integration, localization services, private networks (PN), network slicing, and NR on unlicensed bands (NR-U) for URLLC; Rel-17-and-beyond will cover further improvements in PNs, convergence with industrial networks, network automation, and NR-Light for new IIoT use-cases. (Adapted from Mahmood et al. (2021) with open access (IEEE))

Although 5G becomes a unified connectivity fabric, the seamless introduction to the factory floor raises several questions and challenges about the connectivity gaps 5G can fill, how to integrate it with the TSN, and whether it satisfies the time-sensitive communications (TSC) targets while keeping business-critical data local. Therefore, different businesses and mobile network operators (MNOs) are seeking to understand the 5G architecture's design and optimization (the extending ripple in Fig. 9.5) to expand their business models beyond mobile broadband. In this direction, private (nonpublic) 5G networking models and associated spectrum licensing options need to be explored carefully in realizing dedicated, customizable, and cost-effective services for industrial use cases (Aijaz 2020).

In addition, some standalone 5G deployment models can enable dedicated support for critical use cases while providing necessary privacy and security features. However, for public network integrated private 5G, the solution providers need to reduce the operation and management complexity and provide autonomous solutions to network operation, optimization, on-site and remote functional splits, new service deployment and QoS assurances for multiple scenarios and services. Private 5G wireless brings along many security concerns to the factory floor, especially when sharing resources with a public network. The 3GPP offers solutions to increase the resilience of private networks through the duplication of network functions/resources and service/slice authorization and isolation. Against radio jamming attacks, the solutions can be devised by combining 5G's dynamic spectrum monitoring and allocations and beamforming techniques. Meanwhile, to reduce the additional communication overhead of security measures, new techniques such as physical layer security and federated learning are promising but require realistic suitability analysis for real-time control and analytics (Mahmood et al. 2021).

9.1.5 Smart Health-Care Network Based on 5G

Smart healthcare enables advanced diagnostic instruments to provide advanced treatment for patients and smart healthcare devices to enhance the quality of healthcare by providing vital signs in real time. The aim of smart health care is to facilitate patients through the delivery of information regarding medical issues and their solutions. Smart health care enables patients to take appropriate measures in case of critical circumstances. It enables remote check-up service that results in reducing the treatment expenditures and supports health care providers in broadening their services beyond geographical limits. With the expansion of smart cities, a robust smart health-care system is required to guarantee health services to users. Apart from well-being, one of the significant contributions is reducing health expenditure by timely diagnostics (Ahad et al. 2020).

The IoT will revolutionize health care and reduce the cost of medical devices. 5G networks will play a major role in enabling widespread adoption of the IoT. In 5G networks, smart healthcare is one of the most important applications. Figure 9.6 shows the general architecture of the 5G-based smart health-care network and its main entities. In smart health care, the IoT can improve several applications, including asset management in hospitals, behavioral change monitoring, remote monitoring, treatment compliance monitoring, assisted living, smarter medication, and telemedicine. In 5G networks, machine-to-machine (M2 M) communication and IoT are expected to be the main pillars of smart health care (Ahad et al. 2020).

Globally, healthcare vertical is all set for a paradigm change with an increasing adoption of devices with sensing equipment, technology and telemedicine evolution. The key technology drivers in health care include the following (Rao 2017):

(a) IoMT devices: Management of massive amount of data created by IoMT devices (wearables, implantable medical devices, and smart sensing remote/in house monitoring devices)
(b) Smart wearables: health parameters such as heart rate, basic blood pressure, and breathing rate can be monitored.
(c) Tactile Internet: remote healthcare involving 2-way interaction and robotics places a special premium on millisecond level latency and immediate response times.
(d) Critical communications: remote health monitoring of blood sugar, ECG, temperature interworking with QoS mechanisms to reserve capacity for critical communication.
(e) Emergency medical services: ambulance services communicating back to a nodal hospital with relevant patient information, such as high-resolution images/videos, in a timely manner.
(f) Privacy and Security: Privacy and data integrity of patient data is of paramount importance and necessitates special mechanisms to be adopted. Data handling will be expected to ensure required levels of privacy, preferential network resource handling and guaranteed Quality of Service (QoS).

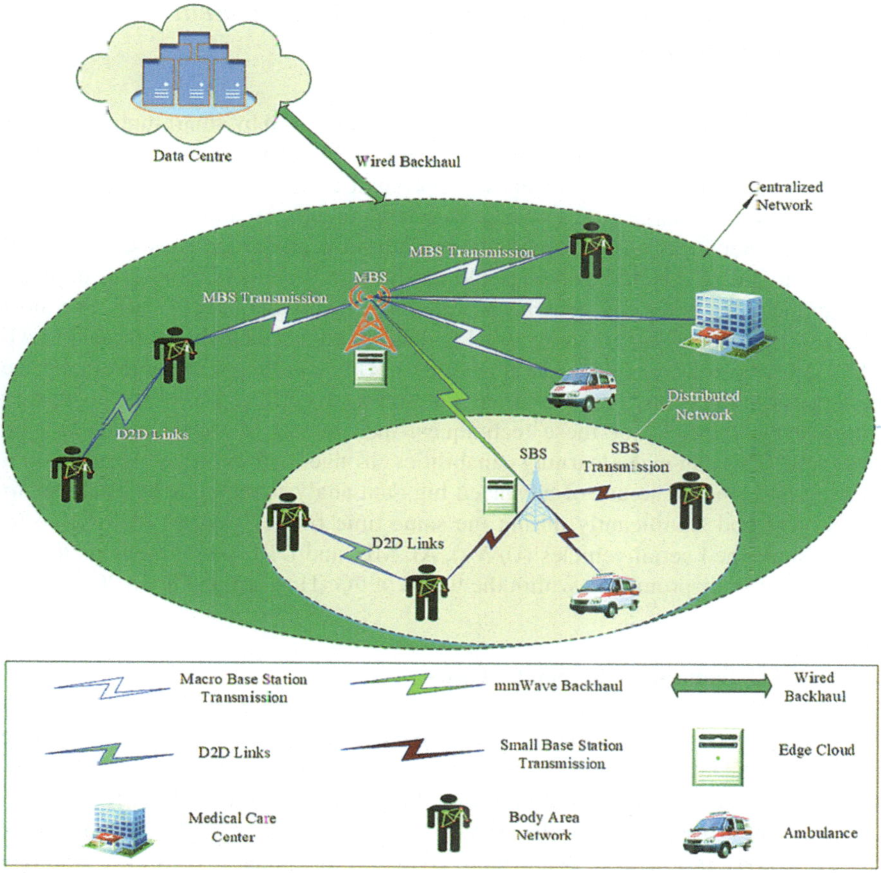

Fig. 9.6 A general architecture of smart health-care network based on 5G. (Adapted from Ahad et al. (2020) with open access (MDPI))

(g) Analytics: With the spurt of UEs and ubiquitous wireless connections in health care, the amount of data processing that must be accomplished exponentially increases. Various parameters, such as user location, services used, signaling information, and applications in use, form a composite big dataset that needs to be processed and handled. This gives rise to new models and use cases that range from data extraction, data formatting, storage and presentation.

There is not a single technology or solution that will be utilized completely by the health care segment. This will be a methodical, stepwise and need-based evolution from existing technologies to the future. Healthcare models are rapidly changing due to demographic and socioeconomic changes from a hospital-based, specialist-focused approach to a distributed patient-centric care model (Rao 2017).

9.1.6 5G Technology for Smart Energy Management and Smart Cities

The development of 5G technology has been largely driven by smart mobile devices and advanced communication technologies. It may thus serve as a technical enabler for a whole new range of business opportunities and industrial applications. Moreover, it may enable different devices to work together seamlessly. In the future, 5G cellular networks will undoubtedly revolutionize global industries, which will have profound effects on businesses and customers. The most crucial part of smart city applications is the IoT, which generates significant amounts of data. The most effective methods to use for analyzing big data are advanced techniques such as AI, ML, and deep reinforcement learning (DRL). This can help companies make the best possible decisions based on their long-term objectives. Moreover, to further improve the accuracy of these techniques, the amount of training data can be increased. This can make learning capabilities stronger and improve their decision-making outcomes. The use of advanced big data analysis methods to create smart cities increased significantly during the same time frame. The use of IoT, blockchain, unmanned aerial vehicles (UAVs), AI, ML, and DRL-based methods to create smart cities is promising within the world of 5G (Huseien and Shah 2022).

9.1.6.1 5G Technology for Smart Cities

5G is a largely communication-based communication network. Moreover, to ensure that the correct layer of control is used for a specific application, 5G can be dynamically programmed. Given the high-speed data transfer offered by 5G technology, many different businesses will be able to use it. As shown in Table 9.2, innovative 5G technology progresses toward more advanced and smarter uses of technology (Huseien and Shah 2022).

For time-sensitive applications, 5G technology incorporates strict latency requirements of under 1 ms; alternatively, it incorporates relaxed latency requirements for nontime-sensitive applications. Additionally, 5G could be used in different settings in which stringent process controls and thus high levels of network reliability are needed. Comparably, for applications that are not process-sensitive, network reliability can be more relaxed. 5G technology can also be used in settings whereby vast amounts of information need to be processed in real time. In settings where only small amounts of information need to be processed, the system can be requisitioned. The IoT has now become one of the most fundamental 5G applications and is used in many industries and industrial sectors. As the IoT continues to evolve, an increasing number of points for accessing and sharing information via the network (the Internet) are needed. The growth of IoT has been largely supported by centralized data storage systems (such as the cloud). Nonetheless, users often have no idea how the information that they have shared on the system is used, thus rendering it a black box. In such cases, a centralized system may be unable to ensure data transparency. Blockchain technology is a possible solution that can be used to improve privacy and security on internet-based networks (Rao and Prasad 2018).

Table 9.2 Advantages of 5G communication technology (Huseien and Shah 2022)

Better network performance	5G networks will be able to give users ultra-high transfers speeds, ultra-low latency, ultra-reliable experiences, ultra-high connection density, ultra-high traffic density and ultra-high mobility access. It will also provide enhanced spectral efficiency for the networks and will lower the costs associated with network operations and maintenance costs whilst simultaneously providing enhanced network energy efficiency.
Smarter network operations	Through 5G technology, intelligent perception and decision-making skills will be significantly enhanced. The innovative features of 5G technology, such as real-time perception and the capacity to analyze user characteristics (such as their preferences, geographic location, network context and terminal status) will largely help companies to develop technological solutions which can ensure that data-driven network functions and resources are efficiently deployed.
Network functions are more flexible	The development of 5G technology is largely based on user-experience. It is thus designed to fulfil mobile internet needs and the requirements of IoT businesses. With regard to access networks, both plug-and-play, as well as self-organising base-station networks will be supported by 5G technology. This will enable a lightweight access network topology to be created, which will ensure that the system is easy to use and maintain. With regard to core networks, the process of using packet-core networks to develop innovative network functions will be redesigned and simplified to ensure that the system and forwarding functions are efficient, and that network control is flexible.
Network ecology is friendlier	The requirements generated by vertical industries and new industry ecologies will also be supported by 5G networks through the provision of a more user-friendly and open network. By enhancing and opening up network capabilities, third parties will be provided with a flexible service deployment service. The purpose of this is to facilitate friendly interactions with third-party applications. Tailored services can be provided on- demand through 5G networks. Moreover, business networks can be enhanced, innovative environments can be developed and ultimately, the value of network services can be enhanced.
Reliability comparison	The data transmission process in industrial automatic control systems is extremely reliable. At present, wired communication systems such as Ethernet, hubs and switches have been around for a while and are highly reliable. In 5G networks, channel coding technology is used to implement verification information. The 5G system is also able to identify any information that becomes lost during transmission and can determine whether any transmitted information is wrong. In such cases, the error will be automatically fixed.
Cost comparison	The information points in 5G communication systems are very flexible, meaning that can be modified and adjusted very easily. No embed pipelines or weak bridges are required, and there is no need to lay cables. This ultimately saves significant amounts of money on network wiring, installation, construction and maintenance costs. Moreover, given the extreme flexibility of the 5G system, it can be adjusted at any time throughout the construction process to suit changed site conditions. The system has the flexibility to react to any changes in demand or issues caused by inter-profession conflicts. The need for any buried pipelines and bridges during the construction process is eliminated and no complex wiring work is required. The process is easy, flexible and cost-effective since there is a significant reduction in project costs.
Comparison of safety	If an unauthorised user attempts to gain access to a wired network, they will first be required to connect to the physical line and then to the network using the network card. Subsequently, they must successfully complete the network's identity verification process. To maximise system security, a firewall is set up on the wired network.

The transformational effects that the IoT has on city infrastructure are becoming more well known, and thus, 5G technology plays a vital role in facilitating this process. This would enable the IoT to be used in all types of business activities across many different vertical industry sectors, including the manufacturing industry and the production of raw materials. 5G will open the door for innovative smart city applications in the future. Moreover, an increasing number of devices will be able to connect to the internet at any time. This would help to ensure that different vertical applications can be integrated (Huseien and Shah 2022).

9.1.6.2 Applications of 5G Technology in the Construction Industry and Infrastructures

Smart home technologies, smart grids, and smarter energy use technologies are spreading and growing in influence. In addition to playing a crucial role in improving cities and their infrastructures, smart buildings can also significantly enhance the comfort of residents. This is because it improves energy efficiency, available services, quality of life, safety controls and general comfort. Many different definitions for 'smart building' have been put forward, most of which are based on energy efficiency and the concept of a "smart grid". Intelligent management systems contain features that enable mass data to be stored and analyzed. Thus, when implemented in buildings, they can significantly enhance energy management. This is because electrical devices on a grid can learn and adapt to new behaviors, which is essentially what makes them 'intelligent' building systems. Therefore, services that incorporate IoT and big data technologies are intelligent because they employ vast amounts of data in analytics and automatic learning. 5G-led ubiquitous sensor networks are a cornerstone of smart city development; meanwhile, the unique ability of 5G networks to meet differentiated smart city needs, smart edge systems built on 5G and other technologies for collaborative intelligence (Daissaoui et al. 2020).

9.1.6.3 Smart Building System Integrated with 5G Communication Technology

Smart buildings have been defined as buildings that integrate advanced communication and technological systems (such as automatic monitoring equipment), organize information resources in the most effective way and receive at least a reasonable return on investment. Such buildings are environmentally friendly, comfortable, efficient and energy-saving. Intelligent building requires a combination of traditional industry factors and new technology. In intelligent buildings, the most advanced communication and control technologies are used in the building construction process to ensure that it is comfortable, convenient, energy efficient and environmentally friendly for those working or living inside (Sukhareva and Vorontsov 2019).

A wide range of devices, systems, actuators and sensors are required in smart buildings, all of which must be interconnected and function together effectively. One key aspect of smart buildings is that a communication network that can be controlled remotely is installed in the building. For instance, a sensor will verify that all windows are automatically closed before the air conditioner can be turned on. This sensor may be triggered by a temperature change. In other words, the sensor will trigger the air conditioner, after which information between the air conditioner and windows is shared. With regard to technical interoperability, the systems implemented to control the air conditioner and windows will most likely be produced by different manufacturers. Thus, it is crucial that an integration process is implemented to ensure that the building can be automatically controlled and managed (Jie et al. 2013).

A 5G network will be a full and comprehensive ecosystem designed to facilitate a society that is completely mobile and connected. It will also allow new business models to be developed that will create value. In the coming years, the traffic volume and data transmission demand on mobile broadband will increase significantly. Moreover, mobile broadband will be used in an increasing number of cases. In addition to supporting the development of already-established mobile broadband services, 5G will also play a significant role in the development of new services that can be more widely applied. This means that a vast range of applications will be able to use 5G, from low-bandwidth applications to services with extensive data transmission and latency demands. A wide range of devices, systems, actuators and sensors are required in smart buildings, all of which must be interconnected and function together effectively to provide inhabitants with comfortable and convenient lifestyles. One key aspect of smart buildings is that a communication network that can be controlled remotely is installed in the building. For instance, a sensor will verify that all windows are automatically closed before the air conditioner can be turned on. This sensor may be triggered by a temperature change. In other words, the sensor will trigger the air conditioner, after which information between the air conditioner and windows is shared. With regard to technical interoperability, the systems implemented to control the air conditioner and windows will most likely be produced by different manufacturers. Thus, it is crucial that an integration process is implemented to ensure that the building can be automatically controlled and managed. Table 9.3 presents this key feature in more depth, along with the key functional objectives and technical requirements of smart buildings (Jie et al. 2013; Blanco et al. 2017).

Therefore, sustainable and smart building is gaining momentum in public opinion, and thus, it is making its way into the agendas of researchers and city authorities all over the world. To move toward sustainable development goals, 5G technology would make significant impacts on building construction, operation, and management by facilitating high-class services and providing efficient functionalities.

Table 9.3 Major goals of smart buildings (Huseien and Shah 2022)

Goals	Descriptions	Technical requirements
Location-based services	Identify building occupants or resources locations and movements for improving convenience of services in building.	Track accurate position of targeted objects.
Energy efficiency	Maximize the use of building energy, with the ideal condition to be a Net Zero Building (NZB), while keep a high level of service at the same time.	Communicate with external elements (building-to building, or building-to-infrastructure).
Facility management	Preventive maintenance and organized operation and control of building facilities and equipment to reduce operations and maintenance time and cost.	Establish communication between building's equipment and devices
Indoor occupant comfort	Optimize ambient environmental conditions according to occupants' preferences for improving health and productivity.	Understand occupants' behavior pattern.

9.2 Perspectives on 6G Wireless Communications

Fifth generation (5G) wireless communication networks are being deployed worldwide, and more capabilities are being standardized, such as mass connectivity, ultra-reliability, and guaranteed low latency. The 6G will have additional requirements of high time and phase synchronization accuracy beyond what 5G can deliver. Additionally, 6G will have to provide near 100% geographical coverage, subcentimeter geo-location accuracy and millisecond geo-location update rate to meet use cases. As 5G networks are still limited to some typical scenarios, remote areas such as villages and motorways are not well covered, which limits some applications such as driverless vehicles. Nonterrestrial and specifical satellite communication networks are needed to complement terrestrial networks for cost-effective, seamless, and ubiquitous service availability. Unmanned aerial vehicle (UAV) communication networks are important for fast response in harsh and difficult environments. A maritime communication network is needed to provide ships with high-quality communication services. While mmWave can provide a Gbps-level transmission data rate in 5G, Tbps-level transmission data rates will be needed for applications such as high-quality three-dimensional (3D) video, virtual reality (VR), and a mix of VR and augmented reality (AR), where terahertz (THz) and optical frequency bands can be candidate bands. Faced with the large datasets generated by using extremely heterogeneous networks, diverse communication scenarios, large numbers of antennas, wide bandwidths, and new service requirements, 6G networks will enable a new range of smart applications with the aid of artificial intelligence (AI) and machine learning (ML) technologies. One automation level is for improving the network performance itself in many aspects, for example, quality of service (QoS), quality of experience (QoE), security, fault management, and energy efficiency (Fettweis 2014; You et al. 2021).

Up to 5G, traffic on the network is dominated by video or streaming applications. Wireless networked control of robotic objects (e.g., automated driving or factory logistics) is a new existing application for cellular technology, but this also generates new challenges. When analyzing the network traffic generated by these applications, many mobile objects must share sensor as well as control information, which overburdens a centralized control system. Instead, distributed control systems using AI are becoming a focus in research and development. In particular, federated learning is a promising approach, where dataset correlation algorithms are distributed over mobile robotic objects and aggregated learning occurs over the cloud. Interestingly, this generates a completely new class of net-work traffic, with large bandwidth and widely varying latency demands. It is highly likely to assume that these and equivalent AI applications will not only overtake but also dominate the network traffic demands of 6G. In comparison with the 5G network, 6G wireless communication networks are expected to provide much higher spectral/energy/cost efficiency, higher data rate (Tbps), 10 times lower latency, 100 times higher connection density, more intelligence for full automation, subcentimeter geolocation accuracy, near 100% coverage, and submillisecond time synchronization. New air interface and transmission technologies are essential to achieve high spectrum efficiency and energy efficiency, including new wave forms, multiple access approaches, channel coding methods, multiantenna technologies, and proper combinations of all these diversity techniques. Meanwhile, novel network architectures are needed, for example, software defined network/network functions virtualization (SDN/NFV), dynamic network slicing, service-based architecture (SBA), cognitive service architecture (CSA), and cell-free (CF) architectures (Saad et al. 2020).

However, softwarization comes at a cost. The use of commercial off-the-shelf (COTS) servers versus domain specific chips in a virtualized radio access network (RAN) implies a large increase in energy consumption and countering measures for improving energy efficiency. This results in the current fact that 5G networks consume more power than 4G networks, but of course at a delivery of a higher bandwidth. Therefore, 6G requires a new computing paradigm to support all benefits of softwarization without bearing the costs of energy consumption (You et al. 2021).

Computing technologies such as cloud computing, fog computing, and edge computing are important for network resilience, distributed computing and processing, and lower latency and time synchronization. To solve the limitations of 5G, including the drawbacks of short packets, provide the delivery of high-reliability, low-latency services with high data rates, system coverage and the Internet of Everything (IoE), and meet the demands of future mobile communications, 6G networks should be human-centric, instead of machine-centric, application-centric, or data-centric. To meet these requirements, 6G wireless communication networks will have new paradigm shifts. As illustrated in Fig. 9.7, 6G wireless communication networks will be space-air-ground-sea integrated networks to provide complete global coverage. Satellite communication, UAV communication, and maritime communication will largely extend the coverage range of wireless communication networks. To provide a higher data rate, all spectra will be fully explored, including sub6 GHz, mmWave, THz, and optical frequency bands. To enable full applications,

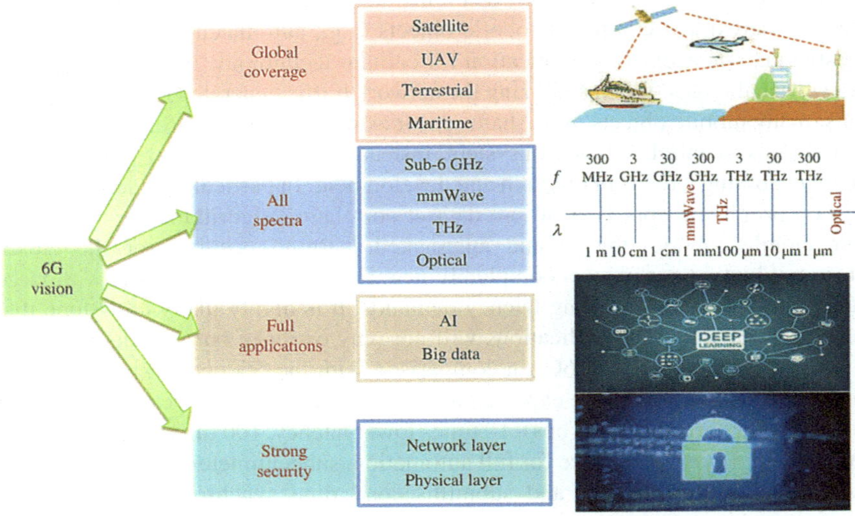

Fig. 9.7 A vision of 6G wireless communication networks. (Adapted from You et al. (2021), Credit: https://www.ncbi.nlm.nih.gov/pmc/articles/PMC7714900/pdf/11432_2020_Article_2955.pdf)

AI and ML technologies will be efficiently combined with 6G wireless communication networks to achieve better network management and automation. Furthermore, AI technology can enable the dynamic orchestration of networking, caching, and computing resources to improve the performance of next-generation networks. The last but not the least trend is the strong or endogenous network security for both the physical layer and network layer when developing it. Industry verticals, such as cloud VR, Internet of things (IoT) industry automation, cellular vehicle to everything (C-V2X), digital twin body area network, and energy efficient wireless network control and federated learning systems will largely boost the developments of 6G wireless communication networks. An overview of 6G wireless networks is shown in Fig. 9.8, where the performance metrics, application scenarios, enabling technologies, new paradigm shifts, and industry verticals are given (Zong et al. 2019; You et al. 2021).

In summation, sixth generation (6G) wireless communication networks are expected to provide global coverage, enhanced spectral/energy/cost efficiency, better intelligence level and security. To meet these requirements, 6G networks will rely on new enabling technologies, i.e., air interface and transmission technologies, and novel network architecture, such as waveform design, multiple access, channel coding schemes, multiantenna technologies, network slicing, cell-free architecture, and cloud/fog/edge computing. There may be four new paradigm shifts from 5G to

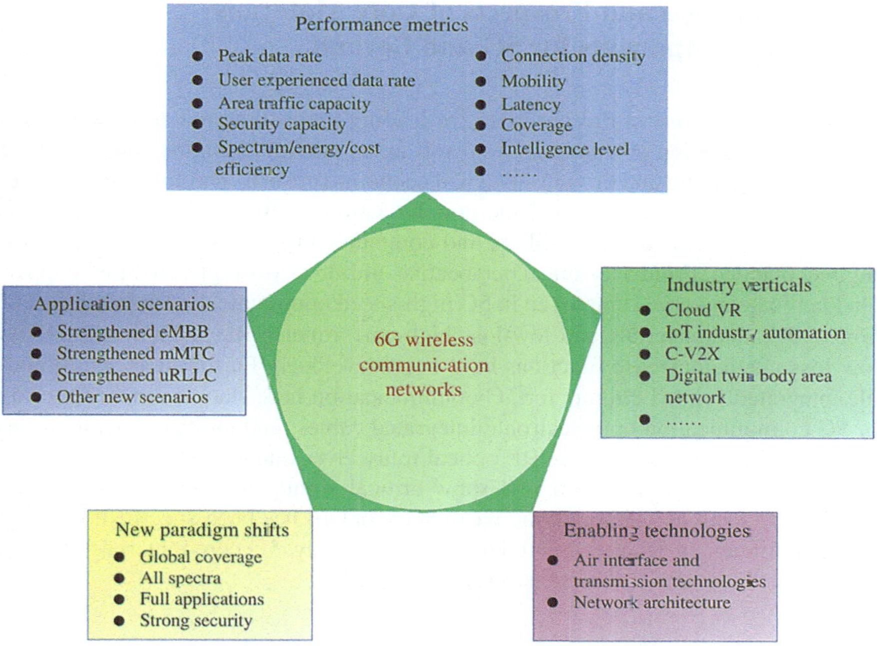

Fig. 9.8 An overview of 6G wireless communication networks. (Adapted from You et al. (2021), Credit: https://www.ncbi.nlm.nih.gov/pmc/articles/PMC7714900/pdf/11432_2020_Article_2955.pdf)

6G. First, to satisfy the requirement of global coverage, 6G will not be limited to terrestrial communication networks, which will need to be complemented with non-terrestrial networks such as satellite and unmanned aerial vehicle (UAV) communication networks, thus achieving a space-air-ground-sea integrated communication network. Second, all spectra will be fully explored to further increase the data rates and connection density, including the sub6 GHz, millimeter wave (mmWave), terahertz (THz), and optical frequency bands. Third, facing the big datasets generated by the use of extremely heterogeneous networks, diverse communication scenarios, large numbers of antennas, wide bandwidths, and new service requirements, 6G networks will enable a new range of smart applications with the aid of artificial intelligence (AI) and big data technologies. Fourth, network security will have to be strengthened when developing 6G networks. Clearly, 6G with additional technical requirements beyond those of 5G will enable faster and further communications to the extent that the boundary between physical and cyber worlds disappears (You et al. 2021).

9.3 Challenges and Prospects of Core Materials and Components for 5G and Beyond

5G has become a novel driving force for leading innovation and stimulating new types of information consumption, as well as an emerging engine for promoting industrial upgrading and driving sustained economic growth. Through the construction of 5G network systems and the popularization of 5G terminals, components such as base station antennas, filters and communication PCB boards are increasingly demanded. From a technical perspective, ultralow-loss high reliability copper-clad laminate materials employed in 5G high-speed communications (56 Gbps, 112 Gbps), 5G metamaterials and low-loss high-performance RF technology, and 5G low-loss magnetoelectric functions have been developed. Furthermore, multimodule integrated printed circuits for 5G communication base stations, manufacturing of 5G communication optoelectronic integrated cables, and multichannel high linearity and large dynamic range RF optical transceiver integrated modules in terms of 5G mobile communication will show critical significance in various 5G and beyond applications. Although the basic 5G standard has been specified, it is still being continuously improved and developed. Certainly, 5G core materials, devices and key technologies are the same (Yang et al. 2021).

9.3.1 Ultralow-Loss High-Reliability Copper-Clad Laminates

The PCB substrate material for high-speed transmission is vital for 5G applications in the future. Ultralow-loss high-reliability copper-clad laminates are the most important 5G PCB substrate material. The dielectric properties, glass transition temperature, thermal decomposition temperature of functionalized polyphenylene ether resin and special crosslinking curing agent, and other performance of the copper-clad laminate should be controlled (Yang et al. 2021):

- Glass transition temperature Tg (DSC) $\geq$180 °C
- Dielectric performance D_k/D_f(RC56%, 10GH) $\leq$3.50/0.0025
- Copper foil peel strength (1OZ HVLP3) > 0.7 N/mm
- Z-axis expansion coefficient Z-CTE < 2.5%

The key technologies used are free radical curing technology, glass fiber cloth selection, development of new filler types and content selection, flame retardant selection, etc. By regulating the proportion of formula resin, the prepared prepreg exhibits high manufacturability and low volatility, which achieves the balance of dielectric properties and heat resistance of high-speed substrates; in addition, it shows good processability in the PCB processing stage, which is suitable for meeting 5G high-speed ultralow-loss electronic circuit substrate for wireless communication.

9.3.2 5G Metamaterials and Low-Loss High-Performance RF Technology

Electromagnetic metamaterials have been processed or synthesized artificially, exhibiting a periodic or quasiperiodic structure and special electromagnetic properties. They have three important characteristics, i.e., a special artificial structure, the physical properties of supernormality, and the electromagnetic properties of supernormality. The extraordinary physical properties make the application prospects of metamaterials very wide, such as metamaterial radar antennas, absorbing materials, electronic countermeasure radars, and metamaterial communication antennas. Metamaterial communication antennas have unparalleled advantages over conventional antennas in 5G and large-scale satellite communications (6G) in the future. Thus, the development of metamaterials that can be mass-produced at low cost is strategically significant to 5G and beyond technologies. Compared with known similar materials, the metamaterial is light in weight, raw materials are easily available, and the process is relatively simple and suitable for automated production. Main technical requirements for 5G metamaterials (Yang et al. 2021):

- The yield rate reaches 95%
- The volume is less than 10%
- The density reaches 45 kg/m^3
- The dielectric constant is 1.04–1.85

For the 5G antenna array, the loss of the existing 141 coaxial cable (in the high-frequency band 3.3–5.0 GHz), for example, is excessively large, approximately 1 dB/m, and it cannot provide a more stable phase. Therefore, waveguide feed technology has been used for the 5G wireless antenna array to reduce losses and provide a consistent phase. Comparably, the metamaterial provides ideal dielectric properties and can be designed with heat dissipation channels for high-power electromagnetic wave applications. The metamaterial also has low loss, and its insertion loss is on the order of 10^{-4}. If 5G uses a metamaterial lens antenna, the power of the main equipment of the base station can be reduced by up to 50%. In addition, the device selection of the main equipment of the base station will be wider, the chips and component devices will be cheaper, and the source channels will be significantly increased.

9.3.3 5G Low-Loss Magnetoelectric Functional Materials and Devices

With the rapid development of 5G communication technology and the blowout requirements for various high-performance and miniaturized and integrated RF electronic components, low-temperature cofired RF ferrite ceramics (LTCCs) are increasingly demanded. To yield the optimal formula for broadband low-loss LTCC materials, the "seawater lava method model" is proposed to optimize the simulation of its doping, nanocrystal implantation, domain-domain transfer pinning and other processes to achieve the low temperature of broadband low-loss LTCC materials (900 °C) co-firing. Based on the loss model, i.e., the grain boundary resistance widening band model and the electric domain pinning model, the LTCC material frequency band is widened up to the microwave/millimeter wave range. Novel effects of multiphase recombination and codoping are developed to form glass-dielectric recombination, i.e., magnetic-dielectric composite material systems, to achieve multiperformance low-loss LTCC material technology breakthroughs in multiple frequency bands. Combined with the original ceramic strip technology, RF and microwave devices have been developed, including RF and microwave filters, antennas, circulators/isolators, switches, waveguides, and high frequency inductors. Therefore, low-loss magnetoelectric functional material technology can be used to develop LTCC gyromagnetic (30–100 GHz) and millimeter-wave low-loss microwave ceramic material systems to underpin the application of materials in future 6G communications and realize multiperformance LTCC composite materials in the radio frequency band, microwave, and millimeter wave band. In the future, the modified LTCC process will promote simpler structures, better integration and lower cost for filters and other combination modules. Moreover, passive and silicon-based active integrated systems or components can be assembled on miniaturized multiple chips or components for microwave communication and radar applications (Yang et al. 2021).

9.3.4 Multimodule Integrated Printed Circuit Boards

The low-frequency medium mixed voltage method is adopted to achieve low-cost multiband transmission design, the development of low-profile copper processing technology to enhance the integrity of high-frequency signal transmission, and the optimization of line etching technology to control single-line impedance errors. Main technical requirements for multimodule integrated printed circuit boards (Yang et al. 2021):

- Number of layers 8 ~ 18, thickness 1.2 ~ 4.0 mm
- Board warpage ≤0.8%; through-hole thickness-diameter ratio ≥ 10:1
- Single wire impedance deviation ±10%; back drill depth error ≤ 5%
- Flatness of the copper block and board surface ±75 μm, copper block thrust ≥300 N (30.6 Kgf)
- Signal transmission insertion loss ≥ −45 dB/m@20 GHz
- 288 degrees 10 s/3 times thermal shock or lead-free reflow soldering 3 times, the product will not delaminate and change color

By locally embedding copper blocks into high thermal conductivity composite materialsto achieve efficient thermal control in the board, high performance multi-module integrated printed circuit boards can be made through developing electrode-position formulations, improving the uniformity of high thickness-to-diameter ratio through-hole plating, developing in-board resistance, improving system integration of functions, as well as building product systems, performance testing methods and evaluation mechanisms.

9.3.5 *Manufacturing Technology of Photoelectric Integrated Cables*

In 5G communication, the long-distance connection between the signal and the power supply requires large bandwidth and high power. Thus, the photoelectric integrated cable is the optimal choice for 5G wireless communication. However, the complex application scenarios of 5G make the manufacture of photoelectric integrated cables challenging with a series of new requirements. Although the 5G communication standards are unified globally, the specific solutions of various communication equipment manufacturers are different. For example, butterfly optical cables have the characteristics of miniaturization and easy construction. Strictly symmetrical and tight cable forming technology is adopted to make the product structure stable with high tensile and compressive strengths and not easy to bend and twist, which can meet various construction conditions. In many complex application scenarios, fiber fusion splicing technology cannot be used; therefore, developing vital technologies for the success of on-site quick splices is critical. By employing precise size control and multilayer tight cladding technology, the optical unit and the quick connector can be perfectly matched and easily joined with high reliability. Additionally, the miniaturization and high power of 5G equipment have caused large heat generation. Therefore, high-temperature resistant materials are needed to make a reliable cable communication system, which can be used under high temperature conditions for a long time.

9.3.6 *Photonics-Assisted Ultrabroadband RF Transceiver Integrated Modules*

To comply with the major requirements of ultrabroadband dynamic analog transmission and low power wireless access in 5G mobile communications, a multichannel High Linearity and Large Dynamic Range RF Optical Transceiver Integrated Module is designed, centering on the new requirements and challenges of the microwave and light wave fusion access system by exploiting changeable and diversified services. The support in the 20 GHz frequency band exceeds 4 Multiband, multistandard wireless signal multichannel (≥4) integrated optical transceiver modules to achieve multiple services, large dynamic, long-distance fiber distribution and multidomain resource centralized management and control of a new generation of microwave optical wave fusion access and ubiquitous spectrum sensing systems.

High linearity and large dynamic range RF optical transceiver integrated chips and modules can be fabricated based on the MZI-PLC silicon-based integrated coupling structure, 42-degree inclined array waveguide coupling method, 3D integrated microwave package design method for electrical design, thermal management and reliability analysis. The 1310 nm band multichannel (≥4) integrated RF optical transceiver module was developed by exploiting the prepared multichannel (≥4) laser array chip and high-saturation-power photodetector array chip. The 3 dB bandwidth of each channel is greater than 20 GHz, and SFDR is greater than 95 dBHZ2/3, effectively increasing the number of indoor coverage users. A multiband, multistandard wireless signal multichannel (≥4) integrated optical transceiver module has been realized, supporting more than 4 services in the 20 GHz frequency band. The new generation of microwave optical wave fusion with large dynamic, long distance fiber distribution and centrally controlled multidomain resources into ubiquitous spectral sensing systems. The key chips and devices of high-performance RF optical modules achieve the overall improvement and practicability of optical link performance indexes; realize broadband, high linear, high power, multichannel analog direct modulation laser technology; and realize broadband, high linear, low insertion loss modulation. Its array technology, high-power broadband photoelectric detection array receiving technology, low noise analog optical amplifier technology and multimode integrated RF optical module have also been realized (Yang et al. 2021).

9.3.7 All-Optical Network and Superlarge-Core Fiber Optic Cables

As impacted by the advantages of high density, easy installation of optical fiber ribbons, convenient offline or branching, and fully dry water-blocking and moisture-proof structure, the skeleton optical cable simplifies the conventional structure of optical fiber extraction, fiber drop, and connection work to the greatest extent. Given the characteristics of the 5G technology's own communication architecture and large data transmission capacity and the optical fiber expansion problem of the last 1 km of the FTTH entrance of the all-optical network, the large core number expansion of the skeleton optical cable is the most urgent product problem to be solved in the construction application of the skeleton optical cable. Technical breakthroughs have been made from fiber coloring technology, skeleton tension control technology, optical fiber in-groove technology, water blocking tape wrapping technology, and optical fiber and tape technology, which have addressed the production bottleneck and flatness quality problems of the coloring process and skeleton cable forming process, as well as the cost of auxiliary materials (e.g., resin). Through optimization and experimental calculation and with the optimization parameters of curing power and coating pressure, the tile-like problem attributed to excessive curing of the optical fiber ribbon, and the problem of excessive flatness of the optical fiber ribbon resulting from uneven coating pressure should be solved.

By optimizing the guide rail and position of the optical fiber mold, the frictional force between the mold and the guide plate and the squeezing force between the optical fibers can be reduced.

When the full factor model of the intelligent production process is constructed for the automatic production line, it needs to realize the rapid learning and design of industrial equipment, as well as the accurate mapping of virtual and real design, resulting in a variety of the production line data and their complex sources. It is difficult to monitor data equipment, product quality, energy consumption and other production line processes. Hence, the relationship between the original design and transfer learning should be determined, and the simulation solution methods are integrated to build a simulation solution interface for modeling and data fusion. This will help to develop high-confidence and effective simulation solutions for production line design. Multicore high-density banded fiber optic cables will become the core product of 5G commercial carrying optical networks (Yang et al. 2021).

As a result, 5G is fully in line with the future trends of human progress, social development, and economic growth, and its large-scale applications will inevitably be achieved. In the 5G era, the door of the Internet of Things will be opened and expanded to all industries and fields, as well as to all aspects of the economy and society. This will change the way people consume, transform the way they produce, and provide a powerful tool for the high-quality development and governance of society.

References

Ahad A, Tahir M, Aman Sheikh M, Ahmed KI, Mughees A, Numani A (2020) Technologies trend toward 5G network for smart health-care using IoT: a review. Sensors 20(14):4047. https://doi.org/10.3390/s20144047

Aijaz A (2020) Private 5G: the future of industrial wireless. IEEE Ind Electron Mag 14(4):136–145

Blanco B et al (2017) Technology pillars in the architecture of future 5G mobile networks: NFV, MEC and SDN. Comput Stand Interfaces 54:216–228

Bouali F et al (2021) 5G for vehicular use cases: analysis of technical requirements, value propositions and outlook. IEEE Open J Intell Transp Syst 2:73–96. https://doi.org/10.1109/OJITS.2021.3072220

Chiasserini C, Magnan A (2017) 5G for the automotive domain. https://futurenetworks.ieee.org/images/files/pdf/applications/5G-for-the-Automotive-Domain030518.pdf

Daissaoui A et al (2020) IoT and big data analytics for smart buildings: a survey. Procedia Comput Sci 170:161–168

Fettweis GP (2014) The tactile internet: applications and challenges. IEEE Veh Technol Mag 9:64–70

Huseien GF, Shah KW (2022) A review on 5G technology for smart energy management and smart buildings in Singapore. Energy AI 7:100116. https://doi.org/10.1016/j.egyai.2021.100116

IEEE 5G Initiative Roadmap Committee Members (2017) IEEE 5G and beyond roadmap whitepaper. https://futurenetworks.ieee.org/images/files/pdf/ieee-5g-roadmap-white-paper.pdf

Jie Y, Pei JY, Jun L, Yun G and Wei X (2013) Smart home system based on IOT technologies. 2013 International Conference on Computational and Information Sciences, pp. 1789-1791, doi: 10.1109/ICCIS.2013.468.

Mahmood A, Abedin SF, Sauter T, Gidlund M, Landernäs K (2021) Factory 5G: a review of industrial-centric features and deployment options. TechRxiv Preprint. https://doi.org/10.36227/techrxiv.17089265.v1

Markakis E, Politis I (2017) 5G emergency communications. https://futurenetworks.ieee.org/images/files/pdf/applications/Emergency-Communications030518.pdf

Rao K (2017) The path to 5G for health care. https://futurenetworks.ieee.org/images/files/pdf/applications/5G%2D%2DHealth-Care030518.pdf

Rao SK, Prasad R (2018) Impact of 5G technologies on smart city implementation. Wirel Pers Commun 100(1):161–176

Saad W, Bennis M, Chen M (2020) A vision of 6G wireless systems: applications, trends, technologies, and open research problems. IEEE Netw 34:134–142

Sukhareva AV, Vorontsov KV (2019) Building a complete set of topics of probabilistic topic models. Intell Syst Theory Appl 23(4):7–23

Warren D, Dewar C (2014) Understanding 5G: perspectives on future technological advancements in mobile. GSMA Intelligence. https://www.gsma.com/futurenetworks/wp-content/uploads/2015/01/2014-12-08-c88a32b3c59a11944a9c4e544fee7770.pdf

Yang R et al (2021) Research on 5G related technologies under the new generation of information technology. J Phys Conf Ser 1907:012056

You X, Wang CX, Huang J et al (2021) Toward 6G wireless communication networks: vision, enabling technologies, and new paradigm shifts. SCIENCE CHINA Inf Sci 64:110301. https://doi.org/10.1007/s11432-020-2955-6

Zhu C, Cheng X, Ye H, Yang J, Xu L, Chao K (2019) 5G wireless networks meet big data challenges, trends, and applications. In: 2019 IEEE SmartWorld, ubiquitous intelligence & computing, advanced & trusted computing, scalable computing & communications, cloud & big data computing, internet of people and smart city innovation. SmartWorld/SCALCOM/UIC/ATC/CBDCom/IOP/SCI, pp 1513–1516. https://doi.org/10.1109/SmartWorld-UIC-ATC-SCALCOM-IOP-SCI.2019.00272

Zong BQ, Fan C, Wang XY et al (2019) 6G technologies: key drivers, core requirements, system architectures, and enabling technologies. IEEE Veh Technol Mag 14:18–27

Index

C. Tong, *Advanced Materials and Components for 5G and Beyond*, Springer Series in Materials Science 327, https://doi.org/10.1007/978-3-031-17207-6

The manufacturer's authorised representative in the EU is Springer Nature Customer Service Centre GmbH, Europaplatz 3, 69115 Heidelberg, Germany. If you have any concerns regarding our products, please contact ProductSafety@springernature.com

Printed and bound by CPI Group (UK) Ltd, Croydon, CR0 4YY
30/09/2026
02225239-0008